The Pesticide Handbook

The Pesticide Handbook

Peter Hurst, Alastair Hay and Nigel Dudley

journeyman
London • Concord, Mass

First published 1991 by Journeyman Press
345 Archway Road, London N6 5AA
and 141 Old Bedford Road, Concord
MA 01742, USA

British Library Cataloguing-in-Publication Data
Hurst, Peter
    The pesticide handbook.
    I. Title II. Dudley, Nigel III. Hay, Alastair
    632

ISBN 1-85172-041-3

Library of Congress Cataloging-in-Publication Data
Hurst, Peter.
    The Pesticide Handbook / by Peter Hurst, Alastair Hay, Nigel Dudley.
        p. cm.
    Includes bibliographical references and index.
    ISBN 1-85172-041-3.
    1. Pesticides—Handbooks, manuals, etc. 2. Pesticides—
Toxicology—Handbooks, manuals, etc. 3. Pesticides—Environmental
aspects—Handbooks, manuals, etc. I. Hay, Alastair, 1947- .
II. Dudley, Nigel. III. Title.
SB951.H89 1990
632'.95—dc20   90-48389
                                                                    CIP

Typeset by Stanford Desktop Publishing Services, Milton Keynes
Printed in Great Britain by Billing and Sons Ltd, Worcester

Contents

To Kathleen and Dorothy, Patrick and Jafa

Acknowledgements

Among the many people who have provided assistance and advice, we would like, in particular, to thank Peter Beaumont, Patrick Bond, Barbara Dinham and Steve Robertson.

We are grateful to New Science Publications, publishers of *New Scientist*, for permission to reproduce the article 'How the Chemical Industry could clean up its act' which appeared in the 12 February 1987 issue of the magazine.

Others also deserve our thanks. They are: Anne Beech who encouraged us at all times as we converted successive drafts into a finished manuscript; Linda Etchart who completed the job and turned it into a book; and Wendy Hay who produced the index.

1

Introduction

Adverts showing us immaculately manicured lawns without a dandelion in sight, or promising us apples free of any blemish, are the stock-in-trade of the pesticide industry. Like manufacturers of cars or chocolate, the producers of pesticides need to sell their products. To do this they need to persuade us, their consumers, that if we do not buy a weedkiller for our lawn, or a fungicide for our apples, life will not be quite as good as it could be. By implication, we deserve better: we too can have a lawn like a bowling green, or apples as smooth as a baby's bottom if only we use what is on offer.

The advertising used to sell pesticides in the UK, Germany, France or the United States is similar to, but far more subtle than, the advertising used to promote products in Brazil, Colombia or Malaysia. In the developing world, governments simply do not have the resources to monitor pesticide use on farms or plantations, nor do they have enough inspectors to ensure that chemical companies are not making extravagant claims about the products they promote. Misleading advertising is much more common in the Third World in spite of international codes of practice designed to stop it.

International codes are necessary because the pesticide industry is a global business. Few realise just how much pesticides intrude into our daily lives: the chemicals are in the food we eat, the water we drink, the grass we walk on, even the air we breathe. Fortunately, for the majority, exposure to the chemicals is limited because regulations limit the quantity of chemicals permitted in food or water. For those involved in the manufacture, application, or use of pesticides, contact is far more extensive and problematic.

The very mention of the word 'pesticides' is calculated to arouse strong emotions in most people. Few are neutral where these chemicals are concerned: people are either for or against them. Sometimes the information people rely on is not as good as it could be. So in this book we have tried to separate fact from fiction, to tease out what is real from what has been exaggerated and, above all, to look at the advantages of using these toxic chemicals against the backdrop of the problems they cause to human health and to the environment. We have strong views about these issues but we have tried throughout to give credit where it is due and not to be too partisan.

Pesticides are a product of the scientific and technological revolution of the last 50 years, the fruits of new methods of synthesis in chemistry and, increasingly, from developments in biotechnology (genetic engineering). In the main, pesticides are produced by large, transnational companies who control production and sales worldwide. These same companies are also investing heavily in other areas of agriculture and food production; many own nurseries and firms that produce seeds for farmers and growers, others control food companies, and many are doing research into genetic engineering. Much of this research aims to develop crops resistant to the chemicals the pesticides companies sell, in order to squeeze competitors out of the market place.

Pesticides are toxic compounds designed to kill, or 'control', living things. Pesticides will kill weeds and destroy bugs and micro-organisms that attack crops, animals, timber and humans. Controlling pests in this way reduces crop losses and food wastage wherever food is kept in storage. On the debit side we need to remember that many pesticides will remain in the body for years; they are also persistent in the environment. In Chapter 2 we review the properties of pesticides, their benefits and drawbacks.

Because pesticides are so toxic they may affect the well-being of other living things which are not their intended targets. The 'acute', or more immediate, effects of exposure to pesticides are well documented and range from obvious poisoning through lung and skin problems, to more vague 'flu-like' symptoms. 'Chronic', or longer-term health problems of exposure are less well documented and much more controversial, and include cancers, birth defects and nervous system damage. Regrettably, many aspects of the long-term effects of chemical exposure are still unknown. In Chapter 5 we review what is known and point out what research still needs to be carried out.

Those most at risk from pesticides are the men and women in direct contact with the chemicals through their jobs. These include workers in chemical plants employed on production or packaging; people employed in spraying on farms, in horticulture and forestry; those who treat timber, control rodents, even those who dispose of waste chemicals and containers. We highlight these high-risk occupations in Chapter 4, and we discuss what workers can do to reduce the risks. We emphasise the importance of knowing what chemicals are being used on a particular job, and the need to find out what health effects the pesticides are known to have. Should anyone be poisoned, we provide some advice on first aid which could help to save a friend's life.

Most of us consume food and drink contaminated with pesticide residues. This is an international problem and is more difficult to deal with when crops and animal produce are exported from country to country. Even if we ban or control the use of certain pesticides in food production in the industrialised countries we will probably still consume many pesticides in the food we import from developing countries where controls are more lax. The health effects of consuming residual pesticides in food is much disputed and a subject where there is often more heat than light. Because of the importance of the residue issue we have explained how different countries and international regulatory authorities arrive at their decisions about whether certain residue levels are dangerous or not.

Pesticides affect the whole environment. Sadly, there are all too many examples where the use of a particular chemical to control a certain pest has resulted in many harmless and even beneficial plants, insects, animals and micro-organisms being killed. Pesticides can even create 'pest' problems by killing off the natural predators, parasites and the bacteria which normally check the spread of pests. With these controls removed, minor pests can spread rapidly and become a major threat to crops. Again, we review these issues and point out where over-use of chemicals could lead to serious problems in some developing countries.

The chemical industry in general is concerned about its public image because many people believe that the chemical companies have caused serious pollution problems. Sandwiched between the desire to improve their reputation and the need to sell their products, pesticide manu-facturers may be tempted to cut corners and allow practices which compromise the welfare of humans and the environment. The ethics of exporting dangerous pesticides to poorer, developing countries have been questioned, and in Chapter 9 we discuss the subject in depth. The story we tell is not pleasant because many of the abuses we have found are so serious. Changes are needed at the international level to reduce the number of poisonings from pesticides and we point out how some consumer groups and trades unions are trying to bring about much needed improvements.

Chapter 10 is a snapshot of the problems associated with the growth of a single cash crop: cocoa. By examining one crop, and the problems cocoa growers face in different countries through the use of pesticides, we show just how complicated the issue is. We also show how exchange of information between cocoa growers can help to improve their cir-cumstances.

Action to reduce the problems pesticides cause must also be addressed at government level. At this stage the role of different government departments can appear, and often is, contradictory. Some government departments will give priority to the promotion of industry, whereas others are more concerned about the effect of chemicals on human health and the environment. In the UK a single government department, the Ministry of Agriculture, Fisheries and Food (MAFF) acts as both judge and jury on pesticides. MAFF decides when pesticides can be used and when the safety of a product is in doubt, but MAFF is also charged with promoting agriculture and the safe use of pesticides, two activities which many people consider to be incompatible with its allotted role as guardian of public safety. In Chapter 11 we explain why we are unhappy with this arrangement and why we consider that changes are still needed. We are most concerned that some of the central decisions about safety are still made without public scrutiny.

Pesticides may be the preserve of the chemical industry, but the problems they cause affect us all. For too long we have waited for governments to give the public the protection it deserves. Even if governments could be made more responsive to some of the more pressing issues the public considers important, vested interests would ensure that this did not happen too rapidly. We provide a guide to some of the organisations involved with this work.

International, European Community (EC) and UK codes and laws are explained in Chapter 11. By providing this regulatory tour we hope that we have shown how the 'machine' works but, above all, we hope that we have armed those who are concerned about pesticides with sufficient information to enable them to bring about changes in their own environment.

In general, we are concerned at the lack of a comprehensive, government coordinated Pesticides Policy based on reducing pesticide use wherever possible. Countries such as Denmark, the Netherlands and Sweden have policies and programmes to reduce pesticide use by 50 per cent or more during the 1990s. Pesticide Policies and reduction strategies are discussed more fully in Chapter 12.

2

About Pesticides

What are Pesticides?

The term 'Pesticides' refers to certain chemical substances and micro-organisms (bacteria, fungi, viruses and mycoplasmas) used to kill and control 'pests' such as weeds, fish, fungi, bacteria, insects, and animals like rats and mice. The term 'pests' is used in a broad sense to include animals, micro-organisms, and plants, as well as insects and insect-type pests.

Pesticides include fungicides, herbicides, insecticides and miscella-neous compounds such as wood preservatives, plant growth regulators, soil sterilants, animal and bird repellents and masonry biocides. Many leaflets and books talk of fungicides, herbicides, insecticides and wood preservatives as though they were separate categories of chemicals, whereas they are all types of pesticides. A summary of the various types of pesticide is given in Chapter 3, and the legal definitions are dealt with in Chapter 11.

Most pesticides are synthetic, manufactured chemicals specifically developed for their toxic qualities and properties. These pesticides consist of the chemical 'active ingredient' mixed with other chemicals, such as solvents, to make up the formulated, ready-to-use, pesticide product. These other chemicals may be as toxic as the pesticide active ingredient itself, or even more so, and can comprise 90 per cent or more of a registered, approved pesticide product. Pesticides are often referred to by other names such as 'agrochemicals', and 'crop protection/plant protection products'.

A small number of pesticides are based on naturally-occurring micro-organisms (bacteria, fungi, viruses and mycoplasmas) and chemical extracts from plants, such as pyrethrum and derris/rotenone. Genetically engineered pesticides produced by manipulating natural micro-organisms are now being rapidly developed as part of the general expansion of biotechnology.

5

In this book, for reasons of convenience only, pesticides are grouped and discussed under the general headings of:

- Fungicides
- Herbicides
- Insecticides
- Miscellaneous

Why are Pesticides Used?

There are three main reasons why pesticides are so widely used:

- to maintain and increase food and commodity production
- to control disease and promote public health
- to assist in the preservation and maintenance of buildings and structures

Food and Commodity Production

The pesticide industry and the government argue that the use of crop protection chemicals, which is the industry name for pesticides, is vital to ensure that food and commodity production is maximized by helping prevent losses during growth, harvest, transport, distribution and storage, and that the quality of food and commodities is maintained and improved, and at an economic price.

Maximizing Production

The International Group of National Associations of Pesticide Manufacturers (GIFAP) has stated that: 'even with modern cultural techniques, at least 30 per cent of the world's potential crop production is lost each year. Crop losses would be doubled if existing pesticide uses were abandoned.'[1]

The British Agrochemicals Association (BAA) – the UK pesticide manufacturers' trade association – estimates that 45 per cent of world food potential is lost to pests, 30 per cent to weeds, insects, vermin and diseases before harvest and a further 15 per cent between harvest and use. The BAA leaflet *Pesticides in Perspective: Why Use Pesticides?* states that:

neither governments nor individual farmers can tolerate the loss of crops which have cost money and effort to grow. Without pesticides UK cereal yields would drop by about 25% in the first year and 45% in the second. Other crops would be similarly affected ... In the tropics, including many of the developing countries, crop losses are far worse than in Britain and far more devastating to the welfare of the people. Denying these countries pesticides would be catastrophic – they need to double their food production over the next 20 years.[2]

The UK-based chemical company ICI, in cooperation with the well-known botanist and environmentalist Professor David Bellamy, have produced a booklet and video entitled *Food for Thought* reiterating the claim that pesticides are vital to feed a hungry world.[3] In particular, David Bellamy looks at ICI's safety testing programme in the development of crop protection products.

Maintaining Quality

The BAA leaflet states that:

> most consumers have an understandable preference for buying food which is unblemished ... the UK's farmers and growers now fill shops and supermarkets with produce which is clean and of first class quality at reasonable prices. This is made possible by the use of pesticides to control the pests, diseases and occasionally weeds, which do the damage.

Quality is also important in the production of commodities – whether home produced or foreign – ranging from flowers and pot plants to cocoa and coffee beans, maize, rice, rubber and cotton.

The Ministry of Agriculture booklet, *Pesticides and Food: A Balanced View*, states that:

> pesticides in agriculture have helped improve yields and protect crops both in developed countries and in the third world. But there's more to pesticides than that. They are used after harvest, thereby extending the life of the produce by protecting them from pests and diseases in store. This makes a wider range of produce available to consumers. They are used to improve hygiene by killing flies and preventing cockroaches and by preventing other food hazards.[4]

Industry statements like these gloss over the large over-production of cereals and other foodstuffs in the US and the European Community (EC) countries resulting in 'grain mountains' and so on. The use of pesticides and artificial fertilisers (nitrates, phosphates, etc.), combined with mechanisation and crop plant breeding, have been major factors in making possible intensive, high-output agriculture. The agribusiness sector has undoubtedly profited from these developments.

The wider human and environmental costs of high-output agriculture are only now being felt and assessed. These include a decline in the number of small to medium-sized farms, widespread job losses among the agricultural workforce, as well as serious water pollution and damage to wildlife. The cost to the taxpayer of subsidising over-production is also high. Under the EC Common Agricultural Policy (CAP) only one-third of the subsidies go direct to the farmers and growers while two-thirds of all the money spent goes on storage and destruction of surplus food.

It is regrettable that industry has not researched the impact of increased use of Integrated Pest Management techniques or organic farming methods on the need for pesticide applications. Pesticide use has to be considered within the context of the type of agriculture practised and the possibilities of developing more economic, safer, and less environmentally harmful methods of food production. Pesticide use should be justified only where there are no alternative methods of pest control available and where the benefits outweigh the costs of using these hazardous chemicals. In fact, there have been very few scientific studies on the benefits of pesticide use.

Disease Control and Public Health

A wide range of animals act as carriers (vectors) in the transmission of disease from humans to humans and from animals to humans. Most diseases are carried and spread by arthropods (insects and their close relations).[5] Arthropod means 'jointed leg', and these creatures are all characterised by an external skeleton and jointed appendages, including legs, mouth parts and antennae. Arthropod species such as lice, mites, mosquitoes and ticks act as vectors of bacterial, protozoal and viral diseases. Other species such as the scabies mite, and bedbugs, are external human parasites, but not normally vectors of disease.

As well as transmitting diseases from humans to humans, arthropods also transmit what are primarily animal diseases, from animals to humans. Diseases capable of being transferred from animals to humans are referred to as 'zoonoses' and examples include rabies, brucellosis, and leptospirosis (Weill's disease). Infection is usually by bites or stings, though transmission by mechanical means, for example on the feet of flies, beetles and cockroaches, also occurs.[6]

In the developing and warmer countries of the world, arthropod-borne diseases are a considerable problem and pesticides often play a central role in public health programmes. Urbanisation, including the spread of shanty towns, is happening rapidly in many of these countries. The consequent high population density, cramped living conditions, and inadequate sanitary facilities result in a highly polluted environment. This background is ideal for the proliferation and spread of disease by arthropods and rodents.

Mosquitoes spread diseases which include malaria, dengue, yellow fever, encephalitis and rift valley fever. Malaria control was the first eradication programme to be undertaken on a truly global scale. Insecticides such as dichlorodiphenyltrichloroethane (DDT), dieldrin and benzene hexachloride (HCH) were primarily used, being sprayed on marshy areas and on the interior walls of dwellings. When sprayed on walls at the proper dose, the insecticides, which were relatively cheap, remained lethal to the mosquitoes for about six months.

At first the programme seemed successful and malaria was being controlled; by 1962 eradication involved 65 countries and a total of 737 million people. About this time problems began to appear. Many mosquito species developed growing resistance to DDT and the other insecticides. These insecticides also proved to be persistent, killed non-target species and had unwanted side effects on human health. The global malaria eradication programme broke down. As a result of this debacle, it is now standard practice to manage the disease with pesticides as only one part of a public health programme, not as the 'miracle' ingredient.

Rodent-borne diseases include leptospirosis, salmonellosis, plague, typhus and haemorrhagic fever. Mice and rats are carriers of these diseases. Specialist pesticides (rodenticides), usually in the form of baits, are used to kill the animals, stopping the transmission of the disease. Rodenticides are commonly used in and around food shops, stores, restaurants, food factories and similar premises, as well as in drains and sewers.

Fumigation control, using pesticides in gaseous form, is widely used to control fungal and insect pests in bulk food and commodity storage premises, such as silos and warehouses, ship and aircraft holds and container lorries. Many of the fumigants can be applied by the general professional pesticide user, for example to fumigate grain bins on the farm with the organophosphorus pesticide pirimiphos-methyl. Other more highly toxic pesticide fumigants such as ethylene oxide, hydrogen cyanide, methyl bromide and even phosphine are applied by specialist fumigation contractors. The treated areas are sealed off and often sheeted for an extended period. These fumigants are 'total' and will kill off all fungal, bacterial, arthropod and rodent species as well as being highly toxic to humans.

Once again pesticide use in disease control is only justified if there are no alternative pest control methods and the benefits outweigh the costs. Very little independent cost-benefit analysis of pesticide use has been carried out, especially in the area of disease control and public health. Pesticide use should be only one factor in integrated public health programmes for disease prevention and control.

Preservation and Maintenance

Pesticide products used in the preservation and maintenance of land-based and marine structures, buildings and equipment include wood preservatives, anti-fouling paints, and surface/masonry biocides. As well as being applied by the professional pesticide user, there is also a lucrative and expanding do-it-yourself (DIY) market, and the home and garden user can buy many of the products available to professional users.

Wood preservatives are the largest group of pesticides in this category and contain fungicides and/or insecticides to control woodboring insects and fungal wood rots. They are used in the pre-treatment of timber in door and window frames, joists, rafters and fence posts, or as remedial treatments on buildings. Pre-treatment of outdoor and indoor timbers is generally carried out in a manufacturing plant with permanent machinery and some degree of automation. The pesticide is often 'injected' into the wood under pressure so that it penetrates to the heart wood to give lasting protection. Fuller details of types of wood preservative are given in Chapter 8.

Remedial treatment of timber in buildings is aimed mainly at eradicating 'woodworm', which is a collective term for the larvae of the common furniture beetle, death watch beetle, house-longhorn beetle, weevils (often found in conjunction with fungal decay in timber) and other less common woodboring insects. Under normal conditions the larval stage of these insects can last from one to ten years, so persistent, slow-release pesticides are used for long-lasting control. Persistent materials are also used for control of dry and wet fungal rots in timber. In hotter countries termites are the major woodboring pests, and ter-miticides are widely used to control them.

In the UK, some 150,000 premises are remedially treated with wood preservatives each year.[7] The treatments are mostly carried out by pest control companies, or smaller contractors who advertise as woodworm and dry-rot specialists. There are no formal entry qualifi-cations for this trade, and only limited training requirements. In practice anyone can set up as a specialist and, while organisations such as the British Pest Control Association attempt to introduce codes of practice and standards for the industry, the quality of 'specialists' is highly variable. Choosing a reputable company is important to ensure that the job is done properly, and to avoid babies, children, families and schoolchildren being directly exposed to the wood preservatives. Some of the persistent products in widespread use have been linked to health problems. Pentachlorophenol (PCP) is suspected of causing a variety of cancers.[8] Lindane is a known cause of aplastic anaemia, a rare and often fatal blood disorder.[9] Approval for use of tributyltin oxide (TBTO) as an anti-fouling paint on marine timbers and boats was withdrawn after scientists discovered it caused sexual deformities in dog whelks, a type of shellfish. There is also an important DIY market for wood preservatives, surface biocides and the like. The DIY individual is directly exposed to these chemicals, so good information and the correct choice of materials is important to minimise risks. The US Environmental Protection Agency (EPA) in its booklet, *Citizens' Guide to Pesticides*, explains how to choose a pest control company.[10] Advice of a similar nature is not available from the UK pesticide authorities.

Wood preservatives, especially if persistent, can also kill bats roosting in buildings. All bats are protected under The Wildlife and Countryside Act 1981. Pesticide operators working on structures where bats are likely to live should consult the Nature Conservancy Council (NCC) beforehand. The protection of bats under the above Act applies to household wood preservatives as well as products intended solely for professional use. The labels of these products contain a phrase referring to The Wildlife and Countryside Act 1981 and any products containing the active ingredients gamma-HCH (lindane) and pentachlorophenol must carry the phrase 'DANGEROUS TO BATS'.[11] The NCC publishes a list of wood preservatives – mainly synthetic pyrethroid-type insecticides – which are less harmful to bats. Chapter 8 has more details.

Properly selected, dried, stored, installed and maintained wood is remarkably durable. Woodworm and fungal rots only damage wood if environmental conditions are favourable. Dry rot, for example, is misnamed – it hates well-ventilated, dry places. It needs timber moisture levels of at least 20 per cent to survive and 30–40 per cent to thrive. The main factors encouraging attack are excessive moisture, humidity and lack of ventilation. Proper building design and maintenance is the sensible way to protect wood and property. Sound, seasoned wood which is well-fitted and maintained, in a well-designed and ventilated building presents little opportunity for insect or fungal attack. Better building design and construction is a key factor in reducing the quantities of wood preservatives used.

Modern building practices, however, often encourage the use of wood preservatives by using inferior timber. The British Wood Preserving Association (BWPA), which represents many specialist contractors, notes in a leaflet the increasing use of soft sapwood in highly stressed sections of timber in buildings. The BWPA goes on to say: 'if untreated, sapwood can be attacked by insects, or fungi then this can lead to potentially hazardous structural failure. Repair or replacement of such timbers can be both costly and inconvenient, so pre-treatment has a vital role to play.'[12]

Where are Pesticides Used?

Chemicals have been used to control pests since ancient times. There are records of the Romans using sulphur in fumigation, and of the Chinese using arsenic to control insect pests in the garden by AD 900. The chemicals employed were mainly inorganic compounds. Copper, mercury salts and sulphur were used as fungicides, while arsenic and cyanide were used as general insect poisons. Organic compounds (which contain carbon as one of their central elements) were used in the form of tar distillates or plant extracts like derris, nicotine and pyrethrum. Selectivity was largely a matter of the timing of application.

Even though some newer chemicals such as 4,6-dinitro-o-cresol (DNOC) were introduced in the 1930s, the modern pesticide industry really dates from the Second World War.[13] War provided the stimulus for the commercial development of existing chemicals and the introduction of new ones. The 1940s saw the introduction of organochlorine pesticides such as DDT, which was used extensively to control human lice in the armed forces and civilian and refugee populations. The organophosphorus pesticides were originally developed for use as human nerve gases, and then applied to commercial insect pest control after the war. Hormone-type herbicides such as 2,4-dichlorophen-

oxyacetic acid (2,4-D) and 4-chloro-o-tolyloxyacetic acid (MCPA) were
first marketed in the late 1940s.

Since 1945 the pesticide industry has developed steadily worldwide.
The number of new pesticide active ingredients and ready-formulated
products has expanded steadily. Currently, over 400 pesticide active
ingredients are approved for use in the UK, and a similar number in
EC countries, while in the US there are over 1,200 basic active
ingredients formulated into more than 30,000 ready-to-use products
and brands. The main pesticide manufacturing and exporting areas are
the US, Western Europe and Japan.

The development and growth of the pesticide industry has been
closely linked to the expansion of modern, intensive agriculture. The
chemical and agricultural industries are closely linked on a global
scale in what is often referred to as the 'agribusiness' sector. Chemical
farming, based on large-scale use of pesticides and artificial fertilisers,
has gone hand-in-hand with developments such as agricultural mechanisation and, reductions in the labour force, plant-breeding and
intensive livestock production.

Developing countries represent a major and expanding market (see
Chapter 9). In 1989 UK pesticide exports totalled £655.5 million, 60
per cent of total production, and much of this was for use in developing
countries.[14] In these countries pesticides are used in plantation, export-
orientated agriculture and by smallholders and small farmers mainly
growing crops for themselves and the local market.

While agriculture and allied industries such as forestry and horticulture
are the main market for pesticides, there are other major areas of use
such as in public health programmes. Pesticides are used in a wide range
of products and situations, many of which are not commonly known.

Agriculture, Horticulture and Forestry

Pesticide manufacture is now a global business. Commercial agriculture
is the main customer, and the more intensive the agricultural production
the greater the use of pesticides. Crop protection by controlling
diseases, insects and weeds is the prime objective, but pesticides are also
important in livestock production. Although cash crops in developed
countries and plantation crops in developing countries form the main
markets, pesticides are being increasingly used in subsistence-type
agriculture (see Chapter 9). The examples given below relate to pesticide
use in intensive agricultural systems in Europe and the US.

US

Pesticide use in the US almost doubled in 21 years, from 540 million
pounds weight of active ingredients in 1964 to over 1,000 million
pounds in 1985. Agricultural use increased from 59 per cent of the total
in 1964 to 77 per cent in 1985. In California, where approximately 250
million pounds of pesticides are used annually, 92 per cent are used
in agriculture. US farmers spent $4.6 billion on pesticides in 1985, nearly

4 per cent of their total farm production expenditure. Pesticides are used on some 2 million US farms, in 75 million households, and by 40,000 commercial pest control firms. In total, these users spent US$6.6 billion on pesticides in 1985.[15]

UK

Agricultural crops: virtually the whole land area of the UK is directly treated with pesticides at some time: 99 per cent of cereals, vegetables, fruit and general crops are routinely sprayed with one or more pesticides. Even pasture land on which dairy cows may graze is sprayed with herbicides; milk is particularly susceptible to chemical residues.[16] Arable pesticide use represents 80–90 per cent of agricultural use. In 1988/89 4 million hectares of cereal were grown in the UK, and 7.7 million hectares were sprayed with herbicide, showing that, on average crops were treated more than once. Fungicides are also used routinely, ranging from seed dressings to foliar sprays. In 1988/89 10.5 million hectares of cereals were treated with fungicides. Insecticide use is more variable but can be intense on some crops. Nearly 2.5 million hectares of cereal were sprayed with insecticides in 1988/89. Plant growth regulators were used on some 2.5 million hectares of cereal.[17] (1 hectare = 2.47 acres = slightly bigger than an average UK football pitch.)

Patterns of pesticide use are changing, as official government figures for arable crops show.[18] Between 1982 and 1988 the volume of pesticide use in arable crops declined while the area treated increased. Usage, measured by area treated, increased 17 per cent, from 19.7 million hectares in 1982 to 23.1 million hectares in 1988. This compares with 1977, when 10.9 million hectares were sprayed. By weight, pesticide use on cereals decreased by 6 per cent between 1982 and 1988, but this still represented an increase of 72 per cent since 1977. Figures are not available for other crops.

As the value of the crop increases so does the frequency, and often intensity, of spraying. Tank-mix applications are generally the rule, with as many as four-way mixes of different pesticides being made at the same time. Average numbers of applications per season are shown in Table 2.1, bearing in mind that each application could consist of a tank-mix of two or more pesticides:

Table 2.1: Frequency of Pesticide Applications per Season

Crop	No. of applications/season
Cereals, oilseed rape	5 to 8
Potatoes	3 to 10
Top and soft fruit	15 to 20 or more
Hops	20 or more

Source: Ministry of Agriculture, Agricultural Development and Advisory Service (ADAS), *Pesticide Usage Survey Report, 1987*, HMSO, London, 1987.

Top and soft fruit production is an area of intensive pesticide use, involving a wide range of tank-mix applications. In England and Wales, apple and pear orchards receive an average of 21 pesticide treatments during the year. It is estimated that 800 tonnes of active ingredients are applied annually to top fruit orchards. Protecting harvested and stored produce from fungal attack, insect pests, and vermin such as mice and rats increases the number of pesticide applications to a crop. For example, commercial grain stores are routinely treated with fumigant pesticides to prevent fungal and insect damage, especially from seed weevils. As they are being put into store, potatoes are often treated with plant growth regulators (classed as pesticides) which inhibit sprouting, extending the commercial life of the crop.

Horticulture uses a wide variety of pesticides, especially fungicides, insecticides and plant growth regulators as well as general fumigants and soil sterilants such as methyl bromide. These chemicals are used on produce ranging from salad and vegetable crops to flowers, seeds and potted plants. The high-value of produce, combined with the need for quality, often leads to intensive, programmed pesticide use and it is no accident that pesticide resistance problems develop most rapidly in glasshouses and polythene tunnels. The use of persistent, systemic pesticides can cause pesticide residue problems in fruit, salad crops and vegetables, especially if the chemicals are applied too close to the harvest date, which is often determined by the market price. Many glasshouse crops such as cucumbers and tomatoes are now grown using 'nutrient film techniques' where the crop is grown without soil in a liquid solution, to which systemic pesticides such as fungicides and insecticides are routinely added. Biological control programmes, as part of an 'Integrated Pest Management' (IPM) approach, are being widely introduced as pest-control techniques in intensive cropping situations (see Chapter 12).

Livestock and poultry production: pesticides play a central role in this area. 'Veterinary pesticides', ranging from general fumigants such as formaldehyde to insecticide sprays and paints, are used in and around livestock and poultry houses.[19] General vermin control with rodenticides is also widely practised. Some pesticides are applied directly to the animal or fish for insect and/or disease control and in the UK they are classed as 'animal medicines' (see Chapter 11). Pesticides used in or on the animal can leave residues which is a problem if the animal is destined for slaughter and eventual human consumption. Examples of pesticides classed as 'animal medicines' include sheep dips, lice/mange treatments and warble fly dressings on cattle, and fish lice preparations on commercially-farmed salmon, which are treated using organophosphorus-group insecticides.

Fish farming: salmon farming in Scotland is a rapidly growing industry. Production rose from 2,500 tonnes of fish in 1984 to 19,500 tonnes in 1988, and this trend is set to continue. The Department of Agriculture and Fisheries for Scotland (DAFS) forecast that the total would reach 54,000 tonnes by the early 1990s. Scientists know very little about the effects salmon farms will have on the environment/ecosystem in the long term, especially the use of chemicals in fish production. The synthetic colouring, canthaxanthin (E161g), banned in the US, is added to the diet as it makes the flesh of the fish pink like that of wild salmon, rather than 'hatchery grey'. The fish and their sea cages are kept clean with calcium oxide, chlorine, sodium hydroxide and iodophors (a solution of iodine). Disease is kept at bay with formaldehyde, malachite green and four common antibiotics. The organophosphorus insecticide dichlorvos (commercial name Nuvan) is used to kill the parasitic salmon louse, *Lepeophtheirus salmonis*. Dichlorvos is also toxic to crabs, lobsters, mussels, shrimps and other marine organisms. Its use has been linked to the sudden increase in the incidence of eye cataracts in wild fish.[20]

Forestry: tens of thousands of hectares of forestry land are treated annually with pesticides. Fenitrothion, an organophosphorus insecticide is routinely sprayed from the air for control of the pine beauty moth. Organochlorine group insecticides such as lindane and synthetic pyrethroid group insecticides like permethrin are routinely used to treat tree seedling roots prior to transplanting. Brushwood killers, as well as general herbicides, are also regularly used in forestry production.[21] Forestry workers have raised fears over links between the use of the herbicide 2,4,5-trichlorophenoxyacetic acid (2,4,5-T) and birth defects.[22]

Aerial spraying of agricultural and forestry land accounts for only 2–3 per cent of UK pesticide applications, but it attracts a great deal of attention because of the number of spray drift accidents. The 1987 Ministry of Agriculture report on aerial spraying stated that there was an increase of 8 per cent in crop area flown compared with 1986, and that 98 per cent of applications were made to cereals, oilseed rape, potatoes, peas and field beans.[23] There was a continued decrease in the application of herbicides and insecticides, but the application of fungicides and molluscicides (slug pellets) increased. The survey also revealed that 3 per cent of total aerial applications in 1987 involved the illegal use of 'pesticides which were not approved for the application in question'.

Aerial spraying is more widely used in other areas of the world for crop protection and the control of the carriers of disease. One important area of pesticide use has been for locust control. Persistent organochlorine group insecticides like dieldrin were originally used, but these were withdrawn in favour of shorter-acting, less environmentally damaging chemicals such as fenitrothion, an organophosphorus group insecticide.

In North Africa in the late 1980s the worst locust plagues for 30 years threatened harvests, in regions already hard hit by years of drought and famine. The United Nations Food and Agriculture Organisation (FAO) called for a re-examination of the use of dieldrin to control the 'hopper' stage of the locust precisely because it is more persistent than other materials and would give longer control.[24] The merits of this approach have to be weighed against dieldrin's greater potential for long-term environmental damage and effects on other forms of wildlife.

Amenity and Industrial Use

Weed control, around industrial estates, factories, docks, harbours, power stations and electricity pylons, airports, defence establishments and along railway tracks, motorways and watercourses, is big business. Long-lasting, residual herbicides are generally used. The chemicals remain active for months, even years and, by reducing the need to spray regularly, help keep costs down. There is evidence that this use of the residual herbicides is a major source of water pollution. Atrazine and simazine, two residual herbicides used along railway lines and to clear weed from pavements, are major pollutants of drinking water in the UK (see Chapter 6).[25]

Pesticides, particularly herbicides, are regularly used to control weeds on pavements, roadside verges and embankments, in parks and gardens and on open spaces on housing estates. Playing fields, including school grounds and leisure and sporting facilities such as golf courses, tennis courts, and bowling greens are routinely treated with herbicides, and even fungicides and insecticides. The British Agrochemicals Association (BAA) leaflet, *Chemicals and the Amenity User*, states: 'pesticides are cost-effective tools for use in amenity areas, on roadways and footpaths for reasons of safety, efficiency, aesthetic value, public health, to meet legal requirements and economics.'[26]

In 1989, some 550 tonnes of pesticide active ingredients were used for non-agricultural weed control in England and Wales.[27] Nine active ingredients accounted for almost 90% of this figure: atrazine (25%); simazine (14%); diuron (12%); 2,4-D (9%); mecoprop (8%); amitrole (7%); glyphosate (5%); sodium chlorate (4%) and MCPA (4%). Non-agricultural herbicide consumption was distributed as follows: the power and industry sector 34%; local authorities 33%; transport 21%; water companies 7% and golf and leisure facilities 5%.

Use in Manufactured Products

Manufactured products, many of them used or found in the home, also contain pesticides, although this fact is often not stated on the product label. Even if we consciously choose not to use pesticides, the chemicals are often in the things we buy. Some examples of ordinary, everyday items containing pesticides are: wallpaper pastes, carpets and textiles, masonry treatments, pet delousing products, fly sprays, wood preservatives, marine anti-fouling paints and medicinal shampoos.

Pure wool carpets often contain pesticides to protect against fungal and insect attack. Two of the main pesticides used are a synthetic pyrethroid insecticide, permethrin, and a halogenated (chlorinated) diphenyl urea derivative (trade name Mitin FF), which is used specifically as a moth-proofing agent. The wool industry claims that both pesticides bind to wool fibres so that only tiny amounts come into contact with the skin. Other pesticides used as general biocides in textile production include pentachlorophenol, pentachlorophenol laurate and phenyl mercuric acetate.[28] The US Department of Agriculture has found that the pesticide active ingredient albamectin – currently used on cotton, oranges and tomatoes in the US – could help protect woollen carpets, furniture and clothing from moths and beetles. Scientists have found that the pesticide keeps moths and beetles off fabrics for five years or more while, by contrast, mothballs made from naphthalene only protect clothes for a year. Tests are currently being conducted to establish if albamectin is safe on fabrics that touch the skin.[29]

Pesticides are also used in medicines. Warfarin, an anti-coagulant rodenticide, is used to prevent blood clotting in patients with heart conditions and those recovering from surgery. Insecticides such as carbaryl, lindane and malathion are ingredients in medicinal hair shampoos for the control of human head lice and scabies. Shampoos containing pesticides can be dangerous to young children if swallowed. Keeping young children away from pesticides is a widespread problem. As part of its public information services, the US Environmental Protection Agency publishes a special booklet entitled *Child-Resistant Packages for Pesticides*.[30]

Other manufactured industrial products such as, for example, heavy-duty rubber-coated cable contain pesticides to protect against mice, rat and termite attack.

Public and Environmental Health

A major area of pesticide use is in public and environmental health programmes to control vectors of human and animal disease, to protect food and commodity storage facilities, and to maintain hygiene in urban and rural areas. Pesticides play a significant role in public health in the warmer regions of the world, especially in disease prevention programmes such as malaria control.[31] In these countries the range and volume of pesticides used can even exceed the amount used in agriculture. Some of the principal public and environmental health uses include:

- Control of tropical diseases such as malaria, schistosomiasis and filariasis. In malaria control, insecticides are used to kill the water-borne larvae of mosquitoes which carry the disease.

- General insect and vermin control in public places, open spaces and especially in and around food shops and stores, cafes, pubs, restaurants, schools and drains and sewers.
- The use of bird repellents to protect buildings from defacement by the likes of pigeons and starlings.
- Public health control in buildings. In addition to general vermin control, wood preservatives are widely used to prevent structural damage. Biocides are used in factory and office water tanks to kill micro-organisms which cause infections such as Legionnaires disease.
- Pesticides are used on aircraft for insect and rodent control. Many commercial airlines use pesticides in, and around, galley and food handling areas. The chemicals typically used are synthetic pyrethroid insecticides and the carbamate insecticide, bendiocarb. Chemicals are applied as wettable powders on walls, around cupboards and doors and, on drying, leave a residue which target species such as cockroaches pick up on their feet; eventually the insects accumulate a lethal dose. Treatments are carried out every six to eight weeks.

If rodents are found or seen on board an aircraft then the whole plane is fumigated. This is because rodents pose both a safety and a health hazard as they can gnaw through cables, wire and even aluminium sheeting. The aircraft is towed to a safe location where a safety barrier is put around it. Specialist contractors then seal off all exits and pump the fumigant, methyl bromide, as a gas into the aircraft where it is left for four hours or so, after which the plane is fully ventilated before returning to active service.

Passengers and aircrew travelling to or from malaria-infested parts of the world are directly sprayed with pesticides to kill off any mosquitoes on board. Trained crew members walk up and down the aisles spraying from single shot pesticide aerosols which deliver a fixed amount of spray. The technique is often referred to as 'block away' treatment. Pyrethroid insecticides, such as difenothrin, are used increasingly. More toxic materials have been used in the past and, indeed, are still mandatory in certain countries. The procedure is regulated by international health rules laid down by the World Health Organisation. Passengers are usually unaware of what is taking place and are given no information on possible hazards.

Marine and Aquatic Pesticides

Pesticides are used as marine anti-fouling paints and surface coatings on ships' hulls, on underwater structures – jetties, piers, and oil-rig platforms – and on nets, floats and other apparatus used in fish farming. The use of pesticides as anti-fouling paints has been linked to commercial damage to oyster and mussel beds as well as general pollution of the marine environment. The use of the pesticide active ingredient, tributyltin oxide (TBTO), has been withdrawn in Britain as a result of this damage.[32]

Herbicides play an important role in aquatic weed control in many countries, including the UK where some 15 products are approved for direct application to water (see Chapter 6).

Direct application of insecticides to water for control of insect and insect-type larvae is particularly important in tropical and sub-tropical countries for disease control. Pesticides are also used to control certain nuisance fish species; these chemicals are called piscicides. No pesticides are approved for this use in Britain.

Garden and Household Use

British gardeners and domestic users can choose from a range of over 600 pesticide products, containing 100 or so different active ingredients (see Chapter 8) and applied at a rate of some million kilos per year. These are listed in the BAA *Directory of Garden Chemicals 1989/90*, which is revised periodically.[33]

Many of the chemicals approved for use are irritants, and suspected carcinogens and teratogens, and are poisonous to fish, wildlife, pets and domestic animals.

The pesticide industry argues that the quantity and concentration of garden and home chemicals used are too small to matter. Others disagree. Norway and Sweden, for example, have banned or restricted some chemicals available in Britain.[34]

Pesticide Exports and Developing Countries

Developing countries represent a major and expanding market for all pesticide producers, with pesticide exports accounting for over 50 per cent of UK production. The United Nations (UN) has called for tighter international legal controls over pesticide exports by manufacturing countries because of the poor human and environmental health and safety record of pesticides in these Third World countries (see Chapter 9).

Who Makes Pesticides?

Pesticides are big business. Global sales exceed US$20,000 million a year, and are still growing. Twelve companies account for over 70 per cent of all pesticide sales. These companies are all major chemical transnationals based in Europe and the US, as Table 2.2 shows, and most rank among the world's top chemical producers.

Table 2.2: 20 Top Agrochemical Companies by Sales in 1989

Company	Country	Sales (US$ million)
Ciba-Geigy	Switzerland	2,271
ICI	UK	2,026
Bayer	West Germany	1,862
DuPont	USA France	1,684
Rhone Poulenc	France	1,646
Monsanto	US	1,558
Dow-Elanco	US	1,485
Hoechst	West Germany	1,090
BASF	West Germany	1,032
Shell	UK/Netherlands	903
American Cyanamid	US	820
Schering	West Germany	740
Sandoz	Switzerland	713
Kumiai	Japan	420
FMC	US	414
Rohm and Haas	UK	367
Sankyo	Japan	344
Nihon Nohyaku	Japan	268
Takeda	Japan	265
Hokko	Japan	265

Source: *Agrow*, No. 112, 1 June 1990.

The UK is one of the world's major pesticide manufacturing and exporting countries and it provides an important base for pesticide research, as most of the major companies have strategic research facilities here. Pesticide manufacturers tend to locate their plants in their home base and near major markets. The pesticide active ingredients are only made at a small number of locations, and are formulated into the required mixture in formulation plants around the world. Sometimes the product will be further processed, or packed into smaller containers when it reaches the end-use market.

For example, Dow-Elanco's main insecticide, chlorpyrifos, is only made at King's Lynn in the UK and Midland, Michigan, in the US. It is formulated at these and other plants, and then shipped to markets all over the world. ICI makes paraquat, the world's second largest-selling herbicide, in only five factories, located in the UK, US, Japan, Brazil and India. It also formulates in Malaysia, and from these six plants it ships to over 130 countries.

The UK transnational, ICI, is among the top three pesticide manufacturers along with the Anglo-Dutch company, Shell. Shell is the one of the leading manufacturers of organochlorine pesticides, which are increasingly under attack because they degrade slowly in the environment.

Of the top 12 pesticide companies, 9 have either active ingredient or formulation facilities in the UK: Ciba-Geigy; ICI; Rhone Poulenc;

Monsanto; Shell; Hoechst; Dow; Schering, and American Cyanamid. Other important producers in the UK are Fisons; A.H. Marks; Wellcome Foundation; Mitchell Cotts, and Rio Tinto Zinc (RTZ). Between them, pesticide production companies in the UK have a workforce of 8,700.

Most pesticides produced in the UK are exported, making it one of the top three pesticide exporting countries in the world. Some 45 per cent of pesticides go to other EC countries and over 30 per cent go to Third World markets.[35]

Safety in Manufacturing

Making pesticides can be a dirty business. A description of the early days of paraquat production at ICI's main plant in Widnes, Cheshire, told how the active ingredient, bipyridyl, and tar-like waste were poured molten from the stills into open drums to wait until the bipyridyl solidified. It was then broken up with sledge hammers, exposing process operators to fumes and dust.[36]

Investment in more sophisticated technology, and advancement in health and safety regulations, mean that these crude forms of production have disappeared in industrialised countries. Where pesticides are manufactured by local companies in Third World countries, however, these conditions may still be common. In spite of advances in technology, pesticide production is still not necessarily safe, and the processes involved are often dangerous. The world's worst industrial disaster at Bhopal, India, in 1984 occurred when a leak of lethal methyl isocyanate gas escaped from the Union Carbide pesticide plant, killing over 2,600 people and disabling more than 100,000. In 1986, a fire at the Swiss pesticide plant of Sandoz in Basle released pesticides and mercury into the Rhine, killing much of the river life.

On a day-to-day level, toxic wastes from pesticide production processes are released into air and water. Some of the very hazardous waste is placed in containers and transported to landfill sites. Companies monitor their discharges, and these are policed by HM Inspectorate of Pollution and the Health and Safety Executive.

Increasing concern for the environment has brought more awareness of the need to further reduce emissions. European regulations following the leak of dioxin at Seveso have helped to lessen plant pollution, and in the UK this is enacted under the Control of Industrial and Major Accident Hazards Regulations (CIMAH), which apply to some pesticide plants. Under CIMAH, certain hazardous chemicals which are stored in large quantities must be registered, the local authority notified, evacuation plans drawn up in case of accident, and local residents informed of the dangers and the escape plans.

Research and Development

The UK is a popular base for research, with ICI, Rhone Poulenc, Dow Chemicals, Schering and Shell maintaining large laboratories there. Some 26 per cent of those employed by pesticide companies in the UK work in the research field.

Pesticide companies both respond to the market and influence it through the products they develop. One of the main trends is for products with greater biological activity, leading to dramatic reductions in the quantity of active ingredient (but not the activity) applied per hectare – sometimes as low as a few ounces. These reductions mean significant savings for the manufacturers in terms of storage, handling and distribution costs.

The top companies spend more than US$100 million a year on pesticide and biotechnology research. The research and development costs for a single major new product are typically US$30–50 million, with an average of 6–8 years from project inception to full commercial introduction. In this time, companies may screen 30,000 products looking for one to sell.[37] The enormous cost of product development and the huge, sophisticated laboratories needed mean that the entrance fee for pesticide producers is very high, and make it unlikely that there will be major changes in the companies dominating pesticide production – although Japanese and Korean companies are expanding.

The high cost of product development leads companies to concentrate on products where they are assured of large sales, particularly with the large cereal crops produced in, or sold to, richer countries. But a significant amount of research is low-key and aimed at sustaining sales of older pesticides. Rhone Poulenc explained this by saying that 'More and more of this money is having to be spent on what is becoming known as product defence, leaving less and less for field development and extension work.'[38] ICI's main product, the herbicide paraquat, is now almost 30 years old and research continues to find new applications.

Agrochemical companies once thought that biotechnology, or genetic engineering, might replace pesticides. However, companies increasingly use it in a complementary manner to develop plants which are resistant to their pesticides, rather than to the pests themselves. Monsanto, with patent rights on the world's leading herbicide, Roundup (active ingredient glyphosate) has used genetic engineering to develop tomato plants tolerant to Roundup.

What are the Concerns?

Pesticides are among the few toxic materials deliberately added to the environment and widely sold for both professional, and home and garden use. Their use in agriculture, horticulture and forestry is a major source of involuntary exposure of the general public. The public is exposed to pesticides through contamination of fresh and processed food by persistent chemical residues most of which cannot be washed off, or degraded by cooking. Workforce exposure is even more direct. Pesticide production and use has to be considered in the context of agricultural over-production in the richer parts of the world such as Europe and the US, and the lack of controls on pesticide use in developing countries.

Traditionally, pesticides have been the province of government and industry. However, other constituencies are now involved. Alternative views on the benefits and disadvantages of pesticides are held by many scientists, environmental organisations, consumer and world development groups, and trades unions.

Different Viewpoints

Expert scientific opinion does not always support the pesticide industry argument that pesticides are essential for increased food production. Leading US scientists, for example, have suggested that from 1945 to 1974, while pesticide usage in the US increased tenfold, pre-harvest losses rose from 7 to 13 per cent.[39]

Environmental groups such as Friends of the Earth have highlighted the over-use of pesticides, linked to over-production and the problem of residues in food and water.[40] Greenpeace, in particular, have emphasised the transport and manufacturing hazards of pesticides as well as disposal problems, especially as pesticides are major items in the toxic waste trade.[41]

Countryside groups such as the Ramblers' Association stress that no one who walks in the countryside should underestimate the dangers of pesticides: 'The sheer volume of pesticides applied to crops throughout the year makes it doubly difficult for walkers to avoid contact with these chemicals.'[42]

Trades unions such as the Transport and General Workers' Union (TGWU), the General Municipal and Boilermakers' Union (GMB) and the National Union of Public Employees (NUPE) have also highlighted the over-use of pesticides and the hazards they pose not only to workers but also to the public and environment. In 1987 these three trades unions published the results of their own pesticide-user survey, entitled *Pesticides: The Hidden Peril*, which documented the hazards to the workforce.[43] The unions also stressed the need for proper legal controls and enforcement of pesticide regulations and emphasised the needless secrecy surrounding the approval of these chemicals. In 1991 they plan to undertake another major campaign to improve the worker, public and environmental safety of pesticides.

Consumer groups such as the International Organisation of Consumer Unions (IOCU) have expressed concern over pesticide residue levels in food. IOCU states that of the 3,350 known pesticide active ingredients worldwide, information to make even a partial assessment of the hazards to humans is only available for 34 per cent and it is campaigning to prevent substances banned in Western countries from being exported to countries in the developing world.[44] The UK Consumers' Association, while recognising that consumers have benefited from a greater range of unblemished foods available throughout the year, has highlighted the aggressive marketing of pesticides as a factor in contributing to the building up of the infamous EC food mountains.[45]

World development groups such as Oxfam point out that much of the developing world uses pesticides to produce export-orientated plantation crops, and not for subsistence agriculture. Pesticides are, therefore, often of limited value to local populations. Oxfam also argues for export controls to developing countries to help reduce the number of easily avoidable deaths and poisonings from pesticides.[46]

The position of bodies like the UK National Farmers' Union (NFU) and County Landowners' Association is harder to assess. In the past they have argued that pesticides were a 'necessary evil' for high agricultural output. Their views are changing with the need to cut back on surplus agricultural production within the EC. However, at the 1990 NFU Annual Conference, farmers passed a resolution backing the continued use of chemicals: 'A succession of speakers at the 1990 NFU meeting dismissed hopes of a large-scale expansion in organic produce. They said that abandoning pesticides would result in food shortages.'[47] Recently, more critical voices have been heard from bodies such as the Small Farmers' Association and the British Organic Farmers' Association.[48]

The British Medical Association (BMA) has reviewed the safety of pesticides to humans and the environment, with special reference to residues in food and water. Their *Pesticides, Chemicals and Health Report*, published in 1991, arose out of the concern of doctors over their profession's lack of knowledge of, and training on, pesticide hazards.[49]

The UK Pesticides Trust is an independent body which provides a forum for discussion of pesticide issues and helps coordinate action and collate information. In particular, the Trust has highlighted the over-use and misuse of pesticides, the lack of public access to health and safety data, the need to develop alternatives, and the lack of controls over the import and export of pesticides.[50]

The Pesticide Action Network (PAN) is a worldwide coalition of citizens' groups and individuals who campaign against the misuse of pesticides. Formed in 1982, PAN now unites over 300 organisations in some 50 countries. PAN aims to raise public awareness about pesticide abuse by campaigning against particularly toxic pesticides as well as unethical corporate marketing practices. It seeks to promote alternatives to chemical pesticides and to encourage effective policies on the manufacture, distribution and use of pesticides.[51]

Specific Areas and Issues of Concern

Global Presence in Humans and the Environment

Pesticide contamination is now global, as monitoring by the US Environmental Protection Agency (EPA) and other regulatory organisations has shown. Pesticide residues are now found in the snow caps of the highest mountains and the arctic and antarctic ice packs. A California study has shown that pesticides even concentrate in fog.[52]

Humans are contaminated with pesticides, especially their fatty tissue which stores persistent chemicals, and slowly releases them into the bloodstream. Many studies have demonstrated that breast milk is contaminated with a variety of pesticides.[53] Many pesticides cross the placenta and so newborn infants are often already contaminated before birth. In the US, for example, high levels of DDT have been found in women in the rural south,[54] while heptachlor, a persistent organochlorine insecticide, was found at levels above the EPA safety limit in women in Hawaii.[55]

Pollution of Water Supplies

Agricultural nitrate and pesticide use, along with industrial pesticide production and use, are the major cause of contamination of ground water, much of which is used for drinking water. In the US, 50 per cent of the drinking water supply is from ground water; in rural areas it is 90 per cent or more. The US Environmental Protection Agency (EPA) estimates that in 38 states pesticides have already fouled the ground water used by some half of the population.[56] There is extensive ground water contamination in the UK and drinking water in many areas contains pesticide and nitrate levels above the EC safety level.[57]

Residues in Food

Pesticides, along with nitrates and other chemical fertilisers, also end up as unwanted residues in food. Concentrations are often low but there is little evidence or research on the long-term health effects, especially on vulnerable groups such as young children and babies. Residues in imported food can contain pesticides banned or restricted in the importing country,[58] highlighting the international nature of the pesticide trade, and the need for international controls and enforcement.

Ill-Health and Pesticides

Acute or immediate health effects of pesticide exposure range from eye and upper respiratory tract irritation and contact skin dermatitis to systemic poisoning which can lead to severe illness and even death. Lower level 'flu-like' symptoms also occur but these are often not correctly diagnosed or linked to pesticides (see Chapter 5).

The exact scale of the problem of acute ill health and pesticides is still unknown. The UK has no central incident reporting or monitoring scheme. Current reporting schemes record as few as 200 or so work-related poisoning incidents per annum, yet the real number is almost certainly higher. Even the Health and Safety Executive (HSE), which is responsible for investigating pesticide incidents involving workers and the public in Britain, acknowledges that there is up to 90 per cent under-reporting of serious accidents in agriculture by employers.[59] Many poisoning incidents are not recorded. Similarly, in the US the number of workers affected by pesticides is unknown but estimated to be up to 300,000 per year.[60] The scale of incidents affecting consumer and public health in the US is unknown, and once again there is no central incident reporting scheme.

Chronic, or long-term, health effects from pesticides, which include cancer, effects on fertility and reproduction, birth defects, neurotoxic and neuro-behavioural/neuro-psychological effects, are similarly under-researched and under-reported.

The UK Parliamentary Select Committee on Agriculture has much to say about the health risks from pesticides and access to information in its 1987 report on *Pesticides and Human Health* (see p. 27).[61]

Risks to Workers and their Families

Pesticides are a regular feature of agricultural, horticultural and forestry work. The crop sprayer is a key piece of equipment on most farms and is in use over much of the year. As well as working directly with these and other chemicals, agricultural workers are exposed to toxic pesticides from a variety of sources – including the crops they grow, harvest and store, the soil they cultivate, and spray drift. Agricultural workers and their families often live in homes surrounded by fields that are heavily and repeatedly sprayed, are likely to consume produce soon after harvesting and they may get higher pesticides residues in their foods than ordinary consumers. In the US, where pesticides are often added to irrigation water, many farmworkers must use this water for bathing and drinking. Agriculture is the only industry in which children comprise a significant part of the workforce, and workplace exposure to pesticides starts at a very young age, as infants and very young children are often taken to the fields with their parents.[62]

Pesticide manufacture and formulation workers are particularly at risk as they work with the concentrated pesticide. A recent survey of UK pesticide manufacturers by the HSE highlighted defects in the setting and monitoring of airborne concentrations of pesticides, poor ventilation, inadequate replacement of personal protective equipment, and gaps in safety training.[63] Women workers may be especially at risk as they are often employed on the packing lines. Exposure is often greater on those jobs where the liquid or powdered pesticide is poured into the can or packet.

Pesticide manufacture can also be hazardous to local communities in the event of fires, explosions or the accidental release of chemicals. The 'Seveso' accident in Italy, when deadly dioxins were released into the atmosphere following an explosion at a local pesticide plant, led to Europe-wide major hazard regulations to protect the workforce and

Committee on Agriculture Report:
The Effects of Pesticides on Human Health, 1987

Concern has now penetrated deeply into the medical and scientific establishment and the government as the following recommendations from the UK Parliamentary Select Committee on Agriculture's Report: *Pesticides and Human Health*, shows:

Having studied the detailed memoranda of the Ministry of Agriculture and heard oral evidence from its officials twice, we have concluded that anxiety can no longer be allayed by merely stating that no harmful effects have been observed in certain pesticides and that therefore they are safe. Those responsible for their clearance must convince the public that they have the resources, knowledge and independence of judgement to investigate potential risk to human health from pesticide use and they must do this in a more open way.

We are not convinced that MAFF has responded seriously enough to the need for a greater effort to be made in the area of reviewing the toxicological data for older pesticides, approved under conditions more lax than today.

In view of the undoubted public concern about possible chronic health effects of pesticide use, we find the lack of epidemiological research into the health of agricultural workers quite unsatisfactory and urge greater effort to be made in this area by the responsible public authorities.

We believe that the public should have knowledge about the potential long-term, as well as acute, effects of the pesticides it intends to use.

We conclude from our extensive discussions in North America that an opening up of access to health and safety data is perfectly feasible. We urge the UK government to take much more of a lead than it has hitherto done in encouraging industry along a path we did not find it unwilling to tread. We also recommended that some systematic programme for the release of health and safety data on pesticides will have to be devised. Merely leaving release of information to chance review in those cases is entirely unsatisfactory.

We recommend that the duty of overall control and coordination of pesticides be transferred from the Ministry of Agriculture to the Health and Safety Executive (HSE) and that the HSE be given adequate resources to ensure that pesticide use is properly monitored and that all regulations are properly enforced.

local community.[64] In developing countries pesticide plants are set up because of nearness to markets, cheaper labour costs, and weaker health and safety/environmental controls.

Pesticide transport in bulk road tankers, and as packages and containers can also be hazardous and there are regular spillages which have to be dealt with by the emergency services. Pesticide production leads to the creation of toxic wastes and safe disposal of such wastes is now a major domestic as well as international problem.

Adequacy of Toxicity Testing

The main method of toxicological testing for short- and long-term toxicity of pesticides, drugs and other chemicals is the bioassay test on live animals, fish, insects and other organisms. The main procedure, known as the Lethal Dose 50 (LD 50) test or Lethal Concentration 50 (LC 50) test (for aquatic organisms), checks the effects of a specific dosage which it is estimated will cause a 50 per cent death rate in laboratory animals and organisms. It is a measure of immediate or 'acute' toxicity. Feeding studies on rats, mice and rabbits may also be carried out to check for longer-term health effects.

The relevance and accuracy of using animal tests to set human safety limits is questioned by many scientists, especially where chronic health effects are involved. Some of the tests are of doubtful scientific value as laboratory animals do not necessarily react to a particular substance in the same way as humans. For example, chemicals which are irritating to humans do not always affect another animal species, and vice versa.

One of the commonest experiments is the Draize Test, which has been used since 1944 to assess whether liquids, including pesticides, are irritating to the eyes. The substance under test is sprayed into one eye of an albino rabbit. These animals are chosen because they do not possess tear ducts, and therefore cannot get rid of any irritating liquid in the eye. The condition of the two eyes is compared for up to seven days. Generally, no pain relief is provided, and the rabbits can suffer severe discomfort and pain.

Draize Test results frequently vary a great deal between different laboratories. Weill and Scala wrote as long ago as 1971 that: 'The rabbit eye [and skin] procedures currently recommended by the Federal Agencies ... should not be recommended as standard procedures in any new regulations. Without careful re-education these tests result in unreliable results.'

The development of new pesticide formulations involves large-scale use of laboratory animals for safety testing. Sometimes animal tests are continued primarily for economic reasons. Reliance on pesticides, and the race to develop new formulations, results in very large numbers of animals being subjected to undignified and inhumane experiments. In other cases, more humane alternatives, such as testing of bacterial cultures, already exist. There is a great deal of replication of toxicity testing on a worldwide scale, involving unnecessary suffering and death for a wide variety of test species. An international or even EC-wide approval and testing scheme for pesticides would cut out much of this unnecessary replication, waste and suffering. The European Commission has announced that it is to drop the need for LD 50 animal tests to classify and register dangerous chemicals; the Commission has agreed to accept data from a different source, known as the fixed-dose procedure.[65] The fixed-dose procedure only requires a small number of test animals, and analysts can finish the evaluation without animals having to die to obtain valid results.

Access to Information

There are genuine concerns over the reliability and validity of acute and chronic toxicity tests carried out on pesticides. This concern is heightened because the pre-market registration and approval of pesticides, or re-approval/review, is shrouded in secrecy in the UK. The public does not have the legal right to see the basic toxicity data or the safety standards on which approval or withdrawal (revocation) of approval is based, and no independent screening of data is possible. Most of the 30,000 to 50,000 products on sale in the UK have not been tested for long-term health effects. Many pesticides in current use would not meet modern safety testing requirements for cancer or birth defects, yet they are still widely sold.

Lack of Independent Research and Monitoring

There has been too little research into the costs and benefits of pesticide use, and the long-term health effects of pesticides, and too few epidemiological studies to research the link between exposure to pesticides and the incidence of a disease.

Responsibility for Pesticides

In the UK the Ministry of Agriculture, Fisheries and Food (MAFF) has a central role in approving pesticides for use. MAFF is also responsible for encouraging food production. There is a conflict of interest in this dual role. It would appear that its main task is the protection and assistance of farmers' interests; very little policing is done.

Enforcement of Laws

Enforcement of laws and regulations is often lacking. In Britain there are about 160 agricultural health and safety inspectors employed by the Health and Safety Executive (HSE) covering some 400,000 premises and a workforce of 750,000 (including self-employed). Routine visits to labour-employing farms take place as infrequently as once every five to eight years. Self-employed farmers, who can be major users of pesticides, can expect a routine visit from an inspector once every 30 years.[66] This is in an industry where the Health and Safety Executive (HSE) acknowledges there is up to 90 per cent under-reporting of serious agricultural accidents and injuries let alone pesticide poisonings and incidents.

In the US, concerned about weak enforcement and conflicts of interest, the US Congress wrested pesticide regulation from the US Department of Agriculture in 1970. Congress handed the job to the US EPA which, in turn, passed it on to the individual states, where most governors gave it to their agriculture department, again raising the problem of poor policing of standards.

Even in California, which the EPA considers the toughest State for regulation, most growers and large agribusiness firms that break pesticide laws are not prosecuted by the State Agriculture Department. The California State Department uncovered 9,287 violations in the year ending 30 June 1988, and investigated 3,122 additional reports of possible pesticide poisoning. Of the violations, only 600 resulted in fines and just 18 were referred to local district attorneys for possible legal action. Agriculture is California's major income-earner, with an annual revenue of over $15 billion, and there is a powerful agricultural lobby. During the 1980s the agricultural lobby defeated three Bills that would have shifted pesticide regulation to a State department whose sole function would be to protect public health.[67]

California's enforcement record shines compared with that of other States, according to the EPA. In the latest major study on enforcement of pesticide laws, the General Accounting Office, Congress's investigative arm, found dubious regulatory practice in a proportion of the cases it reviewed, ranging from 5 to 80 per cent depending on the State concerned. In these instances, agriculture departments either took no action against violators or simply issued a warning letter.[68]

Spray Drift

Spray drift and spray dispersal is a major problem given current pesticide application technology (see Chapter 4). Only some 10–15 per cent of applied pesticides reach the target pests. The remaining 85–90 per cent is dispersed off-target to air, soil and water through drift, volatilization and run-off. Significant concentrations of most pesticides applied by air or ground sprayer can drift downwind for up to a mile or more, even under the best wind conditions. Communities next to agricultural land are most at risk from pesticide drift. While some episodes of illness have been reported in these communities, the problem is essentially poorly documented.

Effects on Wildlife

Pesticide use in agriculture in the US, UK and Western Europe began in the middle to late 40s. By the mid-1950s evidence of widespread contamination of wildlife could no longer be ignored. The organochlorine insecticides, for example, were responsible for the decline of many predatory bird species through a combination of direct poisoning, food chain poisoning and effects on reproduction (see Chapter 7).

Use of Non-renewable Resources

Energy is another area of concern as synthetic pesticides are made from petrochemicals, using oil as the basic raw material. Oil is a non-renewable resource: once used it cannot be replaced. It has been estimated that the total amount of fossil fuel energy used in crop production, either directly in machinery or indirectly as fertilisers and pesticides, is 2.2 per cent of total energy usage in the US and 2.6 per cent in the UK.[69]

As concern over the environment grows, pesticide sales are likely to be subjected to 'green taxation' as part of a 'carbon taxation' to reduce waste of energy. Food produced using chemical pesticides and fertilisers would be more expensive as a result.

Over-use

Pesticides can create pest problems as well as solving them. Many insects, fungal and bacterial diseases and weeds are now resistant, in whole or part, to a wide variety of pesticides, limiting the effectiveness and useful life of many chemicals. The growing, worldwide problem of pesticide resistance is the major factor behind the development of integrated pest management programmes and alternative practices (see Chapter 12).

Pesticides can cause secondary pests to become major pest species. As one pest species is successfully controlled, a previously minor or secondary pest species, no longer held back by the previously dominant species, comes to take its place. The minor pest becomes a major pest in its own right and more and different pesticides have to be found to control it. As many pesticides are not selective, beneficial predators are killed off by the chemicals and natural biological control is reduced, so that pest populations often multiply unchecked.

Agricultural Dependence on Chemical Inputs

Pesticide use can encourage or lead to pesticide dependence. As resistant pest populations develop and natural controls are reduced, farmers and growers can be trapped into using higher dose rates, more tank mixes, more frequent applications and newer pesticide active ingredients to achieve the same, or even reduced levels of control. As the UK Royal Commission on Environmental Pollution noted: 'Farmers feel themselves to be on a treadmill with regards to pesticide usage, compelled by circumstances to depend on chemicals to an extent that they as countrymen, intuitively find disturbing.'[70] This pesticide 'treadmill effect' has been well documented in other countries (see Chapter 12).[71]

Pesticides in the Third World

The great majority of injuries and deaths from pesticide poisoning occur in the Third World, despite the fact that the industrialised countries use far more pesticides overall. There are several reasons for this: poor worker training; operators unable to read instructions on the label, either through illiteracy or because they are printed in the wrong language; a climate which makes wearing protective clothes almost impossible, even supposing the workers could afford to buy them, and antiquated spraying machinery. Also, the impoverished conditions in which many families live mean that farmers are forced to continue to use pesticides to produce larger quantities of food to compete with richer neighbours in selling a cash crop, even when they realise that the chemicals are causing them ill-health (see Chapter 9).

A Pesticides Policy

In response to the problems associated with pesticides, many countries such as Denmark, the Netherlands and Sweden now have policies and programmes to reduce pesticide usage. Sweden, for example, aims to reduce pesticide use by 50 per cent or more during the 1990s.[72]

In the UK, the lack of a national, government-initiated and coordinated Pesticides Policy, based on reducing pesticide use wherever possible, is a major weakness in regulating pesticides and encouraging the development of alternatives. This is true at the EC and international levels as well.

A Pesticides Policy would create the framework within which the central issues associated with pesticides can be tackled. These issues can be placed into two main categories, namely:

(1) Approval, control and regulation of Pesticides – specific issues to be dealt with under this heading include: toxicity testing and research information; approval and registration and re-approval/re-testing; operator/workforce protection; wildlife protection; export controls, and pollution controls.
(2) Development and implementation of alternatives to pesticides – alternatives can be categorised under three broad headings: Integrated Pest Management (IPM) systems; sustainable agriculture, and organic agriculture.

A Pesticides Policy would be based on properly researched and conducted cost-benefit audits weighing the merits of chemical control against risks to human and environmental health. For such a policy to operate effectively there would need to be greater freedom of access to toxicity data and economic and social data.

Until such Pesticides Policies are drawn up and implemented, following the example of countries such as the Netherlands, Denmark and Sweden, a rational, science-based assessment of the future role of pesticides in food production and public health is difficult to make. Pesticides Policies and reduction strategies are discussed more fully in Chapter 12.

3

Information on Pesticides

The range of pesticide products is often bewildering, especially as the same chemical is often referred to by a variety of terms. This chapter looks at some of the main ways of classifying pesticides and examines what is in a can or packet. We look at labelling, product safety data sheets and other sources of information which are vital if you want to find out about a pesticide.

Formulation: What's in a Container or Packet?

Pesticides come in a variety of formulations depending on factors such as the nature of the target species, the persistence desired, and ease of application, to name but a few.[1] Different mixtures or formulations include:

- dry dusts: used mainly in countries where water is scarce
- granules: larger particles used to avoid spray drift from liquid droplets and dusts, now sometimes obligatory for very toxic chemicals
- dry baits: scattered in an appetising form to attract pests to eat poison, for example slug pellets
- wettable powders: diluted to use in a sprayer, with either water or (less commonly) oil, to form a suspension
- emulsions: liquids ready to be diluted, including emulsions of oil-in-water and water-in-oil
- emulsifiable concentrate (ec): a homogeneous liquid formulation which forms an emulsion on mixing with water
- suspension concentrate (sc): a stable suspension of finely ground active ingredient(s) in a fluid intended for dilution before use
- ultra-low volume (ULV): formulations for spraying in a concentrated form in small droplets, usually with an oil-based diluant that resists evaporation
- smokes: pesticides which are burnt in a confined space (such as a glasshouse) and release poisons into the air
- slow release strips and papers: designed for specialist use in livestock houses, food stores and for fly control
- aerosols: used in closed environments or for direct application to animals.[2]

To formulate a ready-to-use product the pesticide active ingredient has to be mixed with a variety of other chemicals, many of which can be hazardous substances in their own right. A container or packet of pesticides is already a chemical cocktail even before tank-mixing with other pesticides. Any container or packet of pesticides, especially liquid formulations, will contain some or all of the ingredients shown in Table 3.1.

Table 3.1: The Make-up of a Pesticide

Pesticide Active Ingredient	• the main component of any pesticide product, as it is this chemical which controls, or kills, the disease, insect or weed • referred to by the common chemical name (which is not necessarily the same as the name of the product) • the product must contain this, and its properties influence the choice of other formulation ingredients
Solvents	• chemicals used to dissolve the active ingredients to make them up into liquids, and even solids • can be dangerous chemicals in their own right, with their own hazard classification, eg: toluene and xylene • such ingredients must now be stated on the label by law
Carrier	• a liquid or solid added to a pesticide to dilute it to facilitate application
Surfactants	• short for 'surface-active agent' • help reduce the surface tension, increasing the emulsifying, spreading and wetting properties of liquid formulations • often referred to as 'wetters', 'spreaders' and 'stickers' • not necessarily stated on the label
Safener	• a chemical which reduces the potential of a pesticide to damage a crop • not necessarily stated on the label
Adjuvant	• a material added to a chemical to increase its efficacy • a chemical in its own right, but inactive without the presence of the pesticide active ingredient • must be stated on the label by law

Pesticide Active Ingredient

The active ingredient is the most important part of the formulation as it is this chemical which controls the target pest; all the other chemicals in the formulation are there to help it do this. The properties of the active ingredient, such as its boiling and melting points and solubility in water and chemical solvents have a major influence on the choice of other formulation chemicals. Stability is another important factor and the active ingredient should normally have a shelf-life of around two years. Stability is especially important with liquid formulations.

The active ingredient will normally have to be ground down into fine particles before it can be formulated. Small particles are more biologically effective but there is a risk of physicochemical interaction with the carrier material during storage if the active ingredient is too finely ground. If the active ingredient is too concentrated the formulation becomes less stable and there is more chance of it breaking down and causing application problems. Stabilizers can be used to remedy this problem.

The pesticide active ingredient is known by a common chemical name. It is important to know the chemical name and if possible the chemical number (known as the 'CAS number') in order to obtain information on a particular pesticide. This information should be printed on the pesticide label. Without such information, especially the common chemical name, it can be hard to trace or identify a pesticide. The common chemical name should be distinguished from the brand name of the product. Brand or product names can be hard to trace as there are tens of thousands of them. To complicate matters further the companies frequently change names to help sales.

The following examples will help illustrate the points made above:

Table 3.2: Identifying Pesticides

Example
Brand name – Marshall 10G
Formulation – granular, ready to use
Classification – insecticide
Pesticide active ingredient – carbosulfan
(using common chemical name)
CAS No. – 55285-14-8

Example
Brand name – Roundup
Formulation – liquid
Classification – herbicide
Pesticide active ingredient – glyphosate
(using common chemical name)
CAS No. – 38641-94-0

Solvents

For liquid formulations, and even granules and powders, the active ingredient has to be dissolved, and the choice of chemical solvents is determined by a variety of factors. Solubility – the ease with which a solvent will dissolve a given pesticide active ingredient – is obviously the main factor. Solvents are often hazardous chemicals in their own right, so their toxic properties, flammability, corrosiveness and toxicity to crop plants (phytotoxicity) are all factors which have to be considered.[3]

Toluene, a very volatile and highly flammable solvent, is used to formulate pesticides. A more widely used solvent, xylene, is less flammable, but still has good dissolving properties. Xylene is the main solvent used in emulsifiable concentrate (ec) formulations, and naphtha is also used. Isopropanol and glycol are two water-miscible solvents used in solution concentrate (sc) formulations and soluble liquids. Some salts of phenoxy acetic acid herbicides are soluble enough to be mixed directly with water without any need for chemical solvents.

Carriers

Carriers are inert, solid ingredients used to dilute pesticide active ingredients. Carriers are used in pesticides normally formulated as dusts, granules and water-dispersible powders.

Absorbing carriers are used if the active ingredient is a liquid. Absorbing carrier materials include attapulgus clay (aluminium magnesium silicate), diatomaceous earth and colloidal silicic acid. Non-absorbing carriers (or diluents) are used to aid the grinding of active ingredients. Examples include dolomite (calcium magnesium carbonate), gypsum (hydrate of calcium sulphate) and kaolin (aluminium silicate).

Problems of compatibility of carriers with active ingredients can arise. For example, many inert carriers are acidic, basic and may even promote certain reactions (catalytically active). Some carriers can absorb appreciable quantities of water which can cause the pesticide to deteriorate during storage.

Surfactants

Surfactants (surface-active agents) include emulsifiers, dispersants, foaming agents, spreading agents and wetters. Surfactants are widely used because plant leaves are often covered with a waxy, water-repelling layer, the cuticle. A solution containing surfactants has a lower surface tension than water, so it sticks to the plant leaves and spreads more evenly. Surfactants can also improve the effectiveness of the active ingredient by increasing its penetration and translocation (migration) within the plant.

Special Additives

(1) stabilizers (or deactivators) are chemicals used to prevent decomposition of the active ingredient during storage. Stabilizers reduce chemical interaction between carrier materials and active ingredients. Chemicals known as 'acid scavengers' are used to counter active ingredients which release acids if they deteriorate, accelerating the decomposition of the formulated product in storage.

(2) synergists are chemicals which boost the biological activity and effectiveness of the active ingredient. For example, piperonyl butoxide is used with pyrethroid group insecticides as the butoxide blocks the detoxification of the pyrethroid in the insect, making the insecticide more deadly.

(3) emulsifiable oils are sometimes added to boost the biological activity of the active ingredient. A variety of types are used including aromatic and paraffinic oils. The oils work in different ways, such as reducing evaporation of pesticides, assisting penetration, and reducing the amount that might be washed off by rain.

(4) defoamers are sometimes added because the use of highly concentrated spray liquid, or poor spray equipment, can result in excessive foaming. Defoamers (or anti-foaming agents) often at concentrations of 0.5 per cent or less, can prevent this.

(5) thickeners or anti-drift agents are used to prevent the formation of very small droplets during spraying, and/or the evaporation of small droplets from sprayed surfaces to reduce the risk of spray drift. Thickeners also enhance the viscosity of the spray liquid, which helps reduce run-off from leaves.

(6) colouring and stenching agents are used to reduce the risk of accidents from swallowing pesticides. Stenching agents give the pesticide an unpleasant smell and/or taste. Colouring agents are also used in seed dressings to distinguish between treated and untreated seed. Granules are sometimes coloured to make them visible on the soil allowing application rates and correct spread to be checked.

What's on the Pesticide Label?

The main sources of information on pesticides are the pesticide label and the product safety data sheet. The quality of information on labels is highly variable, particularly on prevention and control measures. Broadly speaking, the label contains two main categories of information:

(1) How to apply the pesticide effectively – dose rate, water volume, pressure, the range of pests controlled, harvest intervals and so on.

(2) How to use the pesticide safely without harm to yourself, other people, wildlife and the environment – safe handling procedures, technical and engineering control methods, safe working practices, personal protective equipment, first aid and emergency procedures, and safe disposal and decontamination methods.

In the UK every pesticide packet or bottle must, by law, have a label containing the information as shown on p. 39.[4, 5]

Hazard Warning Symbols

Hazard warning symbols on the label of a pesticide container are mandatory in the UK. The labelling and hazard warning symbols are regulated by the Health and Safety: Classification, Packaging and Labelling Regulations 1984; the EEC Labelling Directive 1979, and the Control of Pesticide Regulations 1986. Hazard warning symbols provide safety information in the form of a hazard warning symbol pictogram with a list of precautions which must be followed when using, or storing, the pesticide. Where more than one symbol is applicable to a product only the most hazardous one will appear alongside the warning phrase. The hazard symbol classifications – 'Very Toxic', 'Toxic', 'Harmful', 'Irritant' and 'Corrosive' – mean the pesticide is automatically regulated by the Health and Safety: Control of Substances Hazardous to Health Regulations 1988 (COSHH).

About 20 to 25 per cent of all approved products in the UK are not thought to be dangerous enough to warrant such a classification, but the absence of a hazard warning symbol does *not* mean a product is safe. All pesticides should be treated with care and the safety precautions on the label followed.

Other Sources of Information

Safety Data Sheets

More comprehensive information on a pesticide can be obtained from a product safety data sheet provided by the manufacturer, supplier or importer of the chemical. Data sheets are important because a pesticide label only gives you basic health, safety and control information.[6] A safety data sheet should provide fuller information on toxicological testing, prevention and control measures, medical treatment and so on. Sadly, as with labels, the quality of manufacturers' data sheets is highly variable.[7]

If you work with, or are exposed to pesticides or other substances (chemicals, dusts) in the course of your work, you have a legal right to obtain safety data sheets.[8] Manufacturers of pesticides are required by law to produce safety data sheets but they are not required to supply

---

Information Required on Pesticide Labels

- **Name and concentration of each active ingredient** (common chemical name)
- **Name and concentration of other ingredients** (such as solvents, which could be harmful or toxic)
- **Name and address of the manufacturer or importer/wholesaler** (or other supplier of the substance)
- **Commercial name of the product** (brand or trade name)

User information including:

- directions for use
- mixing instructions
- application rates (including maximum application rates)
- approved tank mixes

Crop information, including:

- pest range (or spectrum) controlled
- tolerant or resistant species
- warnings on possible crop damage
- harvest intervals

Precautions, including:

- safety phrases (for example, 'wash concentrate from skin and eyes immediately', or 'wear rubber gloves')
- hazard warning symbols (one or more symbols may be present on the label, though not all pesticides carry such a hazard warning classification)
- first aid information, including where to obtain medical help

---

them to every single user; that is the legal responsibility of the employer.[9] Manufacturers are obliged to supply the sheets if these are requested. Make sure that the depot, or regional distribution centre which supplies you, keeps a completely up-to-date file of safety data sheets.

Before possible exposure to hazardous substances you should obtain and consult the relevant safety data sheets so that all the substances associated with the product – including breakdown products – can be correctly identified. You need to know about any health problems caused by exposure to the chemicals, as well as what prevention and control measures are recommended.

As well as giving information on acute health effects, the data sheets ought to contain information about a chemical's cancer-causing (carcinogenic) properties. If these details are not available you should ask the company for them. You should also check whether any of the chemicals affect fertility. Some chemicals reduce a man's sperm count, whereas others can damage the sperm or the egg from the ovary and result in malformed children being born. Certain animal studies are designed to find out about these effects.

Because the UK government does not admit to licensing pesticides which could cause cancer or birth defects, information on these effects may be sparse, even on safety data sheets (more information may be found in Chapter 5).

The data sheet should also contain information on how to prevent or control exposure to the pesticide. Simply knowing about the health effects is not enough. You need information and advice on what technical and engineering controls are needed. You also need information on safe working practices designed to minimize exposure to hazardous substances. Finally, you need practical information on what personal protective equipment should be provided, such as disposable masks, rubber face masks or gloves. Personal protective equipment (PPE) should only be used as a last resort to supplement technical and engineering controls, and safe work practices. This information is often weak or absent from data sheets and too much reliance is placed on PPE as the main means of protecting operators. Make sure that good information on prevention and control measures is provided.

Risk Assessments

Employees exposed to pesticides and other hazardous substances can now consult the risk assessment the employer has to carry out under COSHH.[10] The employer must ensure that a properly trained and/or qualified person carries out a risk assessment of work activities or processes involving exposure to chemicals, dusts and disease-causing micro-organisms. As well as specifying the health risks, the risk assessment should state what prevention and control measures the employer should put into practice. Prevention and control measures should be based on technical and engineering controls linked to safe working procedures and systems. The assessment should also state what measures are needed to check that the controls are working. These include planned maintenance, air monitoring and health surveillance. The risk assessment should also state what training of employees and management is required, along with the type of information which should be provided.[11]

Classification

These are some of the main ways of classifying pesticides and the terms used. Categories are not always distinct, and some of the terms may overlap. A single pesticide active ingredient may control a wide variety of pests, have several different modes of action, and therefore be classified in different ways, such as:

- by the category of pests controlled
- by mode of action
- by chemical classification
- by health classification
- by hazard classification

The Category of Pests Controlled: the Purpose Behind Use

Most pesticide active ingredients control a wide range of pests, so they may be referred to by a variety of terms. The main classifications under this heading are:

- fungicides
- herbicides
- insecticides
- miscellaneous compounds.

Each of the main categories can be further sub-divided to give more precise information on the range of pests controlled. For example, a pesticide popularly known as an insecticide may control a wide range of insects and their close relatives and so be referred to as an acaricide, aphicide, nematicide and so on.

The important point is that the *same* pesticide active ingredient, perhaps formulated in different ways and with different brand names, is used to control all these pests, not different chemicals, as might be assumed.

The following examples highlight classification and sub-division according to the purpose behind using the pesticide.

Insecticide is a general, convenient term for pesticides which control arthropod species – insects and their close relatives. (Arthropod means 'jointed leg' and these creatures are all characterised by an external skeleton and jointed appendages, including legs, mouth parts and antennae.) These include 'true insects' such as flies and beetles, and associated species like mites and ticks.

The general term 'insecticide' can be sub-divided into the following convenient terms to give more precise information on the pests controlled:

acaricides – control mites, ticks and spiders
aphicides – control aphids (greenfly, blackfly)
larvicides – control pest larvae
molluscicides – control slugs and snails
nematicides – control eelworms (nematodes)

termiticides – control termites
wood preservatives – control woodboring insects

Herbicides control annual, and perennial, grass and broad-leaved weeds and, in some cases, brushwood and woody species. Herbicidal products can be referred to more specifically as:

defoliants – remove leaves
shrub or brushwood killers – control woody-type plants

Fungicides control bacterial, fungal and viral diseases of plants and animals, and unwanted organisms such as algae and slime-type moulds. Fungicidal products may also be referred to as:

algicides – kill algae
bactericides – kill bacteria
slimicides – control slimes and moulds

Miscellaneous is a general category which includes a variety of chemicals classified for convenience as pesticides. It includes :

avicides – control birds
chemo-sterilants – sterilise insects and related species
dessicants – speed up drying of the plant
marine paints – control marine organisms
ovicides – kill eggs
pheromone attractants – attract insects and related species
piscicides – control fishes
plant growth regulators – inhibit or stimulate plant growth
repellents – repel birds, cats, dogs, insects, mites
rodenticides – control and kill rodents and vermin
sprout suppressants – inhibit sprouting (see 'growth regulator')

Mode of Action

Pesticides may also be classified by the way they kill or control pests (mode of action). The same pesticide may act in more than one way, and knowing its mode(s) of action will help you select the right one to use.

Pesticides may be *selective*: controlling or killing selected disease, insect or weed species but leaving the crop or non-target animal unharmed (in reality, pesticides' selectivity varies greatly); or *total* (non-selective): indiscriminately controlling all species they come into contact with (for some herbicides this is a much-advertised quality).

Herbicides, whether selective or total, may have additional modes of action. They can be applied before the crop has germinated (*pre-emergent*) or after (*post-emergent*).

Contact herbicides only control the part of the plant they land on.

Translocated herbicides are absorbed into the plant through the foliage and then move (translocate) to other parts of the plant, allowing control of parts of the plant which have not come into direct contact with the herbicide.

Residual herbicides act in the soil to control weeds as they germinate. Their effects may last from days to months, or longer, depending on their persistence and weather conditions.

For example, two widely-used herbicides, isoproturon and chlorturon, are selective, foliar and/or soil-acting, can be applied either pre- or post-emergence, and have both a contact and residual action. Paraquat is a total, contact and translocated pre-emergent herbicide.

Insecticides can be selective or total, have contact and/or residual action, and be foliar or soil-applied.

Contact insecticides can kill by direct hitting the target species or by leaving a spray deposit on the plant which the insect eats or walks on. If the insect eats the sprayed plant the insecticide can also be referred to as *stomach-acting*.

Systemic insecticides are absorbed into the sap and move around the plant more freely than translocated materials. Sap-sucking insects such as aphids and capsids are controlled as they feed on the plant juices. Systemic insecticides give protection for 2–3 weeks, or longer, and once absorbed cannot be washed off by rain. Many insecticides also kill by contact action.

Fumigant fumes and vapours given off by the insecticide are breathed in by the insect, for example nicotine.

Fungicides can be used to treat seed prior to planting, put in plant feed, or applied directly to the plant.

Protectant fungicides have to be applied to plant leaves and foliage before the disease is present and they prevent infection by stopping the disease becoming established. These fungicides can be washed off leaves by rain and repeated protectant applications are often necessary if extended control is required, as with potato blight control.

Systemic fungicides, like insecticides, move within the plant and have some ability to control established infection.

The terms 'systemic' and 'translocated' are often used loosely. The important point to note is that the pesticide has some ability to move within the plant, although this movement should be much greater with truly systemic pesticides.[12]

Chemical Classification

Pesticides can be grouped together on the basis of common chemical structure and characteristics.[13,14] This is a convenient way of classifying large groups of pesticides such as the organophosphorus insecticides where there are hundreds of different active ingredients, but it is not always a reliable guide to modes of action or effects on health. The main chemical groups are listed below.

Herbicides
Chloro-phenoxy herbicides consist of chlorine and various organic acids attached to a phenol ring. Pesticide active ingredients in this group include 2,4-dichlorophenoxy acetic acid (2,4-D): 2,4 5-T: 2-2,4-dichlorophenoxy propionic acid (2,4-DP or dichloroprop) and mecoprop (CMPP).[15]

Bypyridilium herbicides are quaternary ammonium compounds containing a free electron radical which are highly destructive of human, animal and plant cells. These chemicals are total herbicides (or dessicants). Pesticide active ingredients here include diquat, paraquat and paraquat dichloride.[16]

Insecticides
Organochlorine insecticides contain chlorine, carbon and hydrogen, and they are sometimes referred to as chlorinated hydrocarbons or halogenated pesticides. These chemicals are persistent in human and animal fatty tissue and in soil and water. Their overpersistence has led to many of these pesticides being banned or severely restricted.[17] Pesticide active ingredients in this group include aldrin, chlordane, DDT, DDE, dieldrin, endosulfan, endrin, heptachlor, lindane (gamma-HCH) and methoxychlor.

Organophosphorus (organophosphate) insecticides poison insects and mammals primarily by inhibiting an enzyme, acetylcholinesterase, critical for the transmission of nerve impulses. This is a large group of pesticides whose toxicity varies from moderate to high. Their mode of action can be contact, systemic, stomach-acting and fumigant.[18] Mevinphos acts in all four modes. Pesticide active ingredients in this group include phorate, demeton-s-methyl, disulfoton, chlorfenvinphos, carbophenothion, malathion, parathion, dichlorvos, azinphos-methyl, pirimiphos-methyl, and propetamphos.

Carbamate insecticides also poison insects and mammals by inhibiting acetylcholinesterase, though the process is more easily reversible than with organophosphorus compounds. Carbamate insecticides are moderately to highly toxic.[19] Carbamate active ingredients include aldicarb, carbaryl, carbofuran, carbosulfan, methiocarb, methomyl, oxamyl and pirimicarb.

Synthetic pyrethroids are a new group of synthetic insecticides based structurally on natural pyrethrum, which rapidly paralyse the insect nervous system giving a quick 'knockdown' effect. Mammalian toxicity is generally low except for certain chemicals such as deltamethrin.[20] Pesticide active ingredients in this group include allethrin, bioresmethrin, cypermethrin, deca- or deltamethrin, fenvalerate and permethrin.

Fungicides

Dithio-carbamates and *thio-carbamates* are primarily fungicides, though some compounds in this group also act as herbicides. While pesticides in this group have broad chemical similarities to each other, subgroups often have quite different effects on human health; some are recognised skin and respiratory irritants, others affect the central nervous system while a few have been linked to cancer.[21] Pesticide active ingredients which act as fungicides include mancozeb, maneb, thiram, zineb and ziram. Pesticides in this group which are classified as herbicides include diallate, EPTC and triallate.

Sterol biosynthesis inhibitors are newer-generation fungicides, some of which can have irritant effects. Compounds in this group include imazalil, nuarimol, prochloraz, propiconazole, triadimefon, triadimenol and triforine.

Health Classification

Pesticides can be classified according to their effects on human health. For a medical explanation of the terms used and general description of the toxic effects of pesticides see Chapter 5. Only limited use is made of this classification in the UK at the moment but the following health warnings may be printed on the pesticide label.

Anti-cholinesterase or cholinesterase-inhibiting compounds refers to carbamate and organophosphorus insecticides which inhibit the enzyme, acetylcholinesterase. Pesticide labels carry a specific warning: 'Do not use if under medical advice not to work with cholinesterase-inhibiting compounds'.

Anticoagulant rodenticides refers to pesticide poisons used for rodent and vermin control which act by preventing the blood clotting, for example warfarin.

In some countries, such as the US, other health classifications may be required on the label by law:

Carcinogenic if it has been scientifically established that the pesticide may cause cancer in humans.

Teratogenic if it has been scientifically established that the pesticide may cause birth defects to the children of those exposed to it.

World Health Organisation: Recommended Classification of Pesticides by Hazard

This classification is based primarily on the acute oral and dermal toxicity of pesticides to the rat. These determinations are standard procedures in toxicology. Where the *dermal* LD 50 value of a compound is such that it would place it in a more restrictive class than the *oral* LD 50 value would indicate, the compound will always be classified in the more restrictive class. Provision is made for the classification of a particular compound to be adjusted if, for any reason, the acute hazard to humans differs from that indicated by LD 50 assessments.[22]

Conclusion

This chapter has been designed to give a grounding in pesticide classification, formulation and labelling to enable you to identify pesticides and their problems, and find out where to get further information. In the next chapter we look in more detail at using pesticides, and the practical safety and control measures which should be taken.

How the Chemical Industry Could Clean up its Act

The Chemical Industry is said to be worried about its image. Recent opinion polls indicate that the industry is held in low public esteem and that two-thirds of the public are convinced that chemicals harm the environment. To polish its image the industry is now engaged on a public relations exercise in schools, colleges and local communities. It would be wrong of the industry to assume, however, that all that is required is a bit more gloss. There is often more substance to the public's perception than mere creation of an image.

Bhopal, Seveso and Flixborough have shaken the industry's complacency. Chemical companies now acknowledge that serious accidents, often involving loss of life, can, and indeed do, happen. But what about the more prosaic types of incidents where workers may suffer from exposure to chemicals that they use every day? There is a noticeable hush when this subject is mentioned. The industry can even be deliberately obstructive when this subject is investigated, as was the case in 1986 when the Pesticides Group tried to conduct a survey of health and safety information on pesticides. The Pesticides Group is a committee of health and safety representatives from several trade unions, including those representing agricultural and chemical workers (TGWU); public employees (NUPE); chemical workers (GMBWU); and laboratory staff (MSF). In an attempt to gather data on a number of chemicals used in Britain, the group sent a questionnaire to 19 chemical companies asking for information about their products.

No company completed the questionnaire. In fact, the companies were specifically advised by the British Agrochemicals Association (BAA) not to complete it, but merely to provide copies of the literature they sent out with their chemicals.

The questionnaire asked for much more than this. It was an attempt to provide a detailed profile of a wide range of chemicals with all the relevant data on toxicity. This information would have allowed the unions to make some sort of assessment of the risk workers faced in their jobs. So much for the plan; the companies and the BAA scuppered it at the start.

As none of the firms completed the questionnaire, I did so on their behalf, by going through the information they provided. The unions asked me to do this. Where a product data sheet, container label, or accompanying literature provided an answer to a question, I credited the company with having responded.

Eighteen of the 19 firms approached sent some information on their products. Ciba-Geigy Agrochemicals did not bother. The company said that it could not complete the questionnaire but that it was conscious of its responsibilities under the 1974 Health and Safety at Work Act and provided information to the users of its products.

Imperial Chemical Industries (ICI) was the only company that indicated where answers to the questions could be found, and it supplied the relevant safety data sheets, user instructions and storage recommendations for its products.

The one section where ICI, and various other companies, provided no information was that concerning tests on animals. So for many products there were no data about the concentration of the chemicals that would be lethal to animals, nor on the amount that would irritate the skin or the eyes. For these chemicals, no information was provided about their effect on fertility, whether they would cause malformation in fetuses or whether they had neurotoxic or carcinogenic properties.

ICI says that all this information is given to the Ministry of Agriculture, Fisheries and Food's Advisory Committee on Pesticides – the body charged with issuing a product a licence.

According to Anne Buckenham, the BAA's company secretary, the principal reason it advised companies not to complete the questionnaire was that companies were already complying with the terms of the Health and Safety at Work Act and supplying data sheets to customers. Furthermore, the relevant animal test data had to be supplied to the Advisory Committee on Pesticides before products would be approved. In doing this, companies were complying with what the law required of them, she said. Buckenham also said that if safety information were made public it would aid the companies' competitors. But this is not an argument that carries much weight. Although a number of companies who responded to the questionnaire adopted ICI's line in not revealing data concerning tests on animals, there were some notable exceptions. Bayer, Monsanto, and Rohm and Haas all give detailed toxicity data on their products quite openly, and discuss the effects on fertility, as well as any potential teratogenic, mutagenic or carcinogenic properties. If these three major companies can release this information there is no reason why ICI, Shell, British Petroleum (BP) and Fisons should not do the same.

Of all the product safety data sheets the unions received, those prepared by Rohm and Haas, an American multinational, were clearly the best. On two sides of A4 paper the company provided considerable information in succinct, readable and easily accessible packages. Stricter laws in the US on the provision of information may have something to do with Rohm and Haas' literature.

Most companies exhort users to avoid inhaling chemical fumes or dust and to avoid contact with skin. They print clear, simple rules of hygiene on labels. They also warn about the dangers of contaminating waterways, or of small children getting hold of a container.

In the event of contamination there is usually some basic advice on first aid. As for more specific advice for doctors, less than half the companies give this. This omission is serious given that only 12 of the 19 companies indicated that they manned an emergency telephone and provided the number. The rest simply recommended seeking medical attention. A general practitioner or junior hospital doctor would be expected to phone one of the regional poisons centres to seek specialist advice.

The BAA encourages its members to send literature about health and safety to the regional poisons centres. In checking this, I found that one regional centre had only patchy information. The Poisons Centre at Guys Hospital in London, a central repository and advice centre, had literature on four of five chemicals I asked about. It seems that the industry is not as assiduous as it could be in sending out literature, or updating old records.

In another instance, three companies producing similar formulations of the herbicide 2,4-D gave conflicting information about protective clothing. For Silvapron D, BP advises the use of full protective clothing where there is risk that spray might drift. Synchemicals doesn't specifically recommend this course of action, but advises the washing of protective gear if contaminated. As for BASF UK, users of its product BASF -2,4-D ESTER 720- are told that there are no protective clothing recommendations.

On the issue of disposing of excess chemicals, most companies recommend that this should be done safely without specifying what to do. Some refer users to other booklets on the subject, but it is extremely unlikely that everyone will have access to these. The advice should be on labels, or in the safety data sheets.

Finally, the survey showed that safety data sheets are not always sent out routinely with chemicals. Synchemicals, for example says that it will issue these to its 'customers on request'. It is a basic tenet of the 1974 Health and Safety at Work Act that safety data sheets be prepared on chemicals. It should be an equally basic tenet that everyone should legally have ready access to these and that the data should be as comprehensive as possible.

Source: An article by A Hay, *New Scientist*, 12 February 1987. Reproduced by kind permission of New Science Publications.[23]

4

How to Use Pesticides Safely

In this chapter we look at the hazards that can occur before, during and after pesticide use, and what practical precautions should be taken to enable you to use pesticides as safely as possible, given the chemicals available today and current legislation.

Pesticides present their greatest danger when they are being used. For most people, this is when accidents could occur and when important safety issues arise. It could mean contamination of someone loading a tractor sprayer, or ruining a neighbour's crop through pesticide spray drift.

Safe use in the UK is governed by two Acts: The Food and Environment Protection Act, 1985 (Control of Pesticides Regulations [COPR], 1985), and the Health and Safety at Work Act 1974 through its COSHH (Control of Substances Hazardous to Health) Regulations. A detailed *Code of Practice for the Safe Use of Pesticides on Farms and Holdings* is available from HMSO.[1] All employers and pesticide users have a duty to be familiar with this Code, which may be cited in legal proceedings for breaches of regulations. The Code is essential reference material for all pesticide users on farms (see Chapter 11).

We recommend some practices which currently are the best option, but which, in future, we would like to see changed. For example, we suggest pesticide waste and packaging can sometimes be burnt. This can be dangerous, and we would like to see proper regional pesticide disposal points set up so that waste is treated as carefully as possible. There is no point in recommending this in a practical chapter until such disposal points are available.

We suggest photocopying the relevant tables and check lists in this chapter, and keeping them pinned up or accessible in the workshed or store-room.

Before Application: Mixing and Handling Pesticides

The law requires employers and operators on farms and holdings to have adequate instruction and guidance in safe use of pesticides. Operators must be trained, and some must have a Certificate of

Competence from an Agricultural College or Agricultural Training Board. Knowledge of relevant legislation is also required. While domestic users do not require these qualifications, this section provides guidance on hazards and precautions.

Pesticides are sold in many forms – liquids, granules, dusts, fogs, smokes, vapours – and all require handling differently to minimize contact. All can create their own hazards if misused, or if the chemicals are particularly toxic. Some pesticides are sold ready to use, while others need diluting, usually with water. The greatest chance of contamination arises during handling, mixing, dilution and loading spraying machines. Operators may become careless as they become more familiar with the process. Contamination occurs by:

- inhaling vapour from the concentrated pesticide
- splashing liquid pesticide on skin or into eyes
- getting liquid or powder pesticide onto the hands, and from there into the mouth.

In practice, even more dangerous incidents commonly occur. Some workers frequently clear the nozzle of a sprayer by blowing through it and they get a concentrated solution of poison directly into their mouth. Anyone smoking whilst handling pesticides inhales the pesticide vapour through a mini-furnace inside the cigarette or pipe. Many workers report being totally drenched by spray in accidents. All of these examples contravene the Pesticide Code. Even though these incidents would be disapproved of by safety experts, and disowned by the chemical industry, they can easily happen in a working environment.

Some newly-designed pesticide packets dissolve, and can be loaded into a sprayer without the operator touching the chemicals. Although more expensive, this method of packaging is by far the best from a health point of view. All pesticide users should ask for the best available dispensers.

Is it Really Necessary?

This is the first point to check before application. Users must ask themselves three questions:

- Has the pest, weed or disease, been correctly identified, and is the degree of damage caused such that it warrants pesticide use?
- Is any other method of dealing with the problem available? For example biological control, integrated pest management or some non-toxic or organic approach based on others' experience?
- Will the pesticide be effective, and is the timing likely to cause problems (for example, is it too close to harvest)?

Establishing Precautions

Having decided that the circumstances warrant pesticide use, the following must also be checked:

- Is the pesticide approved for the intended use?
- Can it be safely prepared and applied by an operator with the equipment available?
- Will it endanger the operator/user's health?
- Does it present unnecessary risks to livestock and the environment (bees, fish, etc.)?
- Will it damage, directly or through leaching, any streams, rivers or other water, whether surface or underground?
- Are adequate precautions established to prevent either livestock, or the public, from coming in contact with the area to be sprayed (for example, warning notices on public footpaths)? During the times of year when bees are foraging, farmers and growers must give bee-keepers 48 hours' notice of spray times.

In addition to specific precautions on each application, large pesticide users, particularly on farms, must prominently display their written plans for the procedures to be followed during emergencies (fire or spillage). *Everyone* on site should be familiar with these plans.

Keeping Records

It is essential that good records are kept of employee exposure and of all operations involving the storage, application and disposal of pesticides. Records must be kept for at least 5 years – 30 years for health surveillance records. Records must both aid stock control and be usable as a reference in the event of the accidental contamination of people, honey bees, other creatures or non-target crops. The Pesticide Code recommendations for records suggests the following headings: operator; date; application site; crop; material or structure treated; reason for treatment; product used; dilution and application rate; hours pesticide used; weather conditions, and any other relevant detail.

Opening, Diluting and Mixing Pesticides

Most contamination occurs at this stage, and extreme care should be exercised, particularly when handling large quantities of pesticides. Before starting, read all labels carefully (see Chapter 3) and follow the manufacturers instructions strictly.

The Pesticide Code supplies an important check list of precautions to follow when filling the application equipment (see p. 53).

Precautions When Filling Equipment

DO NOT

- use bare or gloved fingers to break the seal on a container
- open more than one container at a time
- make a direct connection between a domestic water supply and a spray tank
- take water from a stream without preventing run-back
- decant pesticides between containers and spray equipment unless you have to
- lift containers above shoulder height
- clamber up a sprayer with an open container
- cause foaming by sucking air into the tank when using an induction probe
- return a probe to its holster without washing it
- cause glugging
- mix two or more concentrates before loading them into the tank
- let fine particles of dry pesticide become airborne

DO

- use any purpose designed device which is fitted or available
- replace the cap/close the container
- ensure there can be no run-back of pesticide into the water supply
- use an intermediate tanker or system
- measure out pesticides only in an appropriate vessel and rinse it immediately, use scales dedicated to the task for powders
- make sure of good foothold if you have to pour directly into a tank, preferably on the ground or a platform at the right height
- use filling attachments such as chemical induction probes and low-level filling and mixing tanks
- use a probe to rinse containers if you can
- pour slowly with the container opening positioned so that air can enter
- if two compatible pesticides are to be mixed together follow the correct procedure
- add them to water separately in the recommended order
- measure out powders in still air conditions

During Application:
Using Pesticides on the Ground and in the Air

Pesticides also pose great hazards during application. Problems occur in three main areas: ground spraying, aerial application and aquatic use. Users should check all technical and engineering safeguards, and consult the guide on protective clothing.

Ground Spraying

Ground spraying is the commonest form of application because it is a cheap and rapid way of treating large areas. The procedure usually involves use of a tractor-drawn sprayer which releases liquid pesticide through a series of nozzles arranged along a boom held at right angles to the tractor, and the chemical is sprayed in a swathe behind the vehicle. Factors which influence pesticide spraying are the type of nozzle used, the length of the spray boom, and the pesticide mixture. Ground spraying can also be done using a backpack sprayer.

The Right Equipment

Good spraying equipment is essential; it should be clean, not leak, and must be calibrated to apply the correct quantity of pesticide.

A crucial requirement for successful spraying is equipment which will create the right size droplet. Larger drops tend to bounce straight off a target crop; if they remain on the plant, they may overdose the spot where they land. The smallest droplets simply float away, creating problems of spray drift. Composition of the spray depends on three factors:

- orifice size – smaller holes produce finer spray
- nozzle angle – wider-angled produce a thinner sheet of liquid which tends to make smaller droplets
- pressure – higher pressure creates finer spray and thus smaller droplets.

The majority of nozzles used in Britain are fairly old-fashioned hydraulics, which force pressurised liquid out through a narrow hole, creating a mass of droplets. The nozzles have the advantage of simplicity, but create droplets of variable size. A typical hydraulic sprayer can produce drops between 50–500 microns ($\mu m = 1/1,000mm$) in diameter – the volume of the latter is 1,000 times greater than the smaller droplet. If the 50 micron droplet is sufficient to kill the pest, the large drop will waste 99.9 per cent of the chemical, even if it hits the target. It is estimated that only 0.02 to 1 per cent of insecticides applied are effective in killing pests, and large improvements in efficiency are possible.[2] The average rate of application for conventional sprayers is between 225–450 litres per hectare (20–40 gallons per acre).

Other types of nozzles have been developed and many of these involve Controlled Droplet Application (CDA), where the spray operator can calibrate the sprayer to select the size of droplet required depending on use, weather conditions, and so on. This is sometimes known as the 'dial-a-droplet' system. CDA has now become virtually synonymous with one particular system, the spinning disc nozzle, which produces much more regular droplets by spinning them off a rapidly turning disc. In general, droplets become smaller the faster the disc spins.

Although this system is still being perfected, some farmers have found that the greater efficiency of CDA allows them to reduce pesticide use to one-fifth of normal. However, as CDA can also form very small droplets, incorrect use could result in even more drift. In addition, the pesticide is usually applied in more concentrated form with CDA, making drift more serious if it does occur.

Another application method being developed is electrostatic spraying, which can be used with both hydraulic and CDA systems. Spray droplets are given an opposite electric charge to their earthed targets, which makes them stick more firmly to plant leaves and target surfaces, reducing drift.

Electrostatic spraying is also supposed to enable the use of smaller droplets which, without an electric charge, would drift. The optimum size produced is about 30 microns, about one-fifth the thickness of human hair. The proportion of droplets which will drift in an electrostatic system is still unknown, and levels may still be unacceptable.

The length of boom is also critical: longer booms mean that the tractor needs fewer runs up and down the field to apply the pesticide. However, a longer boom also tends to bounce up and down and causes a larger proportion of the pesticide to be lost as spray drift – reducing efficiency and increasing the environmental hazard. In general, today's booms are considered to be too long for safe spray application.[3]

Manual Spraying

Manual spraying with portable sprayers worn on the back, and using both hydraulic and CDA systems, is widely practised in and around buildings, in forestry plantations and for general amenity use in parks, on gardens, flowerbeds and pavements. A survey commissioned by the Health and Safety Executive has highlighted the poor design of hydraulic knapsack sprayers and the potential for pesticides spilling onto the neck, back and arms of the operator.[4] Sprayers are often heavy and unwieldy to lift and carry, causing neck, shoulder, back, and general muscular problems for the user. The survey aimed to encourage manufacturers to improve the design and safety of knapsack spraying equipment, and to help introduce a new British Standard.

A study of spray operator safety by the British Agrochemical Association (the UK pesticide manufacturer's trade association) in 1983 found that the risk of exposure to the spray operator was significantly greater with a manual sprayer than a tractor-mounted boom.[5] Hands, arms and legs were especially at risk from leaks. The study concluded that there was 'considerable' risk of dermal (skin) exposure from knapsack spraying. Workers also walk through treated areas, further increasing exposure.

Using Granules, Dust and Other Forms of Pesticides

Pesticide granules are solid and vary in size from something like fine sugar to the size of fertiliser granules. The granules are usually a mixture of the chemical and an inert carrier. Granule application is usually via modern fertiliser applicators, which place the product in bands between rows of crops. Some granules are also applied directly below soil level by crop-drilling machinery. Granules are often applied by hand around plants, in glasshouses and on lawns.

Granules may appear safer, but can be especially hazardous, as they are sometimes used as a carrier for chemicals judged to be too hazardous for liquid application. Examples include aldicarb (Temik) and phorate. With these and other toxic pesticides it is important to:

- wear protective clothing
- incorporate the granules into the soil, or water them in to avoid release of toxic fumes
- observe the minimum re-entry period when using in an enclosed space like a greenhouse

Other application methods require particular precautions. Dusts are especially dangerous to operators when opening packages, as they become airborne and can be inhaled or get into the eyes. Smoke generators and vaporizers produce lasting pesticide control in glasshouses, and the minimum re-entry times must be observed. Baits can be extremely dangerous to wildlife if used incorrectly or carelessly.

Aerial Spraying

Although only about 2 per cent of pesticide is applied by aircraft in Britain, aerial spraying accounts for a far larger proportion of accidents and complaints.[6] Aerial application is used when: the ground is too wet to take a tractor-mounted sprayer; a large area has to be sprayed in a very short time, or the terrain is unsuitable for tractor-mounted sprayers (for example, a conifer plantation). During the 1980s the number of aircraft operating in Britain reached 100 or more at times; now it is far less. Soft-wheeled land vehicles mean that aerial spraying is no longer needed in arable farming areas.

Aerial spraying uses much the same system as ground sprayers – a row of nozzles suspended along a boom. Both fixed-wing aircraft and (less commonly) helicopters are used. Opinions differ as to which is safest. Helicopters are more manoeuvrable, but the turbulence from the rotors can increase drift (although some authorities maintain that it forces the pesticide down onto the crops). Fixed-wing planes have the disadvantage that, in turning, they can easily spray beyond the field. Pilots are supposed to turn off the sprayer when turning, but it is claimed by activists in heavily-sprayed areas that this instruction is frequently ignored in practice. For example, in 1984 a crop-spraying plane overshot the village of Blackhall, near Durham, resulting in some 15 people being treated at hospital for burns, rashes, headaches and skin irritations. Villagers were advised not to eat garden vegetables for at least four weeks.[7]

Aerial spraying is more likely to cause contamination than ground spraying. Aircraft travel faster, making over-runs more likely, and pilots are insulated from the weather and less able to judge wind conditions in the target area. To minimize weight, low water volumes are used in aerial spraying, so that more concentrated mixtures are applied. Planes frequently fly too low over houses and there are many examples of gardens, playing fields and houses being sprayed by mistake. The relatively short spraying season means that pilots are under pressure to do as much work as possible in the time available, making mistakes more likely.

Although only certain chemicals are passed for aerial application, these include some known to be hazardous to health, such as benomyl, captan, chlorpyrifos, 2,4-D, dichlorvos, malathion, metaldehyde and 2,4,5-T. At least eleven pesticides cleared for use by aircraft are suspect because they may cause cancer, birth defects or simply because they are intensely poisonous.

Pilots are at less risk of poisoning than ground crews who mix and load the pesticide; flaggers who direct the aircraft or helicopters from the ground run the greatest risk because of over-spraying or exposure to drift.

Many controls in Britain, including the nearest distance that a crop-spraying aircraft can approach someone's home, are far more lax than in most other European countries, despite the fact that new regulations were introduced a few years ago.[8] The specific legal controls are dealt with in a booklet available from the Civil Aviation Authority.[9] (See also Chapter 11.)

Aquatic Pesticides

Only a few herbicides are cleared for use against both aquatic weeds and on the banks of rivers and canals, and permission must be given for this by the National Rivers Authority (in Scotland, the local River Purification Board). In general only workers specifically involved in waterway management are allowed to use pesticides in water; use on river and canal banks is much more common. Herbicides used include: asulam, chlorthiamid, 2,4-D amine, dalapon, diclobenil, diquat, fosamine ammonium, glyphosate, maleic hydrazide and terbutryn. Many other pesticides are toxic to freshwater life and it is illegal to use them near watercourses.

Guidelines for the Use of Herbicides on Weed in or near Watercourses and Lakes, published by the Ministry of Agriculture, Food and Fisheries tells you what you can and cannot do in this area.[10]

Hazards During Application – Spray Drift

If spray is applied carelessly, pesticide drifts onto non-target areas causing environmental problems. Spray drift incidents can ruin entire crops, kill livestock, and damage people. Growers in Lincolnshire who lost thousands of pounds' worth of vegetables when spray drifted across their farm in 1980 are still fighting for compensation.[11] A smallholder in the Quantock hills lost a ram, 25 ewes and 59 lambs after 2,4,5-T spray drifted across from a nearby conifer plantation.[12] And a farmer near Cambridge was confined to a wheelchair after being contaminated when she and her cattle were caught in drift from an aircraft.[13]

There are three types of spray drift:

- *Droplet drift* is caused by spray drifting during application. Care when using spraying machinery and improvements to sprayer technology can control the problem. Proper training and greater controls are needed to prevent careless spraying, particularly in unsuitable weather conditions. CDA sprayers cut out most of the small droplets, which will drift, thus reducing the 'chronic' impact of spray drift; more of these sprayers are needed.
- *Vapour drift* occurs when particularly volatile pesticides evaporate after landing on their target, and drift onto neighbouring plants. A few years ago in England, herbicides sprayed on cereal crops volatised and drifted onto adjacent fields of oilseed rape, preventing some of the rape from flowering. Today, vapour drift has been reduced by a voluntary ban on some of the most volatile herbicides, organised by the National Farmers' Union because of crop damage. The problem has not stopped, as some volatile herbicides are still on sale. There is a need for a formal ban on very volatile pesticides, and proper testing for volatility before new pesticides are introduced.
- *Blow* is caused when strong winds blow granules or dusts out of fields into neighbouring areas. Although blow can be serious (because the most hazardous pesticides are usually prepared in granular form), it is less of a problem in Britain than droplet or vapour drift.

Droplet drift is a serious problem. Very small droplets from hydraulic nozzle sprayers are extremely likely to drift, even if the wind is light. Up to 20 per cent of droplets (those less than 100 μm in diameter) are liable to drift. If water is used as the dilutant, it can evaporate and the droplets become particles of pure chemical which stay in the atmosphere indefinitely causing serious pollution. In windy conditions, a far greater proportion of spray is lost.

Droplet drift problems include:

- Damage to crops when herbicides drift, typically from cereals onto vegetables. This is commonest where cereals and vegetables are grown together, such as the Vale of Evesham and parts of Lincolnshire.

- Illness to livestock when hazardous insecticides drift onto grazing areas or into stables. Sickness and death in sheep, cattle and horses have all been caused by spray drift, which also kills a large number of bees every year.
- Destruction of wildlife in hedges, copses and other sprayed areas. Partridges and butterflies both tripled in population within a year when hedges were left unsprayed on a Hampshire farm.[14]
- Poisoning of people, caused by inhaling spray. Acute poisoning can occur, with individuals claiming that they have developed serious and sometimes irreversible health problems as a result. In addition, many people suffer short term problems such as headaches and asthma attacks.

Aerial spray drift has harmed many people. Twelve part-time firemen were hospitalised after being sprayed by two planes while fighting a haystack fire in Hertfordshire. Fifteen people in a village near Durham had to receive hospital treatment after a plane oversprayed their homes. And Conservative MP Sir Trevor Skeet was sprayed in his own back garden by a plane near Bedford.[15]

Cabless, airblast sprayers used in orchards, nut and fruit groves cause some of the worst pollution and pose special problems – as research by the UK Health and Safety Executive has shown – for operators, neighbours and the environment.[16]

Orchard Spraying – a Major Hazard

Commercial apple and pear orchards in Great Britain receive an average of 21 pesticide treatments per year, involving an estimated 800 tonnes of pesticide active ingredient. Many pesticides are applied to 'run off', in excess of 1,000 litres per hectare.

Most orchard spraying is now carried out with blowers or air-assisted sprayers which use an airstream to carry droplets into the trees: an arc of spray is thrown 27 feet vertically – just above tree height – at a rate of 10–40 metres per second. Up to 5 per cent of the active ingredient applied will drift more than 20 metres downwind – with low-volume application rates drift can be as much as 16 per cent because of the high proportion of small droplets. Reduced-volume spraying using the same quantity of pesticide active ingredient is being developed.[17]

Spray drift in orchards is a particular hazard for the operator, as much spraying is done without tractor cabs to enable the machine to get among the trees. When high volumes of water are used the operator can be drenched by spray drift and run-off from the trees. At low volumes the higher concentration of pesticide poses an additional risk.

One study concluded that orchard spraying is potentially more hazardous than ground crop spraying. As much as 15 per cent of the applied pesticide was lost as spray drift. With the conventional spraying

Controlling Spray Drift

- Plan spraying well in advance and inform neighbours.
- Avoid chemicals which damage neighbouring crops or wild plants, and, if possible, those which damage wildlife and beneficial insects.
- Select the correct nozzle for the operation.
- Calibrate the sprayer with respect to the amount of pesticide required to cover the area to be treated.
- Ensure that protective clothing is available up to the standards required and check the pesticide label to see if special precautions are needed.
- Check the dilution rate of the pesticide.
- Check the weather forecast on the day, and also check wind speed and direction in the target field. It is safest at about wind force 2 blowing away from susceptible crops or wild areas. Totally still air may well mean turbulence later during the summer months. Spraying is not advisable at force 4 and over (moderate breeze, moving small branches).
- Switch to a coarser spray if conditions are marginal.
- When spraying keep the spray boom as low as possible.
- If possible, leave strips unsprayed at the edges of fields and in headlands.
- If conditions are unsuitable, DO NOT SPRAY.

method, when wind conditions were unfavourable, the protective clothing worn by the spray operator was maximally contaminated after spraying less than one hectare (2.5 acres).[18]

Controlling Spray Drift

There are a number of ways to minimize spray drift and its effects during pesticide application. Sprayers should follow the rules given above.

After Application: Storage and Disposal

Once pesticides have been used, remnants must be cleared up: any remaining chemical stored; ready-mixed formulations stored or disposed of, and containers disposed of safely. All these stages can result in contamination if they are not carried out properly.

Poor disposal of pesticide wastes can have catastrophic effects on wildlife. In June 1985 effluent containing highly toxic pentachlorophenol from a mushroom factory farm at Quernmore,

Lancashire, poisoned 200 metres of the River Condor so severely that all life was killed.[19] A few months later, empty cans of the highly toxic and persistent pesticide lindane were found dumped in a lane in the same area.

Storage

Poorly stored pesticides can cause a number of problems, such as seepage into watercourses, fire, or contamination of workers. The provisions for storing pesticides on farms are laid down by law, and further advice is set out in 'HSE Guidance Note CS19'.[20] Guidance on construction of a storage shed is given on p. 62 (see also Chapter 11).

Disposal of Unwanted Pesticides

It is an offence under the Control of Pollution Act 1974 to abandon poisonous or polluting pesticide waste where it may create an environmental hazard. If pesticides are unwanted, they must be disposed of very carefully. When disposing of unwanted chemicals, the following points must be noted:

- Pesticides washed from the sprayer must never be allowed to run into public sewers or where there is a danger of seepage into any form of watercourse. Drainings should run into a specially constructed and maintained soil soakaway.
- If possible, use a flushing device (rather than filling the spray tank with water and pumping it through) to clean out equipment; this will reduce the volume of washings.
- Large quantities of unwanted pesticide, or any pesticides used in urban areas, should be disposed of after consultation with the local official agricultural adviser or (in a town) the local Health and Safety Executive officer.
- Rodenticide baits and bodies of poisoned rodents should be buried or burnt.
- Sheep dips should be disposed of either by using a soil soakaway near the dip bath (being careful to avoid water pollution), or by spreading over the nearest suitable area of level soil.
- Fruit dips should be disposed of in the same way as other crop pesticides. More hazardous or difficult pesticides, including grain, soil and other fumigants, bulb dips, potato dips, seed steeping processes and cyanide gassing powders should not be disposed of without seeking the advice of local official agricultural advisers.
- Large quantities of pesticides should be disposed of through a reputable specialist waste disposal contractor.

Storage Sheds

A special storage shed for pesticides must be constructed, and it must:

- be built well away from houses, livestock, other inflammable materials and water
- have walls, floor and roof which are fire and corrosion resistant, impervious to liquids and insulated, and the floor should be slatted to avoid contamination from spills
- include a raised sill so that the building can contain any spillage prior to mopping up, and also the shed must have an external watertight tank to store spilled liquids
- be provided with suitable entrances and exits
- be fitted with strong and durable shelving
- include natural ventilation or extractor fans to avoid build-up of fumes
- have sufficient lighting so that all labels can be read, while avoiding direct sunlight because of the risk of spontaneous combustion if chemicals became too warm
- be heated well enough to avoid frosts, especially when storing liquids
- be theft-, vandal- and animal-proof with a clear sign displayed giving warning that poisons are stocked within
- include washing facilities and a place to store protective clothing

In addition, care must be taken to:

- store items carefully, and categorise them where possible into poisons, flammable materials etc.
- to store single steel drums no more than three high, and single plastic drums no more than two high.

Disposal of Pesticide Containers

All too often pesticide containers are left in fields or dumped in water-courses. Ideally, containers should be returned to the manufacturer for disposal, and in some cases this is possible. Legal guidelines and codes of conduct are laid down for the disposal of all kinds of containers. A number of stages are involved:

Pre-disposal: all containers (including paper packages) should be rinsed out and the washings disposed of as for tank washings described above. Protective clothing, especially face-shield/goggles, should be worn if necessary.

Storage of containers prior to disposal: must be in a well defined and marked area, preferably under cover or with a special waterproof container provided for empty paper sacks if these are stored outside. Containers should NEVER BE RE-USED for diesel, or any other purpose.

Disposal: should be carried out carefully depending on the container material:

Glass and metal (not aerosols) should be either crushed in a sack (for glass) or flattened, and buried at a depth of at least 450 mm (18 inches) in an isolated place, with a record kept of where burial takes place and what is buried.

Paper and plastic should be punctured or crushed to make unusable. Labels should not be disfigured. Containers can be buried, with the same precautions as glass or metal, and in some cases can be burned. Advice on products suitable for burning should be obtained from the Air Division of HM Inspectorate of Pollution and the Environmental Health Department of the local authority. If burning, it should be on a roaring fire away from buildings, at least 15 metres (50 feet) away from public highways, and not where drifting smoke could reach livestock, people, houses or covered crops. AVOID BREATHING the smoke and/or fumes.

User Safety

All pesticide users must be well informed of the law and what constitutes good practice. Operators on farms and holdings must have the correct training for the pesticides and equipment they handle, or be supervised by someone with a recognised certificate. Good washing facilities are essential, and farm employers are obliged to provide ready access to these. All users should wash after spraying.

Health surveillance is particularly important to ensure that maximum exposure limits are not exceeded, and is vital when the product used could have a serious effect on health if control measures fail (for example with organophosphorus pesticides and carbamates). Outside help may be needed in monitoring.

The pesticide and agricultural industries have relied far too heavily on the use of personal protective equipment (PPE) as the main means of protecting the spray operator. In practice, PPE is the main means of operator protection from pesticides during mixing/filling, use and disposal/decontamination. In reality, PPE is the *least effective* means of protecting the operator, and should supplement technical and engineering controls and safe working methods (see also Chapter 11). These principles are requirements of the newly introduced COSHH regulations. It is hoped that these regulations will move agricultural industries away from dependence on PPE to control at source.

Technical and Engineering Controls

If pesticides have to be used, as part of an integrated pest management approach, then technical and engineering measures are of primary importance.

The approach adopted in forestry after protests from forestry workers affected by lindane is shown on p. 65.

Here again, redesign of equipment can help. A pilot study in Sweden which tested two new planting devices: a planting stick, and sponge clamps, reduced workers' exposure to permethrin when treating conifer seedlings. Permethrin was measured in air and urine samples and the men were examined for allergic reactions. The study was a success and there were far fewer complaints about ill-health when compared with traditional dipping practices.[21]

Dissolvable Pesticide Sachets

Mixing, filling and agitating the spray tank are dangerous operations for pesticide users, given the poor design of many spray containers and packets. Existing containers are often difficult to pour and pesticide can easily spill onto exposed skin and clothing, or splash into eyes. One advance here has been the development of water-soluble plastics made of polyvinyl alcohol film and similar materials. Many pesticides are now being formulated and marketed in these soluble sachets which are simply placed whole into the spray tank. The tank cover is replaced, the mixture agitated by the sprayer pump, and the packaging dissolves inside the spray tank, releasing the pesticide. There are no containers to dispose of afterwards.

Personal Protective Equipment

PPE should be available in good condition, uncontaminated from previous use. In practice, clothing is often poorly maintained and not replaced often enough. It can be uncomfortable, and users often resist wearing it for long periods.

The term PPE includes gloves (which should *always* be worn when handling pesticides), respirators, goggles, disposable dust masks, rubber, long-sleeved shirts or coats, or, when very hazardous chemicals are being used, complete protection suits. Additional clothing is sometimes required while preparing pesticides for use.

Most equipment only offers limited protection and is often unsuitable for the job in hand. It is common to see operators wearing 'nuisance' dust masks when spraying in the mistaken belief that these will prevent them breathing in the fine spray droplets. A list of the PPE required by the Pesticide Code for workers on farms and holdings is given on p. 66.

Example: Engineering Controls Protect Forestry Workers

The UK Forestry Commission dips young tree roots in either lindane or permethrin to protect them against beetle attack. Workers either dipping or planting trees have suffered lindane and permethrin contamination above safe levels.

After a number of incidents in Ennerdale and Keilder forests in the UK, the Health and Safety Executive began monitoring forestry workers' health. Forestry workers who dipped young trees before planting them had absorbed lindane through the skin and had raised blood concentrations. Those affected reported non-specific symptoms similar to flu – headaches, dry mouth, sore throat and diarrhoea.

The forestry workers discussed methods of improving the safety of dipping and transplanting techniques with their union, the Transport and General Workers' Union. This led to the Forestry Commission – acting together with the chemical company ICI – developing a safer method of applying pesticides to coniferous tree plants.

The new system is known as the 'Electrodyn Sprayer Conveyor Treatment'. Young trees are individually positioned on a series of powered support wires and carried by conveyor into an enclosed spray chamber where they are treated with pesticide, using electrostatic principles (referred to earlier on page 64). Operator exposure is greatly reduced as a result.

This is a good example of the type of engineering controls required under the COSHH regulations, but it has still not afforded forestry workers full protection. Contamination can still occur from pesticide treated roots when the trees are carried up hillsides and planted.

Source: HSC Agriculture Industry Advisory Committee (AIAC), *Forestry Commission: Engineering Control Development. AIAC Paper 89/6,* 1989.

Pesticide labels give little concrete information on what types of PPE should be worn and what degree of protection they offer. This is made worse by the fact that with many types of PPE there is no standardisation of equipment, and vital safety information such as protection factors and durability is not given. Gloves provide a good example: they come in a confusing array of types and materials. The purchaser has little or no safety information on which to make a rational choice. The Health and Safety Executive (HSE) is trying to force PPE manufacturers and suppliers to provide improved safety information with their products.

One study showed that personal protective equipment worn by tractor drivers when spraying crops can be a safety hazard in its own right, by exposing the wearer to heat exhaustion and subsequent loss of concentration. The environmental conditions in the cab, and the skin and deep body temperature of the driver, were monitored over a 2.5-hour period to measure heat stress. The driver was wearing non-absorbent, plastic-type protective clothing. Results showed that even

Protective Clothing Required for
Workers on Farms and Holdings

Protective: The term 'protective' applies to items made of a material or substance which impedes the passage of pesticide such that a wearer is protected from contamination in the normal circumstances of use.

Coverall: A protective garment, or combination of garments, offering no less protection than a single garment, close-fitting at the neck and wrist and which:

(a)   covers the whole body and all clothing other than that which is covered by a good, face-shield, respiratory protective equipment, footwear and gloves, and which minimizes thermal stress to the operator when worn;

(b)   when required to be worn in connection with the use of a pesticide in the form of a granule, dust or powder, has all its external pockets covered and has its sleeves over the tops of gloves being worn;

(c)   is white or of a colour which produces a clearly noticeable contrast if contaminated with pesticide.

Gloves: Protective gloves not less than 300 millimetres in length measured from the tip of the second finger to the edge of the cuff.

Boots: Protective boots extending to at least immediately below the knees.

Face-shield: A transparent shield covering the whole of the forehead and face and so designed as to protect the forehead and face from being splashed.

Apron: A protective apron covering the front and sides of the body from immediately below the shoulders to at least 70 millimetres below the tops of any boots that are being worn.

Hood: A hat or other covering to the head so designed as to protect the forehead, neck, and back and sides of the head from contamination by pesticide in the circumstances in which is is being used.

Respiratory Protective Equipment (RPE): Any respirator of a type approved by, or conforming to a standard approved by, the HSE, including breathing apparatus.

Source: HSE, *Protective Clothing for Use with Pesticides.*[22]

on a moderately warm day, heat exhaustion sufficient to impair judgment, was likely to occur. The authors concluded that for tractor drivers, the wearing of a plastic suit was dangerous, 'when the operator needs to exercise skills, and seems most undesirable when handling potentially harmful chemicals.'[23]

Conclusion

While all efforts should be made to find alternatives to pesticides, particularly those which are hazardous and harmful to the environment, we recognise that in practical terms, and until more research is directed to integrated pest management and organic farming, pesticides will continue to be used. Recent legislation in the UK and other countries will reduce pesticide exposure of users, the public and the environment. We recommend obtaining the codes and guidance documents on pesticide safety from the relevant agencies for more detailed advice.

5

Health Risks

Working with pesticides can be dangerous. The chemicals chosen to kill pests are selected because their toxic properties make them efficient at killing unwanted plants, insects, rodents and so on. More often than not, these chemicals are potentially harmful to humans.

The effects that pesticides have on humans vary considerably, depending on the type of chemical being used. Exposure to a toxic chemical can sometimes result in an immediate (acute) effect, and/or a longer-term (chronic) effect.

Acute reactions usually occur while the chemical is being used, or shortly afterwards. The effect is often visible and, even if not, the cause can usually be identified without much difficulty. However, some acute effects produce symptoms like a cold or hay fever, and the cause of these is often overlooked. Most acute reactions last only a short time, and most victims recover completely without long-term complications. A few people may suffer permanent damage of some kind.

The sort of acute symptoms that might occur while a pesticide has been used, or shortly after, include headaches, nausea, tiredness, dizzy spells, vomiting, and even difficulty in breathing. Symptoms will vary and depend on the chemical being used.

Chronic effects of exposure are more difficult to recognise: unlike acute effects they are not immediately obvious. Symptoms may take years to appear and are not necessarily severe at first. There is a risk that they will be put down to someone simply being 'off colour', 'run down', or just 'getting old'. Establishing the cause of a chronic effect is always difficult, particularly if it does not become obvious until 20 or 30 years after the exposure occurred.

The most serious long-term effects are cancer, and genetic or birth defects. The time between exposure to a carcinogen (cancer-inducing substance) and cancer occurring can be anything from 10–40 years. This delay is known as the 'latency period'. Exposure to mutagens and teratogens (substances which induce genetic mutations or birth defects) pass the problem on to the next generation and pose particular risks to women of child-bearing age.

Different pesticides pose different types and degrees of health risk. Table 5.1 at the end of this chapter lists 20 different categories of pesticide. It gives examples of each chemical type, describes the common routes of entry into the body and lists the physical signs of exposure and the symptoms of poisoning.

Protecting Your Health

If you work with chemicals there is always the risk of something going wrong and someone being poisoned. It is important to know what to do if this happens, including what to do if you are poisoned yourself. This section describes some simple first aid procedures for dealing with various sorts of pesticide poisoning. If you follow these you could save someone's life.

First Aid

If someone is poisoned by pesticides there are a number of steps which should be taken immediately to minimise the risk of serious injury. If someone becomes ill while using pesticides, you should:

(1) Get them away from the spraying area.
(2) Take them to hospital as soon as possible. When you get to the hospital, tell the doctor what chemical was being used or, better still, bring along the label from the drum or canister. This will contain information about treatment. Tell the doctor exactly what the person was doing when they became ill, as this may affect the type of treatment chosen. If there is no hospital nearby, or transport is difficult, or the poison victim seems to be acutely ill, there are a number of steps you should take yourself to alleviate pesticide poisoning:
(3) Keep the victim still.
(4) Loosen any tight clothing.
(5) Remove any clothing which is contaminated with chemical, being careful not to get any on yourself.
(6) Wash any contaminated skin well with soap and cold water; washing on its own will help.
(7) If eyes are affected, rinse them continually with cold water for about 15 minutes.
(8) If the patient complains of feeling hot, keep them cool by fanning, or sponging them down with cold water.
(9) If vomiting occurs, make sure that the victim does not choke (and do not be tempted to make a person vomit as this can sometimes be dangerous – leave it to the doctor).
(10) If breathing stops, or weakens, give artificial respiration immediately. Remove any false teeth in case convulsions

start. Lie the patient down on their back, tilt their head back to make sure the airway is clear and that they can breathe, and start heart massage and artificial respiration straight away. The local St John's Ambulance Brigade can teach you how to do this. (In a few cases, involving corrosive or highly toxic pesticides, artificial respiration is dangerous; information about toxicity should always be checked before starting use.)

First aid is only a stop-gap measure. It is better not to have to use it at all. The rest of the section discusses what information you should ask for to help you assess the dangers of using a particular pesticide.

What You Should Know About the Pesticides You Use

Every pesticide you use should be sold with a list of instructions telling you how dangerous it is and how to protect yourself. This information is very important. Read it carefully and follow the instructions. Do not take any short cuts. If you do not have the protective clothing or the equipment recommended for use with a particular product do not handle it until the correct equipment is available. If the list is on a leaflet rather than attached to the container make sure this is kept safe and accessible so that it can be consulted quickly.

In the field there is a serious risk of contamination when pesticides are being prepared for use. This may involve dissolving a powder in a liquid; further dilution of a concentrated solution, and, finally, pouring the 'ready-to-use' pesticide into a spray bottle or tank. Splashing can occur during any of these steps, therefore it is important to protect hands, feet and face. If clothing gets wet with the pesticide, the clothes should be changed to avoid getting the chemical onto the skin.

A great deal of attention is rightly given to the dangers of inhaling the pesticide. However, for most users skin contamination presents a greater risk, and this is not so widely discussed. Although the skin acts as a partial barrier to foreign substances, many chemicals can pass through it fairly easily. The more chemical you get on your skin the greater the amount that can penetrate and get into your blood. Poisoning becomes serious once the pesticide is in the bloodstream. Chemicals take some time to get through the skin, and this is why symptoms of poisoning take longer to appear after skin contamination than after inhalation or swallowing.

Protective clothing will reduce the risk of skin contamination. However, the vapour from some of the most toxic chemicals can pass fairly easily through some 'protective' clothing, so wearing PPE should not be an excuse to take less care.

Good hygiene reduces the risk of swallowing the chemical. By not smoking, eating or drinking when you use pesticides, and washing your hands after you have finished work, you will greatly reduce your chances of swallowing anything dangerous.

If it is recommended that you use breathing equipment, such as a respirator, it is important that you do so. Some chemicals vaporize very easily. With others the way they have to be used means they are in a form that can be breathed in easily. Good ventilation and breathing apparatus are your best means of protection with chemicals which are dangerous when inhaled.

Long-term Health Problems Associated with Pesticide Exposure

Most of the toxicological information we have on pesticides comes from animal studies, or other *in vitro* (outside the body) tests using bacteria, yeast or some cell culture system. A single long-term animal health study on a pesticide costs in the region of £500,000.[1] These animal tests are required by law and in the absence of information from human studies they provide the best guide to the likely effect of the chemical on humans. For new pesticides they are the only guide we have. However, for older pesticides, which have been in use for many years and for which there is information from both human and animal studies, it is not unusual to find results do not agree. For example, lead arsenate used on vines is reported to cause lung and skin cancers in orchard workers and vine growers,[2-5] yet the chemical rarely causes cancer in animals. In this case the animal model does not help us to predict cancer in humans. The human evidence collected from agricultural and industrial surveys suggests that lead arsenate does indeed cause cancer of the lung and skin, but that individuals need to be exposed to significant amounts of the chemical before they are likely to contract cancer.[6]

For many other cancer-causing chemicals there is continuing debate about which ones pose the greatest risk and ought therefore to be restricted or banned, and which should be ignored because the risk of contracting cancer is so small. This trade-off between lives saved and the cost of controls is a feature of safety regulations everywhere. In the US, if the increase in cancer following exposure to a chemical is less than one case in a million, regulation is unlikely. Regulation is also unlikely (in the US) if the cost is reckoned to be more than US$2 million per life saved.[7] Figures of this kind are not available in Britain.

In general, pesticides would be considered a carcinogenic risk to humans if there was evidence that the chemical damaged the genetic material in a cell (genotoxic); if there was positive evidence of cancer in animals, in addition to it being genotoxic; and finally, if there is positive evidence from studies in humans. In the absence of any of this information it is important to be circumspect about the evidence on carcinogenicity and review each pesticide case-by-case. The following is a summary of some of the information from studies on humans and animals.

Aldrin/dieldrin

On the basis of five studies it is clear that aldrin/dieldrin caused cancer in the livers of mice and rats; dieldrin alone caused cancer in the lungs, and other sites, in several studies on mice.[8] The evidence in humans is less clear-cut. In a study of workers exposed to aldrin and dieldrin at a chemical plant for at least six months between 1964 and 1976, there was an increase in diseases of the lung and deaths from pneumonia. An excess of liver, oesophageal, rectal, lymphatic and haematopoietic (blood-producing) system cancers, was also noted but the increase was not statistically significant.[9] A second study of workers (average exposure 11 years) exposed to these two pesticides (as well as endrin and telodrin) revealed a lower than expected incidence of cancer, and no liver cancers.[10]

Amitrole

This chemical causes cancer in mice and rats; both species develop thyroid and liver tumours on exposure.[11] Humans exposed to amitrole have also been exposed to other herbicides, making assessment of amitrole's risk difficult. One study in Sweden reports a slight excess of cancer (at all organ sites) in railroad workers. [12]

Anilides and Phenylureas (Linuron, Diuron and Neburon)

These chemicals are broken down by plants and soil micro-organisms to give 3,4 dichloroanaline. One review suggests that all the chemicals in this family have the potential to cause cancer.[13]

Aramite

This chemical was once used as an acaricide on fruits and nuts. The chemical is no longer made in the US and was no longer registered after 1975. Aramite causes liver cancer in rats and male mice, and gall bladder and biliary duct cancers in dogs. There is no information about cancer in humans.[14]

Carbamates, Dithiocarbamates, Benzimidazole, Thiocarbamates

This is a large category of chemicals including insecticides, herbicides and fungicides. Some examples include the insecticides sevin and propoxur; the triazine herbicide atrazine and simazine; and the fungicides ethylene thiourea, maneb and zineb. These chemicals may be converted to breakdown products which are themselves carcinogenic;

they may form n-nitroso compounds in the stomach of a mammal. N-nitroso carbaryl is a potent carcinogen which causes cancer of the mouth. Ethylene thio-urea is a breakdown product of bisdithiocarbamates (such as maneb and zineb) and produces liver cancer in mice. Ethylene thiourea also produces thyroid cancer in rats and mice. Thyroid cancer in these animals is likely to have been caused by a disturbance in the function of the thyroid gland, leading to increased production of thyroid hormone. This type of chemical is called an epigenetic carcinogen. It is another way of saying that the chemical causes cancer but not by interfering directly with the genetic material in a cell.[15] The risk to individuals occupationally exposed to chemicals causing cancer in animals in this way is still not clear.

A recent study in Italy tentatively linked an increase in cancers of the epithelial tissue of the ovaries with exposure to triazine herbicides, including atrazine. Only 7 of the 65 cancer cases reviewed were definitely exposed to triazines. For those exposed for less than 10 years, the risk of getting cancer was 2.3 times greater than normal; for those exposed for more than 10 years, it was 2.9 times.[16] Although this investigation was carried out on a small number of cases, the results are nevertheless worrying.

Carbon Tetrachloride, Haloalkanes, and Alkenes

Carbon tetrachloride produces liver cancer in mice, rats, hamsters and trout. Other compounds similar in structure, such as chloroform and tetrachloroethylene, are also suspected of being carcinogenic.[17] In humans exposed to this family of chemicals there have been some cases of liver cancer in individuals who also had liver cirrhosis. Many of these chemicals have been used in dry cleaning, and an excess of leukaemia, as well as of cancers of the respiratory system and liver has been reported in dry cleaners.[18]

Chloracetamides (Alachlor), Lasso, Butachlor, Propachlor

Alachlor is perhaps the most widely used commercial herbicide in the US; it is a suspected animal carcinogen.[19] Alachlor has been restricted in the US since June 1988.[20]

Chlordecone (Kepone)

Chlordecone was used as an insecticide in the US between 1958 and 1975 for controlling leaf-eating insects, ants and cockroaches. It was also used as a larvicide for killing flies at the maggot stage. In 1975 chlordecone was discovered to be the cause of serious disorders of the nervous system and it was banned in the US. Chlordecone produces liver cancer in rats and mice.[21] There is no information on whether it causes cancer in humans.

Chlorinated Terpenes and Related Compounds (aldrin, chlordane, chlordecone, dieldrin, endrin, heptachlor, mirex, and toxaphene)

All of these chemicals are insecticides and they are toxic to the liver. In mice they cause cancer of the liver and some cause thyroid cancer. Most of the studies in rats have been negative.[22]

Several investigations have been carried out on workers exposed to chlordane and heptachlor, either in the course of manufacture of the pesticides or during their use in agriculture. The evidence is not clear. For example, in one investigation of men employed for at least three months between 1946 and 1976 and exposed to chlordane at one chemical plant, and to heptachlor and endrin at another, there was no overall excess of cancer deaths. A slight increase in the number of lung cancers was noted.[23] In a follow-up investigation of workers exposed to chlordane for at least three months between 1946–1985, the overall death rate was less than expected. Surprisingly, the incidence of cancer decreased the longer people were employed. An unexpected finding was the high number of strokes in retired workers and those no longer working at the plant.[24]

A study of workers exposed to chlordane, heptachlor and endrin between 1964 and1976 indicated no overall increase for all cancers, but an increased rate of stomach cancer in those exposed to chlordane. With so few cases it is not clear whether the stomach cancers were due to exposure to the chemical, or a chance finding.[25]

Two further investigations have been carried out on workers using chlordane and heptachlor for controlling termites. In the first, workers employed for three months between 1967 and 1976 were assessed for cancer trends. These showed an increase in bladder cancers only.[26] Nine years later this group no longer showed a high incidence of bladder cancer, but it did reveal an excess of lung cancer, a finding which had not been observed earlier. These lung cancer cases occurred in men employed for less than four years, none of whom had been exposed to high levels of chlordane, or heptachlor. It would seem unlikely, therefore, that the lung cancers were due to exposure to the pesticides.[27] Yet in another study of 4,411 pesticide applicators in Florida, some of whom used chlordane or heptachlor, deaths from lung cancer in those using pesticides for 20 years or more was nearly three times that of the general population. Three deaths from acute myeloid leukaemia (a rate slightly higher than expected) were also noted.[28]

Chlorophenoxy Acids (2,4-D and 2,4,5-T)

Both 2,4-D and 2,4,5-T have been tested for their carcinogenic effects in animals. To date the studies have been far from ideal, either because too little information was given in the subsequent report, or else too few animals were used in the trials, so although both 2,4-D and 2,4,5-T have shown increased incidences of cancer in mice, it has not proven possible to evaluate the cancer risk accurately.[29]

2,4,5-T is usually contaminated with 2,3,7,8-tetrachlorodibenzo-dioxin (dioxin) in varying amounts. The amount of the contaminant present is always tiny. Animal studies with dioxin have shown that it is a *potent* carcinogen causing liver and thyroid cancers in rats and mice.[30, 31]

Ever since the first reports of an increased incidence of cancers of soft tissue (soft tissue sarcomas) in workers exposed to 2,4-D and 2,4,5-T in Sweden, there have been many investigations of the effect of these herbicides on humans. Some of the information on the effects of exposure to these pesticides has been set out in Table 5.1. By and large, the Swedish studies are the only ones that report an increased incidence of soft tissue sarcomas in workers exposed to phenoxy herbicides.[32,33] Studies in the US and New Zealand do not agree with those done in Sweden.[34–40] However, investigations in the US confirm another claim made by Swedish scientists that the incidence of a lymph gland cancer (non-Hodgkin's lymphoma) is increased in workers exposed to phenoxy herbicides, particularly 2,4-D.[41, 42]

DDT Group: TDE, DDE, Methoxychlor, and Perthane

There is sufficient evidence that DDT causes cancer in animals. Rats and mice fed DDT in their food develop liver cancer. Mice also develop cancers of the lymph glands and of the lungs.[43] DDT has not caused cancer in hamsters, and feeding studies in dogs and monkeys are inconclusive.[44]

DDE and two of its breakdown products, chlorbenzilate and dicofol, produce liver cancer in mice. These chemicals are all suspected of belonging to a family of cancer-causing chemicals called non-genotoxic carcinogens. Non-genotoxins are regarded as being less dangerous than genotoxic carcinogens which start the cancer process by damaging DNA. DDE produced liver tumours in rats after they had also been exposed to another chemical, n-nitroso diethylamine.[45]

From the evidence available it does not seem that methoxychlor or perthane are carcinogenic.[46] In humans the evidence on DDT is difficult to interpret because workers exposed to the pesticide have usually also been in contact with other chemicals. There are few situations where the effect of DDT can be examined on its own. Four studies have been reported which show that tissue levels of DDT are higher in cancer patients than in individuals who died from other causes.[47] Other investigations report increased numbers of soft tissue sarcoma and non-Hodgkin's lymphoma cases in workers exposed to DDT. The increases were small; some of the individuals were also exposed to 2,4-D, 2,4,5-T and chlorophenols. An excess of chronic lymphocytic leukaemia has been reported in workers exposed to DDT and other chemicals.[48] Three cancer deaths where only one would be expected were recorded in forestry foremen exposed to DDT, 2,4-D and 2,4,5-T; this is a small excess.[49]

In two other studies of men involved in the manufacture of DDT there was no overall increase in cancer deaths[50, 51] but in one of the investigations deaths from respiratory system cancer were slightly increased.[52] Excess lung cancer deaths have also been recorded in agricultural workers and pesticide applicators using DDT and other pesticides.[53] In licensed pesticide applicators using DDT in addition to other pesticides, the risk of lung cancer increased the longer the person held a licence;[54] the rate was three times higher in those with a licence valid for 20 years or more.[55] There is one study – of orchard workers – which has not shown an increase in lung cancer deaths.[56]

1,2-dibromo-3-chloropropane (DBCP)

This chemical was used as a fumigant in soil to kill nematodes. Its use in the US was suspended in 1977.[57]

Orally administered, DBCP produced squamous-cell carcinomas of the forestomach in rats and mice, and adenocarcinomas of the mammary gland in female rats. In inhalation studies, the chemical produced cancer of the nasal cavity and lung in mice; the nasal cavity and tongue of male and female rats, and adrenal cortex adenomas (cancers) in female rats.[58]

A group of 550 chemical workers exposed to a variety of chemicals, including DBCP and arsenic, had an increase in deaths from cancers of all types (12 were seen, 7.7 were expected). Most deaths were from respiratory tract cancers. When those workers exposed to arsenic – known to cause lung cancer – were excluded from the investigation, there was still a small excess of lung cancer deaths.[59] Another group of 3,384 workers handling a variety of pesticides between 1935 and 1976, of whom some 1,034 were exposed to DBCP there was a slight increase in all cancer deaths. Nine cancers of the respiratory system were also seen whereas only 5 would have been expected.[60]

1,2-dibromoethane (ethylene dibromide)

This chemical is a definite carcinogen in animals. Given orally it produces squamous-cell carcinomas of the forestomach in rats and mice; alveolar/bronchiolar lung tumours in male and female mice; liver cancer in female rats; haemangiosarcomas in male rats, and cancer of the oesophagus in female mice. After inhalation, (EDB) produced adenomas and carcinomas of the nasal cavity; haemangiosarcomas of the spleen; mammary gland tumours; subcutaneous mesenchymal cancers (cancer of the tissue under the skin surface); alveolar/bronchiolar lung tumours in rats and mice, and peritoneal mesotheliomas (cancers of the lining of the lung) in male rats. EDB also produced skin and lung tumours in mice after application to the skin.[61]

1,2-dichloroethane

This is an insect fumigant used on grains and in mushroom culture, and also as a soil fumigant in peach and apple orchards. In female mice, 1,2-dichloroethane (DCE) causes cancer of the mammary glands, uterine tissue, and lungs. DCE also produces cancers of the lymph gland in male and female mice, and liver cancer in male mice. Rats develop haemangiosarcomas after exposure to DCE, with females contracting cancer of the mammary gland, and males cancer of the forestomach.

. The evidence is inadequate for an assessment of DCE-linked cancer in humans, even though there has been plenty of opportunity to study this given that the chemical has been in use since 1922.[62]

1,3-dichloropropene (Technical Grade)

The technical grade of 1,3-dichloropropene contains 1 per cent of epichlorohydrin, a known carcinogen. When 1,3-dichloropropene was introduced directly into the stomach of mice and rats, it produced cancer of the bladder, lung and forestomach in mice, and of the liver and forestomach in rats.[63]

Two cases of lymph gland cancer have been reported in firemen exposed to 1,3-dichloropropene six years earlier; the two men died of malignant lymphoma and histiocytic lymphoma respectively. In another incident, a farmer exposed continuously to 1,3-dichloropropene for 30 days developed acute myelomonocytic leukaemia one year later.[64] Given that this chemical is mutagenic (in bacteria and human cells),[65] it must be monitored closely because it could cause cancer in humans.

Dimethylcarbamoyl Chloride

This is an intermediate chemical formed during the production of substituted ureas and carbamate pesticides such as Tandex, Dinetalon and Pyremor (all US pesticides). Apparently there is no information available about the concentration of dimethylcarbamoyl chloride in these pesticides, and the extent of human exposure is unknown. In one study of 107 workers involved in dimethylcarbamoyl production for periods ranging from 6 months to 12 years, there was no reported death from cancer. In animals, the chemical produces localised tumours; it causes skin cancer when applied to the skin of mice, and nasal tract carcinomas in rats and male hamsters inhaling the chemical.[66]

Hexachlorobenzene and Chlorinated Benzenes

Hexachlorobenzene was used previously as a fungicide to control wheatbunt, and slut fungi on other grains. It is present as an impurity in several other pesticides including dimethyltetrachloroterephthalic acid and pentachloronitrobenzene. In animals hexachlorobenzene causes cancer of the thyroid and liver in hamsters, and of the liver in mice and female rats.[67]

In the late 1950s there was an epidemic of poisoning in Turkey caused by the consumption of grain treatment with hexachlorobenzene. Twenty-five years after the incident, the population was surveyed but no excess of cancer was reported.[68]

Lindane

This insecticide has been covered at some length in Table 5.1. Lindane produces liver tumours in mice.[69–71]

Malathion and Malaoxon

Malathion causes an increase in benign and malignant cancers at all (but not the same) sites in two species of rats (the Osborne-Mendel and Fischer-344); males are more susceptible than females. Malathion also caused an increase in the incidence of liver cancer in male mice (B6C3F$_1$ species). Malaoxon – a breakdown product of malathion caused by exposure to light – causes an increase in both benign and malignant cancers at all sites, including endocrine (hormone-secreting) organs, in Fischer-344 rats.[72] Cancers were also produced in the adrenal gland, liver and the haematopoietic (blood-producing) system.[73]

Mirex

Between 1962 and 1978 mirex was used in the south-eastern US to control fire ants and other ant species. There was also limited use of mirex as a flame retardant. In soil mirex degrades to chlordecone (kepone).

Mirex produces liver cancer in rats and mice; and reticulum (bone marrow) cell cancers in the males of two strains of mice.[74] There is no information about mirex's carcinogenic risk to humans.

Nitrofen

This is a contact herbicide used for pre-emergent and post-emergent control of annual grasses and broad-leaf weeds which are problems on land used for the cultivation of various food crops and flowers. Most nitrofen stocks were recalled in the US in 1980. Technical grade nitrofen caused liver cancers in mice and cancer of the pancreatic gland in female rats.[75] There is no information about its carcinogenicity in humans.

N-nitroso-di-n-propylamine (NDPA)

NDPA is a contaminant found in pesticides such as Treflan and isopropalin (used in the US).

NDPA produces cancers of the liver, kidney, oesophagus and respiratory tract in rats and hamsters. A breakdown product of NDPA also produces cancer of the liver and respiratory tract in rats and hamsters. There is no information about the carcinogenicity of NDPA in humans.[76]

N-nitroso Derivatives

Many nitrogen-containing pesticides (substituted ureas, carbamates, quaternary salts, and the triazine pesticides) could have a nitroso group added to them during manufacture or at some stage thereafter. This addition makes these compounds more of a carcinogenic risk. 80 per cent of the n-nitroso derivatives are known to be mutagenic, or carcinogenic. Many of these chemicals are further metabolised in plants and animals. Di-alkylnitrosamines (an impurity of dinitroanaline-based herbicides) are among the most potent animal carcinogens. Cyclic nitrosamines, produced from compounds containing what are known as tertiary amines, are also proven carcinogens.[77]

Organophosphates (dichlorvos, DDVP, dimethoate, parathion-methyl, parathion, phosphamidon, tetrachlorvinphos, trichlorphon)

These organophosphates are in a family of generally non-persistent insecticides. Many organophosphates are mutagenic in cell culture systems (mainly bacteria). However, the chemicals have not, as a rule, caused cancer in animals (but see malathion, Table 5.1, p. 78). These negative results may be because the organophosphates are both extremely toxic and break-down very quickly in the body.[78] Their extreme toxicity means that animals cannot be exposed to too high a concentration before they begin to suffer symptoms of poisoning.

Looking at some of the organophosphates individually the evidence suggests that great care is needed when dealing with them.

On the basis of its chemical structure, dichlorvos would be suspected of being a carcinogen. The chemical is also a mutagen in bacterial tests using *Salmonella*. But dichlorvos does not cause cancer in animals and therefore it is not considered to be a carcinogen.[79]

Similarly with parathion-methyl, this chemical's structure would alert one to the possibility of its being a carcinogen. In bacterial systems this chemical is a mutagen. However, it does not cause cancer in animals.[80]

Parathion would also be considered carcinogenic, both because of its structure and the fact that it is a mutagen in bacteria. The results from carcinogenicity tests in the rat are equivocal and difficult to extrapolate to humans.[81]

Phosphamidon should also make us pause, again because of its structure, and because it is a mutagen in bacteria. Like parathion however, animal test results are equivocal and difficult to use for predicting the risk to humans.[82]

Tetrachlorvinphos would again alert us because of its structure, but it is not a mutagen. It does, however, cause cancer of the pancreas in rats and mice.[83] Tetrachlorvinphos must, therefore, be considered to pose a potential carcinogenic threat to humans. The other organophosphates must also be considered in this light, although the fact that they do not cause cancer in animals suggests that the carcinogenic risk to humans is much less than their mutagenic activity would lead us to believe. The danger from the organophosphates is more from their acute toxic effects than their carcinogenic potential.

Paraquat

Some 760 employees involved in the manufacture of paraquat by ICI in the UK have been monitored for cancer risks. 79 deaths have been recorded in this workforce. The overall incidence of all cancers in the group was slightly raised and more lung cancer cases occurred than were expected (13 observed against 10.5 expected). Workers with the higher incidence of lung cancer were involved in a process using either high-temperature sodium between 1961–1963, or low-temperature sodium from 1966 to the present day. From the information available, the increase in lung cancer is not related to an increased exposure to paraquat. Smoking is not the cause of the excess lung cancer cases either.[84] A previous study (see paraquat, Table 5.1, p. 100) shows that workers involved in the manufacture of paraquat had a higher incidence of skin lesions and skin cancer.[85]

Sulfallate (di-ethylthiocarbamic acid cnd 2-chloroallyl ester)

Sulfallate is a pre-emergent selective herbicide used against certain annual grasses and broad-leaf weeds in the US. The herbicide has also been used on shrubbery and ornamental plants. Sulfallate causes mammary gland tumours in female mice, and cancer of the forestomach in male rats, and of the lung in male mice. No information exists about its carcinogenic risk to humans.[86]

Toxaphene

In the US following the elimination of DDT, the use of toxaphene (a mixture of chlorinated xanthines) has increased. The chemical is used mainly as an insecticide against pests on cotton, soya beans, sorghum, peanuts, livestock and poultry. The evidence would suggest the chemical is an epigenetic, or non-genotoxic, carcinogen (it causes cancer but doesn't interfere directly with the cell's DNA genetic material) as the technical grade of the insecticide causes liver and thyroid cancer in rats. There is no information about its carcinogenic risk to humans.[87]

Trifluarin

Trifluarin is a pre-emergent herbicide which may be contaminated with a potent carcinogen called dipropylnitrosoamine. In female mice the pesticide causes liver, lung and forestomach cancers.[88]

Others (ethylene oxide, formaldehyde, N-(2-hydroxyethyl) hydrazine and beta-propiolactone)

All these chemicals are genotoxic carcinogens (they act by interfering directly with DNA in cells). All the chemicals cause cancer in animals and could cause cancer in humans. The evidence that they have done so is equivocal. However, given the nature of the chemicals, and the way in which they exert their carcinogenic effect, they must all be used with great care.[89]

Permethrin

Permethrin is not reported to be mutagenic in bacterial systems (or others) nor is it a mutagen in animals. Of the three long-term cancer studies in mice, there was evidence that the insecticide caused lung cancer in one strain of mice at the highest dose of the chemical tested. Permethrin did not cause cancer in rats. On the basis of this information the chemical is probably an epigenetic carcinogen with a low carcinogenic risk to humans.[90]

Cancer and Occupation

Many studies have been carried out to assess the risks farmers face from pesticides and other chemicals. Although many of the studies suggest there this is a link between the chemicals and cancer, it is usually difficult to make a firm pronouncement as most individuals affected have been exposed to a wide variety of substances, including viruses.

Farmers were more likely than others to die from the lymph gland cancer, non-Hodgkin's lymphoma, according to a review of 774 deaths in the US State of Wisconsin; no individual chemicals were singled out by the study. Farmers were also at increased risk of developing reticulum (marrow cell) sarcomas. The risk of developing reticulum sarcomas was linked to the production of grains, soya beans, corn, animals, fertiliser use, and the acres treated with herbicides and pesticides.[91] Farmers were also prominent in a second study of 411 people who died from multiple myeloma in Wisconsin, and they were at greater risk if they were older than 65.[92]

Yet another study reports farmers as more likely to contract cancer of the stomach and prostate gland, non-Hodgkin's lymphoma, and multiple myeloma. Like the previous study the risk of developing multiple myeloma was linked with exposure to chickens, herbicides and insecticides. The risk of developing non-Hodgkin's lymphoma was linked with similar risk factors, whereas prostate cancer had no specific link with any of these. Stomach cancer was linked with the quantity of milk products sold, the number of cattle on the farm, and the quantity of corn per acre.[93]

A third study on multiple myeloma (in Sweden) found that agricultural workers were in a high-risk group, but this study had no information about the quantity of chemicals that workers were exposed to, making it difficult to draw definitive conclusions.[94]

Leukaemia is one cancer where there is good evidence to show that farmers are at risk of contracting the disease, and the risk is linked with exposure to insecticides, and exposure to animals, particularly chickens, pigs and cattle.[95-97]

Cancers of the lymph gland have also been shown to be more common in farmers than in other groups. One investigation of 25,945 Italian male farmers, of whom 631 had died from cancer, recorded a far higher proportion of lymph gland and skin cancers than would have been expected.[98] Although the overall incidence of cancer was lower than expected in 5,923 Icelandic farmers, there was a small but definite increase in the number of cases of skin cancer, Hodgkin's and non-Hodgkin's lymphoma, and leukaemia.[99] This link between non-Hodgkin's lymphoma and pesticides was not found in a study of 501 cases who died from the disease in North Carolina. Farming was not one of the high-risk occupations and there was no apparent link between pesticide use and the incidence of cancer.[100]

Stomach cancer is reported in several studies to be linked with working in agriculture; 1,214 pest control workers in one study had a significantly increased incidence of cancer of the stomach, oesophagus and skin. The likelihood of dying from stomach cancer also increased with the length of time workers were licensed to use pesticides, although no specific chemical was implicated in the cancer deaths.[101] Another investigation of the causes of 266 stomach cancers in Chile reported that the cancer risk increased if the individual worked in agriculture, or else had lived in the area for the first part of their life.

It is suggested that there may be a relationship between exposure to nitrate fertilisers and stomach cancer given that Chile is a producer of nitrates. Further investigation of this cohort would be helpful for interpreting the stomach cancer risk.[102] A third study showing that farmers were at increased risk from stomach cancer was carried out in the Canadian State of British Columbia and involved 28,032 male farmers over the period 1950–1978. Analysis showed that there was an increased risk of farmers developing cancer of the hip, stomach and prostate gland, leukaemia and aplastic anaemia.[103]

A small increase of the incidence in cancer of the testes has been noted in a study of 20,245 Swedish pesticide applicators. Individual pesticide exposure was not known, so the incidence of cancer was assessed according to how long people had been in contact with the chemicals; this showed that testicular cancer did not increase with increasing contact with chemicals, a finding which might have been expected had there been a real link between exposure and the cancer. The overall incidence of cancer deaths and of specific cancers of the liver, lung, pancreas and kidney were significantly lower than would have been expected in this group of applicators.[104]

Two investigations by Italian scientists suggest that workers in agriculture are at risk of developing brain cancers (gliomas). In 1982 61 glioma cases had been identified and it was noted that the cancer was more likely to occur if the person was employed in agriculture.[105] A further study of 240 brain glioma cases diagnosed between 1983–1984 in two hospitals in Milan confirmed that farmers were at increased risk of developing the cancer. The risk of a farmer developing the tumour was 60 per cent greater than for other workers. Farmers exposed to insecticides, fungicides, herbicides and fertilisers were the ones at risk. Of the chemicals involved exposure to insecticides or fungicides carried the greatest risk (twice that of the other groups). Many farmers reported that the fungicides they used were commercial compounds based on copper sulphate. Some copper compounds contain methyl urea, itself a known carcinogen of the nervous system in animals. The brain gliomas may therefore be related to the presence of the methyl urea, and not to copper. Only one patient had used a herbicide containing methyl urea; most of the glioma cases were exposed to methyl urea through the use of fungicides.[106]

Workers involved with forestry, or in contact with wood, are reputed to be at increased risk of developing a variety of cancers. Exposure to specific chemicals has not been investigated. The suggestion is that exposure to pesticides in general may be the reason for the increased incidence of cancer. In one of the earlier investigations, a review of the death certificates of 1,549 white males aged 25 years, or older, and who had died of Hodgkin's disease in New York State between 1940–1953, and 1957–1964, suggested that those who had been in contact with wood (for how long was not known), and whose country of birth was given as Italy, were at increased risk of developing Hodgkin's disease.[107] An earlier investigation had shown that wood workers were at increased risk of developing cancer of the nasal cavity and accessory sinuses (exposure to hard woods is linked with nasal cancer).[108]

Another study refers to the high number of woodworkers in a group of 1,577 cases of Hodgkin's disease; woodworkers were 1.6 times more likely than others to develop the cancer.[109] A large study of some 36,622 white workers of the US National Furniture Workers' Union, employed for 20 years or more, showed the men to be an increased risk of developing cancers of the lymph gland and haematopoietic system (marrow). The incidence of leukaemia, and non-Hodgkin's lymphoma, was twice what would have been expected in the normal population. The deaths from acute myeloid leukaemias were particularly high, and 4.7 times higher than would have been expected.[110]

A more recent study again points to an increased risk of people in contact with wood developing non-Hodgkin's lymphoma. Workers involved in forest conservation in the US were said to be at increased risk for non-Hodgkin's lymphoma and colon cancer. Occupational exposure to specific chemicals, family history details, or exposure to drugs that might affect the immune system, are not referred to by the authors of the study. They do say that smoking habits or socio-economic factors were not the cause of the cancers and that the increased incidence of non-Hodgkin's lymphoma in people in contact with wood in their study is similar to that observed by others.[111]

Allergies

Many chemicals can cause allergic reactions and pesticides are no exception. True allergies may show up when someone develops asthma after exposure to extremely low concentrations of a chemical. Immuno-logical tests can also be used to demonstrate a hypersensitivity to chemicals. Often it is a change in the blood level of the specific immunoglobulin IgE that is associated with hypersensitivity.

The organophosphate insecticides are widely used and have been introduced because they degrade in the environment far more rapidly than the organochlorine compounds they have replaced. Poisoning with organophosphates is common. Parathion-methyl, a very toxic organophosphate, is responsible for many deaths in developing countries. Exposure to small concentrations of organophosphates can affect the nervous system (see Table 5.1). In the blood, a fall in the activity of the enzyme cholinesterase below an individual's baseline value is indicative of exposure to organophosphates. This is a sensitive test that is widely used. A fall of 30 per cent in the enzyme activity is evidence of serious exposure, and all workers must be removed from their work environment if cholinesterase activity falls by this amount.

Allergic reactions to organophosphates can occur at concentrations of the pesticide 20 times lower than the amounts regarded by the World Health Organisation (WHO) as an acceptable intake; at these levels there will be no fall in cholinesterase activity. Two cases were reported recently where asthma developed in individuals exposed to small quantities of organophosphate insecticide, but in whom no alteration of cholinesterase levels in blood were seen. The cases involved an abattoir worker found to be sensitive to fenthion – used to protect sheep skins during storage – and a cat owner affected by dichlorvos in a 'pet collar' worn by the cat to deter fleas. In the former case the amount of insecticides that was estimated to have caused the asthma was one-twentieth of that considered an acceptable intake by the WHO.[112]

Fungicides caused allergic skin reactions in 46 individuals out of a group of 652 who were tested. Contact dermatitis occurrs in many agricultural and farm workers exposed to pesticides such as copper sulphate, while allergic contact dermatitis is caused by the likes of bis-dithiocarbamates and benomyl. It would appear from one investigation that many patients who worked, or had worked, on the land were particularly sensitive to chemicals.[113, 114]

Pesticides and Immunity

Many people exposed to pesticides describe their symptoms as similar to those caused by 'flu'. The runny noses, muscle aches, rheumy eyes and tiredness suggest that exposure to pesticides has affected the immune system. Very little research has been done in this area and it is one where considerable research effort will have to be directed in future. From what evidence there is it is clear that chemicals and heavy metals will affect the immune system.[115, 116]

If vague 'flu-like' symptoms develop after using pesticides it is important that GPs be contacted and made aware of the problem. Many GPs have little or no idea of the symptoms likely to occur in people exposed to pesticides. If you suspect that your symptoms have been caused by pesticides make sure that you tell somebody.

Conclusions

From the evidence we have presented it is quite clear that exposure to pesticides is not without risk. In some situations exposure can lead to dire consequences, as evidenced by the number of deaths linked to both acute and longer-term contact with the chemicals. Deaths from acute exposure are far less common in the developed world than they are in the developing world (see Chapter 9). However, it is clear that there is still a disturbing lack of information on the longer-term risks of exposure to pesticides. Where we could, we have presented information about the outcome of exposure to particular chemicals. In many circumstances, however, workers have been exposed to a bewildering array of substances and it is difficult to extract the information about the risk individual chemicals may pose. In this situation it is difficult to apportion blame. It is equally difficult to undertake investigations to identify the cause of an increased incidence of a single cancer. If people are exposed to a variety of chemicals, then studies will be extremely complicated and may not reveal the information we are seeking.

A disturbing feature of all this is the absence of accurate information on the extent of acute poisoning in developing countries. Results from some of the investigations that have been carried out indicate that there is a major problem in the developing world. Deaths from acute poisoning would appear to be extremely high, a pattern that is not seen in the more developed nations. Of equal concern is the absence of any information about the effects of long-term exposure to chemicals in the developing world. We have virtually no information about this. Given that many workers in developing countries are exposed to exceedingly high concentrations of pesticides, information about the health of these workers over the longer term is urgently needed. Mortality studies and investigations to determine the incidence of cancer in workers in the developing world should be undertaken.

In the developed world, it is clear that there are far fewer deaths from acute pesticide poisoning than there are in the developing nations. Workers everywhere, however, are still concerned about the cancer risks. Regrettably, no guarantees can be given that there is little or no cancer risk from pesticides. This is why we feel that workers or consumers need to be provided with all the information about the toxic effects of particular chemicals so they can make an informed choice about the product they use.

We have concentrated on the effects of acute poisoning and also the longer-term, chronic problem of cancer. We have not said a great deal about the effects of exposure to particular chemicals on fertility. There only two chemicals where there is definite evidence that repro-duction is affected following exposure: di-bromochloropropane (DBCP) and ethylene dibromide (EDB). Workers exposed to DBCP have a reduced sperm count and in the majority of those affected the sperm count does not return to normal.[117-121] An investigation of the expected births in families of workers exposed to DBCP showed that the exposed group had a lower number of children than would have been expected. In one population exposed to EDB, men on one chemical plant showed a significant decrease in fertility.[122] Apart from these two chemicals, there are no other pesticides that are linked with a failure to reproduce. However, that is not to say that other pesticides do not have this effect, it is simply to note that the information is not good enough to make an assessment.

The herbicide 2,4,5-T is said to have caused spontaneous abortions in women in the US.[123] Vietnamese and US veterans of the Vietnam War who were exposed to the 2,4,5-T based herbicide Agent Orange are reported to have fathered a disproportionately high number of children with birth deformities.[124-126] The evidence, when reviewed carefully, does not lend itself to the conclusion that 2,4,5-T (or the dioxin contaminant in it) is to blame because none of the investiga-tions gives a clear-cut result. Perhaps in a few years, when 2,4,5-T is such a less controversial herbicide, it will be possible to assess its danger a little more objectively.

Table 5.1: Pesticide Poisoning and Symptoms

Type and Example	Entry, Physical Signs, Symptoms and Consequences
(1) Dinitro dinoseb, dinoterb, DNOC (dinitro-o- cresol).	*Entry into Body* Through the skin. Inhaling spray. *Physical Signs* Yellow stain on skin/hair indicates overexposure. *Symptoms and Consequences of Poisoning* Tiredness, excessive thirst, excessive sweating. Serious illness likely if patient is anxious, restless, breathing rapidly and feels hot.[127] Dinitro compounds interfere with the production of energy in the body (by affecting oxidative phosphorylation).[128] Loss of weight and inability to sleep suggest longer-term chronic exposure. The symptoms of chronic exposure to DNOC may resemble the medical condition of hyperthyroidism. A detailed occupational history will help to establish that the pesticide is the likely cause. Thyroid function tests will exclude hyperthyroidism. Blood and urine tests for DNOC are helpful. Blood levels of DNOC above 10 parts per million (ppm) represent considerable intake of the chemical. DNOC is cleared from the bloodstream extremely slowly and anyone with a blood level of 20 ppm 8 hours after exposure should not be in contact with the pesticide for at least 6 weeks.[129] DNOC is a very toxic chemical and responsible for a number of deaths. Between 1946 and 1959 there were 9 deaths from DNOC poisoning whereas organophosphorus compounds killed 2 people.[130] DNOC is reported to be a mutagen[131] but so far it has not been listed as a potential human carcinogen.[132] Dinoseb is less toxic than DNOC and the deaths reported to have been caused by this pesticide up until 1981 were suicides.[133]
(2) Uncouple Oxidative Phosphorylation bromoxynil, ioxynil	*Entry into Body* Through skin most common. *Physical Signs* Compounds not brightly coloured. *Symptoms and Consequences of Poisoning* The symptoms of acute and chronic poisoning

Type and Example	Entry, Physical Signs, Symptoms and Consequences
	are similar to those above; they include headaches, thirst, sweating, restlessness, rapid breathing and anxiety in acute episodes, and loss of weight and inability to sleep after chronic poisoning.[134]
	Bromoxynil and ioxynil are toxic at the same concentrations.
	Poisonings are reported in workers making these chemicals in industry. All of those affected recovered after they were taken off the site.[135]
	The chemicals in this class also affect energy production in cells.
pentachlorophenol	*Entry into Body* Inhaling spray mist for dust for long periods. *Physical Signs* Staining of skin. Unreliable as proof of exposure. *Symptoms and Consequences of Poisoning* Pentachlorophenol is a wood preservative and it has been the cause of many deaths over the years in workers either manufacturing or applying the chemical to timber. Most deaths have occurred following skin contamination with 1–2% solutions of the chemicals, or through powder settling on the skin and subsequently getting wet and being absorbed.[136–140]
	Pentachlorophenol is also reported to have caused a breakdown of red cells in the body,[141] and more seriously to have poisoned the bone marrow in 4 workers, causing aplastic anaemia. In 2 other workers the chemical was said to have caused a failure in the production of red blood cells; one of these workers subsequently developed leukaemia.[142]
	Pure pentachlorophenol causes liver cancer in mice,[143] but it is far from clear whether it causes cancer in humans. Apart from the case of leukaemia referred to above, penta-chlorophenol has been linked with other chlorinated phenols and phenoxy herbicides as the cause of cancer of the lymph gland in forestry workers.[144] Some other chlorinated phenols, such as 2,4,6-trichlorophenol which is known to cause cancer in animals,[145] may be

Type and Example	*Entry, Physical Signs, Symptoms and Consequences*
	the more likely cause but it will probably be years before we know what to blame.
(3) **Organo-chlorine**	*Entry into Body* Through skin – most important. *Symptoms and Consequences of Poisoning* Headache, nausea, vomiting, tiredness and dizziness can occur. A few hours after exposure the patient becomes anxious and over-excited; muscle twitching and convulsions follow. Breathing becomes rapid – this later slows down. • **Danger**. Onset of symptoms may be delayed for 2 days.
DDT	*Symptoms and Consequences of Poisoning* Although many workers have been exposed to significant concentrations of DDT there have been few fatalities reported from acute poisoning with this pesticide.[146] Most cases have exhibited the symptoms listed above with some exhibiting paralysis in the hands and feet but recovering after 48–72 hours.
aldrin, dieldrin	*Symptoms and Consequences of Poisoning* Aldrin is converted to dieldrin very rapidly in humans. Poisoning with the pesticide has occurred in workers using the chemical as well as in families consuming grain treated with aldrin. Most people have recovered after exhibiting a range of the symptoms described above. Serious poisoning by aldrin usually involves convulsions and is similar to poisoning by dieldrin. An exception was a 3-year-old girl who, after poisoning, was briefly excited and unable to walk before she collapsed, fell into a coma, and died.[147] Dieldrin was primarily used as an insecticide in agriculture although in tropical countries it was sprayed with the intention of leaving residues on walls and ceilings of homes to control the malaria mosquito. Most occupational exposure has occurred among sprayers involved in malaria control although some farmers have been poisoned as well.[148] The overwhelming majority of those poisoned have recovered although there was

Type and Example	Entry, Physical Signs, Symptoms and Consequences
	the tragic death of a child who drank some dieldrin.[149]
endosulfan	*Symptoms and Consequences of Poisoning* Several suicides and accidental deaths are attributable to endosulfan. One 70-year-old woman died a few hours after taking a few drops of the pesticide for a stomach complaint. Four others who consumed far more endosulfan died much quicker – under an hour in one case.[150] Many workers seriously poisoned by endosulfan have had convulsions and in one case the fit was so violent that two vertebrae in the spine were fractured.[151] In another case a man who had been loading spray planes for a month developed a dull headache, weakness, stomach pains, a feeling of warmth and lost his appetite. He thought the symptoms were the start of 'flu'. However, on returning to work the following day he collapsed, had mild convulsions and remained unconscious for 2 hours; recovery was uneventful except that his memory was completely blank for about 2 hours before and after his convulsion.[152]
chlordane	*Symptoms and Consequences of Poisoning* Apparently the majority of established cases of poisoning with chlordane have occurred through gross overexposure to the pesticide. In most instances convulsions started half an hour to 3 hours after the chemical was swallowed, or absorbed through the skin. Convulsions were often the first indication of poisoning.[153,154] A case of megaloblastic anaemia,[155] and one of aplastic anaemia leading to acute leukaemia,[156] have been linked with the exposure to chlordane although in both instances it is possible that other factors were to blame.
lindane	*Symptoms and Consequences of Poisoning* Dermatitis and irritation of the nose and throat are also symptoms of exposure to lindane, the irritation occurring particularly after exposure to the pesticide in vapour form.[157,158] One serious condition linked with exposure to lindane is that of aplastic anaemia.[159–163]

Type and Example	*Entry, Physical Signs, Symptoms and Consequences*
	Many cases of this serious blood disorder have been reported in people exposed to the technical preparation of the insecticide known as BHC. BHC is sometimes referred to as benzene hexachloride or hexachlorocyclohexane. Lindane is a particular form of hexachlorocyclohexane and the only one efficient at killing insects.[164]
endrin	*Symptoms and Consequences of Poisoning* Some of the chlorinated insecticides cause cancer in animals and should be treated with great care; they are: DDT, aldrin, dieldrin, chlordane, endrin and lindane.[165–169]
(4) Organo-phosphates demeton, dichlorvos, mevinphos, parathion, malathion, triorthocresyl phosphate (not a pesticide)	*Entry into Body* Through skin inhaling or swallowing. *Symptoms and Consequences of Poisoning* Exhaustion, weakness and a sense of confusion can occur during exposure and may even be delayed up to 12 hours. These symptoms can even occur days after exposure. Vomiting, cramp-like pain in stomach, cold sweats, dribbling from the mouth and diarrhoea may follow.[170] Eye and chest symptoms may just indicate limited, local exposure of the eye and lungs. Muscular twitching indicates severe poisoning. The organophosphate insecticides (OPs) vary a great deal in their toxic effects with some being much more toxic than others. In general there has been a tendency to replace the most toxic chemicals with others which are less toxic.
dichlorvos	*Symptoms and Consequences of Poisoning* Deaths have occurred from exposure to many different types of OPs. Two workers in Costa Rica died after splashing their arms with a concentrated solution of dichlorvos and failing to wash it off.[171]
demeton	*Symptoms and Consequences of Poisoning* Demeton is also responsible for a number of deaths,[172] and in particular for some fatalities in workers exposed to the chemical at work.[173]

Type and Example	Entry, Physical Signs, Symptoms and Consequences
parathion, parathion-methyl	*Symptoms and Consequences of Poisoning* Parathion and parathion-methyl are responsible for many poisonings and deaths particularly in developing countries.[174–175] Much of our knowledge about the medical effects of organophosphate poisoning has been collected from people poisoned by parathion. The chemical has been used in countless suicides and even as a poison in various murders. In Denmark between 1951 and 1963 parathion was used in 344 suicides, 5 murders and one suspected murder.[176] The organophosphates are probably the one family of pesticides that we know most about. We know for example that they act by stopping the message being transmitted from the nerve to the muscle. This blockage, brought about by inhibition of the enzyme acetylcholinesterase, causes muscles to go into spasm and is why victims of poisoning often have abnormal twitching, or even convulsions. The steps involved in poisoning are also well understood and provided treatment can be started rapidly there is usually a good chance that it will be successful. However, prompt treatment is not always available and for this reason it is essential that some of the more toxic OPs are replaced by others that are less toxic. In developing countries where doctors and hospital treatment are thin on the ground, replacing parathion and parathion-methyl by less toxic alternatives should be a high priority. For most OPs there are usually no long-term complications from poisoning. But this is not true for chemicals such as triorthocresyl-phosphate (TOCP). TOCP is not an insecticide. Acting through a different process TOCP causes direct damage to the nerves which could lead to paralysis. As many as 20,000 people may have been paralysed in the US in the 1930s after consuming an extract of Jamaica ginger fortified with TOCP. These were the years of prohibition in America when drinking alcohol was against the law. Because alcohol could not be obtained legally, many illegal products were made and sold outside the law; Jamaica ginger was one of these. Some of the victims of the ginger extract poisoning recovered fairly

Type and Example	*Entry, Physical Signs, Symptoms and Consequences*
	promptly, but others were paralysed for life.[177,178]

OP poisoning can be confirmed by a blood test. The laboratory will measure an enzyme in the blood called cholinesterase. If the enzyme is less active than in 'normal' individuals this suggests poisoning by an OP. However, the activity of cholinesterase varies between individuals and it is not always possible to infer from a low value that a person has been exposed to the pesticide. Ideally, 2 blood tests several weeks apart should be done to check that the enzyme has returned to the higher normal value after the incident. If this increase occurs it is near certain that OPs caused the poisoning.

Some of the OPs have caused cancer in animals during testing. However the way that the tests have been carried out has not always been satisfactory so it cannot always be said that the same pesticides will cause cancer in humans.[179]

(5) Carbamates
aldicarb, aminocarb, carbofuran, propoxur

Entry into Body
Through skin, breathing, or swallowing; also local effect on eyes.
Symptoms and Consequences of Poisoning
The symptoms of carbamate poisoning are similar to those described in the previous section for organophosphates. Carbamates poison in a similar way to the OPs but they can act more rapidly.

Recovery from carbamate poisoning is also generally more rapid.

The speed with which poisoning occurs, the severity of the poisoning and the length of time that symptoms persist depend on the chemical structure of the individual pesticide.[180, 181]

carbaryl

Symptoms and Consequences of Poisoning
Carbaryl appears to have caused episodes of poisonings in the past but the one death recorded with this pesticide occurred in a 39-year-old man who, while drunk, swallowed half a litre of an 80% solution of the pesticide. The man died 6 hours later.[182]

aldicarb

Symptoms and Consequences of Poisoning
Aldicarb-contaminated cucumbers poisoned a

Type and Example	Entry, Physical Signs, Symptoms and Consequences
	number of people aged between 6–80 in Nebraska in the late 1970s. The source of the poisoning was traced to a greenhouse that had been treated on two occasions with the pesticide. After eating the cucumbers most people had symptoms some 15–60 minutes later, although in some individuals they occurred more than 2 hours after eating. The poisoning caused diarrhoea, nausea and vomiting, stomach pain, sweating, blurred vision, headache, muscle twitching, difficulty in breathing and temporary paralysis in the hands and feet. Age did not affect how severely people were affected; young and old had the same range of symptoms. Everyone recovered without needing any specific treatment.[183]
propoxur	*Symptoms and Consequences of Poisoning* Propoxur has been used extensively throughout the world in malaria control programmes and subjected to extensive field trials in El Salvador, Iran and Nigeria. Episodes of poisoning occurred in those individuals carrying out the spraying and in some of the homes that had been sprayed. Symptoms similar to those described above were reported and when investigated it appeared that most of those affected had absorbed some pesticide through their skin. Those who washed it off rapidly were the least affected. Some of the householders were affected by vapour because they were said to have gone back into their homes too soon after the spraying.[184]
	Carbamate poisoning can be established by a blood test. As with OPs the laboratory will measure the enzyme cholinesterase.
(6) Fluoroacetic acid derivatives fluoroacetamide	*Entry into Body* Only major concern is ingestion by mouth. *Symptoms and Consequences of Poisoning* The delay in onset of symptoms is anything from 30 minutes to several hours. Symptoms include feeling sick and anxious; these are followed by muscle twitching and shaking of the arms and legs. Convulsions and unconsciousness will occur if poisoning is severe. Fluoroacetic acid compounds are used to kill

Type and Example	*Entry, Physical Signs, Symptoms and Consequences*

rodents. In the UK their use is restricted to ships' holds and sewers.[185]

Fluoroacetamide has been responsible for the deaths of several young children in Britain and the US.[186] In one case an 18-month-old girl died 4 days after drinking some of the rodenticide from a bottle kept in a kitchen drawer which she was able to reach. Although the little girl was given something to make her vomit, proper treatment was not started until nearly a day later when she was found unconscious. By then it was too late, and she died 3 days later.[187] Like many other pesticides, prompt treatment is necessary for cases of poisoning with fluoroacetic acid compounds.

(7) Organic mercury, Aryl compounds
phenyl mercury-acetate

Entry into Body
Through skin or by inhaling.
Physical Signs
Skin burns which redden and blister after 6–12 hours.
Symptoms and Consequences of Poisoning
Stinging of eyes and throat. Phenyl mercury fungicides have been used extensively with very few reported problems.[188] There is some suggestion, however, that these compounds may cause damage to the kidneys. Mild kidney damage has been reported in one man who was heavily exposed, however as he also had an acid burn on his skin it may have been this that caused the kidney damage.[189]

a) Organic mercury, Alkyl compounds
ethylmercury phosphate, methoxyethyl-methyl chloride

Entry into Body
Through skin or by inhaling.
Physical Signs
Skin burns which redden and blister after 6–12 hours.
Symptoms and Consequences of Poisoning
Many symptoms are the same as those caused by aryl mercury compounds.

In addition there are effects on the nervous system. Long-term exposure leads to tiredness, loss of memory, an inability to concentrate, and a tingling and numbness in lips, fingers and toes. Shaking of arms and legs, and difficulty in walking, are early signs of more severe damage

Type and Example	Entry, Physical Signs, Symptoms and Consequences
	to nerves.[190] Some, but not all, workers exposed to alkyl mercury complain of a metallic taste in the mouth and slight stomach upsets.[191] Severe poisoning can also cause a loss of side vision – as if someone were wearing blinkers – and can make people extremely irritable and bad-tempered. Mental retardation is not uncommon and has been reported in children who have been poisoned.[192] The symptoms frequently become worse after the illness has been diagnosed and exposure to the chemical has stopped. Mental disorders are not uncommon after organic mercury poisoning and some patients have been admitted to psychiatric hospitals.[193]
	Recovery from poisoning is extremely slow and if poisoning is severe an improvement may not be seen. Over half the children poisoned after eating contaminated seed grain in Iraq were still physically and mentally handicapped 2 years later.[194]
	Children exposed to methyl mercury in their mother's womb may be severely and permanently injured even though their mothers show no symptoms.[195]
	In the Iraqi poisoning incident mentioned above 6,148 patients were admitted to hospital, of whom 452 died.[196]
(8) Inorganic mercury mercuric chloride, mercuric oxide	*Entry into Body* Accidental or intentional ingestion by mouth, much more likely than entry through skin. *Physical Signs* Grey colour in mouth and throat. *Symptoms and Consequences of Poisoning* Burning or metallic taste in mouth. Excessive dribbling (salivation).
	Ashy colour in the mouth and throat. Intense thirst, pain in stomach, vomiting of blood, mucus, and bloody diarrhoea are recognisable symptoms.[197]
	About 1 gram of mercuric chloride, if swallowed, is enough to kill someone.
	If the patient does not die of shock, they will experience the symptoms described above. Teeth will also become loose and the kidney progressively damaged until urine is no longer

Type and Example	*Entry, Physical Signs, Symptoms and Consequences*

produced. Bloody diarrhoea will continue.
Treatment for inorganic mercury is well-known
and relies on agents that bind the mercury and
allow it to pass out in the urine.[198]

(9) Copper fungicides
copper oxide, copper
sulphate, copper
hydroxide

Entry into Body
Mainly through swallowing.
Physical Signs
Staining of tissue in the mouth.
Symptoms and Consequences of Poisoning
Many copper compounds are not very soluble
in water. The most soluble is copper sulphate
which dissolves readily and it is this compound
which has caused most poisoning. It should be
noted, however, that copper compounds are
said not to be an important source of
poisoning.[199]

Copper is essential for life and without it we
would not be able to transport sufficient
oxygen in our blood or develop healthy bones.
But copper in excessive concentrations is clearly
toxic. Swallowing of copper solutions will cause
pain in the stomach and intestines, nausea, and
diarrhoea, which may turn bloody.

Death has occurred 2–3 hours after a large
dose of copper was swallowed. In some patients
who survived poisoning, the copper damaged
the red blood cells causing anaemia. Liver and
acute kidney failure are common, too, in severe
poisoning.[200, 201] Two children are reported to
have died, 10 and 24 hours respectively, after
eating several bunches of grapes heavily
sprayed with copper fungicides.[202]

Those spraying copper solutions in the field
may develop an itchy dermatitis with raised
spots on the skin. There is also a danger if
copper sulphate is rapidly absorbed through
the skin. Some patients with skin burns have
had these treated with copper sulphate but the
chemical was absorbed by the damaged skin
causing haemolytic anaemia, the same
condition as in those poisoned by copper.[203]
Copper solutions – if they get into the eye –
can cause conjunctivitis and may lead to ulcers
developing on the cornea of the eye.[204]

Bordeaux mixture is a fungicidal spray
containing copper sulphate. Cases of

Type and Example	*Entry, Physical Signs, Symptoms and Consequences*
	pneumonoconiosis have been reported in men who sprayed Bordeaux mixture in vineyards for several years.[205] The symptoms of vineyard sprayers' lung, as the illness has been called, included shortness of breath, weakness, loss of weight and thick yellow phlegm on coughing. The men recovered a little when they stopped spraying but relapsed when they were in contact with the spray again.
	Some individuals who were also exposed to Bordeaux mixture did not have breathing difficulties but experienced chills, fever, muscle and joint pains, weakness and loss of weight that became so severe that hospital treatment was necessary. It appears that the workers suffering from vineyard sprayers' lung are much more likely to get infections and this was why many died. Many individuals with lung disorders following exposure to copper sulphate solutions also have severe liver damage.[206] In some cases the liver was so damaged that blood flow through it was restricted; this restricted blood flow is referred to as portal hypertension and is commonly found in alcoholics. It may have been alcohol, rather than copper, that was to blame for the severest liver damage.[207]
	It is known that some copper fungicides – particularly copper sulphate – also contain another type of chemical known as alkyl-ureas which are known to cause brain cancer in rats.[208] The presence of the alkyl-ureas in the fungicides may explain the results of a recent investigation which found that farmworkers exposed to fungicides were 4.7 times more likely to develop brain gliomas (a particular type of cancer) than the general population.[209]
(10) **Arsenicals** sodium arsenite, potassium arsenite	*Entry into Body* By mouth, inhaling and through skin. *Symptoms and Consequences of Poisoning* These compounds are highly toxic – prompt diagnosis is vital. Symptoms – similar for all routes of entry – occur most rapidly after ingestion by mouth. In people exposed to sufficient arsenic dust the first signs of poisoning are usually difficulty in breathing,

Type and Example	*Entry, Physical Signs, Symptoms and Consequences*

and pains in the chest. These symptoms are followed by a feeling of nausea. After about 30 minutes vomiting occurs, together with severe diarrhoea. Repeated exposure to low amounts (chronic poisoning) can cause loss of appetite, weight loss, diarrhoea, skin rash, hair loss, and tingling in the hands and feet.[210]

Some people may also have rough, raised areas on their skin. On occasions the skin may become pigmented.[211] The tingling in the hands and feet is caused by damage to the nervous system. Those affected in this way may also experience problems with their sight, taste, smell and the control of their bladder. Damage to the nerves is said to be greater in alcoholics.[212]

Arsenic causes skin and lung cancer, and all arsenic compounds (including pesticides) are suspect.[213–215]

Evidence to support the claim that arsenic causes cancer has been obtained from people treated with arsenic compounds such as potassium arsenite as well as from those exposed to arsenic in a sheep-dip factory and at a chemical plant where exposure was to lead arsenate and calcium arsenate.[216]

(11) Pyrethrins (synthetic)
decamethrin, permethrin

Entry into Body
Through the skin and inhaling; occasionally by mouth.
Physical Signs
Dermatitis may occur after skin contact; exposure to sunlight can make it worse. Severe swelling of the face including lip and eyelids can occur.[217]
Symptoms and Consequences of Poisoning
Sweating, fever, anxiety and rapid heartbeat may occur. If swallowed, symptoms are likely to include feeling sick, vomiting, diarrhoea, twitching of arms and legs, convulsions if poisoning is severe.

A health survey of 199 workers who re-packed pyrethroid insecticides into boxes by hand showed that about two-thirds of the people had a burning sensation and tightness and numbness on the face; one-third had sniffs and sneezes. Abnormal sensations in the face, dizziness, tiredness and red rashes on the skin

Type and Example	*Entry, Physical Signs, Symptoms and Consequences*

were more common in summer than in winter. Workers did not wear protective gloves in summer because of the heat and sweated more. The symptoms usually occurred 30 minutes after exposure to the pyrethroids and rarely lasted more than 24 hours.[218]

(12) Bipyridillium compounds
diquat, paraquat

Entry into Body
Small quantities may penetrate through the skin or be inhaled. Accidental or deliberate intake by mouth is the most dangerous.
Physical Signs
Splashes in the eyes or on the skin – if not washed off – will cause irritation and reddening and blistering of affected area.
Symptoms and Consequences of Poisoning
Poisoning by mouth will cause vomiting, upset stomach and diarrhoea.

Stinging of the mouth and throat and difficulty in swallowing are also likely. The symptoms of poisoning are similar for paraquat and diquat. Onset of symptoms may be delayed 2–3 days after exposure. Large intake will cause death in 24–48 hours.[219] There are reports where the interval between swallowing paraquat and death occurring has been 26 days,[220] and in one case 102 days.[221]

By 1977 some 564 deaths from suicide, or accidental poisoning, had been caused by paraquat.[222] In Ireland over the 10-year period from June 1967–May 1977 there were numerous deaths from paraquat. Of the 77 cases who tried to commit suicide with the pesticide 56 succeeded and died, whereas 17 of the 40 people accidentally poisoned, died. In the 19 cases where the cause of poisoning was unknown, everyone died.[223]

Paraquat is also responsible for a significant number of deaths in Sri Lanka. Over the 17-month period from January 1985 to May 1986 paraquat accounted for 25 deaths, or 21% of all poisonings by pesticides.[224]

Some recent investigations in Canada have linked paraquat spraying with a number of cases of Parkinson's disease.[225, 226] Some scientists are not convinced by the evidence.[227] However, there does appear to be an

Type and Example	*Entry, Physical Signs, Symptoms and Consequences*

environmental link, for the illness is more common in rural than urban areas of Canada.

Although neither paraquat nor diquat cause cancer, workers involved in the manufacture of paraquat have developed skin cancer through contact with one of the chemicals used in the process. The process in question is one in which pyridine is converted to a 4'-bipyridyl which in turn is used to make paraquat. 20 workers out of a population of 550 who had been in contact with the bipyridyl, and tarry by-products, between 1961–1980 developed skin lesions. Most of the lesions were pre-cancerous (and likely to become cancerous at a later date); 6 men had cancers on the skin.[228]

(13) Phenoxy acetates and related compounds
2,4,5-T, 2,4-D, MCPA

Entry into Body
Through the skin and by inhaling, occasionally by mouth.
Symptoms and Consequences of Poisoning
If taken by mouth – burning of the mouth and throat, dribbling of saliva, cramp in stomach, vomiting and diarrhoea will occur.

Penetration of 2,4-D through skin can cause muscle weakness, mental confusion and difficulty in walking. Unconsciousness occurs in severe poisoning.[229]

In recent years concern about the phenoxy herbicides has centred around the contamination of 2,4,5-T by a toxic by-product called dioxin. Although there are 75 different types of dioxin – their full technical name is chlorinated dibenzodioxins – the one found in 2,4,5- T is particularly toxic and a very potent cancer-causing chemical.[230–232] Fortunately the concentration of dioxin in 2,4,5-T is extremely small and in recent years it has been reduced even more. The dioxin in 2,4,5-T also causes birth deformities in animals,[233] however, it is unlikely that it will do so in humans as it is generally present in concentrations that are exceedingly low. It is far more likely that 2,4,5-T itself will cause problems before its dioxin contaminant does.[234]

The chlorophenoxy herbicides have been implicated as the cause of certain types of cancer in agricultural workers in Sweden and

Type and Example	Entry, Physical Signs, Symptoms and Consequences
	the US. Initial findings suggested that forestry workers exposed to phenoxy herbicides like 2,4,5-T and 2,4-D were 6 times more likely to develop soft tissue sarcoma than their unexposed colleagues.[235] Further studies in Sweden suggested that there was a higher risk of non-Hodgkin's lymphoma (a cancer of the lymph gland) occurring in workers exposed to phenoxy herbicides.[236] But work in New Zealand,[237] and the US[238] has not confirmed the link between spraying and soft tissue sarcoma. However, the studies in the US have added weight to the evidence suggesting that exposure to 2,4-D may cause lymphoma. Farm workers exposed to 2,4-D in the mid-West of the US have a higher than expected incidence of non-Hodgkin's lymphoma.[239]

In view of the studies in New Zealand and the US casting doubt on the link between phenoxyacetic acid herbicide exposure and soft tissue sarcoma, the Swedish scientists have reassessed their original work. This reassessment has not revealed anything different and the Swedes say that they stand by their earlier conclusions that forestry workers exposed to these herbicides are at greater risk of developing soft tissue sarcoma.[240]

Type and Example	Entry, Physical Signs, Symptoms and Consequences
(14) Organotin triphenyltin (fentin), tributyltin oxide (TBTO)	*Entry into Body* Absorbed through skin or from gut. *Physical Signs* Severe skin irritation can occur. *Symptoms and Consequences of Poisoning* Feeling depressed, tired, unsteady on their feet, and weak, are the more common symptoms reported in those affected. Some diarrhoea and loss of appetite can occur. If swallowed, organotin compounds can be lethal. After swallowing, symptoms can include headache, vomiting, dizziness, disorientation when exposed to light, abdominal pain, weight loss, feeling cold, slowed heart rate and an inability to pass urine. Changes in mood, convulsions, coma, and paralysis may also happen.

Trimethyl compounds are less toxic than triethyl, or triphenyl.[241]

The principal problem for anyone using this

Type and Example	Entry, Physical Signs, Symptoms and Consequences

family of chemicals is skin irritation, which can be quite pronounced. Other occupational problems with the organotin compounds appear to be rare.[242] Where poisoning has occurred it was through the use of a particular preparation of diethyltin di-iodide (which also contained 10% triethyltin iodide and slightly less ethyltin-triiodide) for treating bacterial infections of the skin, infections in bone, anthrax, and acne. This 'cure-all' medicine called 'stalinon', was used by about 1,000 people of whom it poisoned at least 224, and killed 103.[243] TBTO suppresses the immune system in test animals making them less able to fight infection.[244]

(15) Nicotine

Entry into Body
Rapidly through skin by inhaling, or by mouth.
Symptoms and Consequences of Poisoning
Symptoms start with the victim feeling sick, and this nausea is followed by dizziness, vomiting, headache, rapid breathing, fast heart rate, sweating and dribbling. In severe poisoning the person, after experiencing these initial symptoms, will collapse, have convulsions, fall unconscious and may have a heart attack. Death can occur in minutes or a few hours.[245]

Between 1930–34 there were 106 accidental deaths and 182 fatal suicides from nicotine insecticides in the US.[246] In New Zealand between 1957–60 nicotine was used in 24 of 49 suicides and was responsible for 4 of the 14 accidental deaths from pesticide poisoning.[247]

During the 1946–47 fruit season in the US some 30,000 pear trees were sprayed with nicotine insecticide in one area. Many of those doing the spraying on a large scale soon showed signs of poisoning. The most seriously affected complained of having strange dreams, a heavy feeling in the chest, difficulty in breathing, and violent nausea that continued in some cases for up to 10 hours. Those who experienced the symptoms could not work more than 1 or 2 days without becoming sick. When they recovered they were unable to work with nicotine again without becoming sick. Most of the workers had no warning that they were poisoned. Everyone

Type and Example	Entry, Physical Signs, Symptoms and Consequences
	complained of a burning sensation when they were in contact with the insecticide. It was penetration of the insecticide through the skin, rather than inhalation, that caused most of the poisonings.[248]
	Symptoms can persist. Three weeks after a man sat on a chair on which nicotine had been spilled he still experienced weakness, sweating, lightheadedness on changing position, nervousness and poor sleep. These symptoms took a further 3 weeks to disappear.[249]
(16) Dithiocarbamate fungicides maneb, zineb, nabam	*Entry into Body* Minimal absorption from the gut after intake by mouth. *Physical Signs* Irritation of the skin or mouth after local contact. *Symptoms and Consequences of Poisoning* Feeling sick and vomiting occurs only after considerable over-exposure and is usually only seen during the manufacturing stages of these pesticides.[250]
	The major occupational health problem from exposure to the dithiocarbamates that has been reported appears to be dermatitis.[251] There is some evidence that exposure to these fungicides could cause cancer, but the general consensus is that more information is needed to confirm these suspicions.[252]
	Notwithstanding this the US EPA is now deciding whether to ban the use of these fungicides on food crops,[253] whereas in the UK MAFF says they can still be used.
(17) Triazine herbicides simazine, atrazine	*Entry into Body* Through skin and by inhaling. *Symptoms and Consequences of Poisoning* These are not very toxic compounds. Large intakes are needed to cause symptoms, which would be delayed, and might include feeling sick, loss of weight, and diarrhoea.[254]
	Cases of severe dermatitis have been reported in workers exposed to atrazine and simazine. A few hours after a farmer (who had a history of dermatitis) had cleaned the nozzle of

Type and Example	*Entry, Physical Signs, Symptoms and Consequences*
	a sprayer used to dispense atrazine, his hands began to blister. The man had worn no protective clothing when cleaning the sprayer. During the afternoon, and evening, the condition of his hands became worse; they were red, swollen, blistered and painful. It took 4 days of treatment with pain killers, steroids and antibiotics before the pain stopped, and a further 24 days before his hands and arms could be said to have recovered, although even then they were still slightly affected. Subsequent tests showed this man to be allergic to atrazine and his dermatitis returned after skin contact with small amounts of the herbicide.[255–256]
	124 workers are reported to have developed contact dermatitis in the Soviet Union after exposure to simazine and propazine. The milder cases had reddening and swelling of the skin lasting 3–4 days. More serious cases had more pronounced skin changes lasting 7–10 days.[257]
amitrole	*Symptoms and Consequences of Poisoning* Amitrole is a herbicide that is related to the triazines, although its chemical structure shows some distinct differences when compared with the triazines. Amitrole has been implicated as a possible cause of cancer in Swedish railroad workers. In a study of 348 Swedish railroad workers exposed to amitrole, 2,4-D, 2,4,5-T and other organic herbicides (for example monuron and diuron) for 45 days or more, there was an excess of cancers in the exposed group compared with what would be expected.[258, 259] 17 cases of cancer occurred in the workers exposed to the herbicides, whereas there should only have been 11.9 cases. (The 11.9 figure is a statistical result based on analyses of the number of cases in the Swedish population as a whole.)
	There is no doubting amitrole's ability to cause cancer in animals. Amitrole causes cancer of the thyroid and liver in male and female mice, and thyroid cancer in male and female rats.[260] In view of this animal evidence there is little doubt that the chemical could cause cancer in humans.

Type and Example	Entry, Physical Signs, Symptoms and Consequences
(18) Chlorates sodium chlorate	*Physical Signs* Skin has a blueish colour. Irritation of the skin, eyes, nose and throat *Symptoms and Consequences of Poisoning* If taken by mouth the person will have stomach pain and feel confused. More severe intakes may cause convulsions.[261] Most deaths from sodium chlorate have been suicides,[262, 263] and death can occur anywhere between 4 hours and 36 days after consuming the pesticide.[264] Besides the symptoms listed above, poisoning may also cause liver and kidney damage and the spleen may become enlarged. Blood becomes brownish in colour and the urine (if there is any) is a characteristic brown or black, due to the damaged blood cells in it.[265] Although there is no evidence available at the moment some thought needs to be given to the possibility that this pesticide could cause cancer. Sodium chlorate is a powerful oxidising agent. Use of a similar oxidising agent – potassium bromate – in bread, where it was used to improve the quality of dough, has been stopped in the UK because the compound is a mutagen and causes cancer in animals.[266–270] Potassium bromate and sodium chlorate are likely to act in similar ways and may cause cancer through their ability to form free oxygen radicals in the body. Until we know more, sodium chlorate should be treated with considerable care.
(19) Anti coagulant rodenticides hydroxy coumarins, indandione derivatives, warfarin	*Entry into Body* Swallowing is the main risk. *Physical Signs* Bruising *Symptoms and Consequences of Poisoning* Only small amounts of chemical used in baits. Risk of poisoning is small. Blood in urine and faeces indicates overdose.[271] Warfarin is used in medicine to stop blood-clotting. Swallowing the chemical will cause a person to bleed internally. Only one case is reported of a man who had extensive skin contact with warfarin when preparing baits. A 23-year-old farmer used a solution of warfarin

Type and Example	*Entry, Physical Signs, Symptoms and Consequences*

10 times over a 24 hour period, getting his hands wet each time and not bothering to wash them for several hours. Two days later the man had blood in his urine and the following day bruising was noticeable on his arms and legs. He also complained of a dull pain in his groin. After 3 days in bed he returned to work only to have his nose start bleeding and blood again appear in his urine. Hospital care was prescribed and the man responded well to treatment with vitamin K.[272]

Sadly, warfarin has also been used to murder people. One 32-year-old man was killed after being fed warfarin over 13 days.[273] But the most tragic report is that from Saigon, in Vietnam, where an estimated 250 infants were killed after someone put warfarin, deliberately, into talcum powder. The talc was put onto the cut umbilical cord of newborn babies by their mothers to stop bleeding and irritation. But instead of stopping the bleeding the warfarin in the talc made the children bleed all the more, causing many to die.[274, 275]

(20) Fumigants
(i) methyl bromide

Entry into Body
By inhaling in the main, although skin contact can also cause poisoning.
Physical Signs
Burns on skin, followed by blistering.
Symptoms and Consequences of Poisoning
Very toxic. Early symptoms include headache, stinging of eyes, feeling tired and sick, followed after some hours by difficulty in focusing the eyes and walking, and an inability to sleep. Convulsions may occur.

Some symptoms such as depression, irritability, change of personality and unable to sleep may persist for months.[276]

Methyl bromide is an extremely toxic chemical. By 1955 there were some 47 fatal and 206 non-fatal cases of poisoning linked with this chemical. Most of these cases occurred during manufacture of methyl bromide, or through its use as a fumigant, or fire-extinguishing fluid. Eleven of the deaths, and 44 of the other poisonings were due to its use as a fumigant.[277]

If the victim does not die he/she may take days, or weeks, to recover. In some cases there

Type and Example	Entry, Physical Signs, Symptoms and Consequences
	may not be a full recovery and the person could remain severely disabled for life.[278]
Fumigant (ii) carbon tetrachloride	*Entry into Body* By inhaling. *Physical Signs* Direct contact causes dry, broken skin. *Symptoms and Consequences of Poisoning* Headaches, dizziness, vomiting, feeling confused and sick, are the likely symptoms. Repeated low exposures may cause feelings of sickness, tiredness and loss of appetite.[279] Liver and kidney injury occur in carbon tetrachloride poisoning with some cases experiencing more severe symptoms than others. The reason for the differences between people is not understood. Many cases of poisoning from carbon tetrachloride have been reported, most involving inhalation of the chemical when it was used as a cleaning fluid, and some through deliberately swallowing cleaning fluid. A few cases of poisoning were recorded following use of carbon tetrachloride as a pesticide.[280] Carbon tetrachloride causes liver cancer in rats, mice and hamsters.[281] It could cause cancer in humans.
Fumigant (iii) ethylene dichloride (often used with carbon tetrachloride)	*Entry into Body* By inhaling. *Physical Signs* Direct contact causes dry, broken skin. *Symptoms and Consequences of Poisoning* The symptoms of poisoning are similar to carbon tetrachloride. Some deaths have been recorded 17–24 hours after deliberate swallowing of ethylene dichloride.[282, 283] Ethylene dichloride was often mixed in a ratio of 3:1 with carbon tetrachloride.[284]
Fumigant (iv) ethylene dibromide (EDB)	*Entry into Body* Mainly by inhalation. *Physical Signs* Blisters on skin after direct contact. *Symptoms and Consequences of Poisoning* Stinging of eyes and throat. If ethylene dibromide comes into contact

Type and Example	*Entry, Physical Signs, Symptoms and Consequences*

with any skin it will irritate; the nose and throat are particularly susceptible. However, if concentrations reach the point where irritation occurs, the level of exposure is dangerous.[285]

It is not ethylene dibromide's irritating properties which cause concern but its cancer-causing properties. The chemical is mutagenic in various cell culture test systems using bacteria and moulds.[286, 287] There is some evidence that ethylene dibromide causes cancer in man. In a 1976 investigation, of 161 men exposed to ethylene dibromide from the mid 1920s until 1942 there were 7 deaths from cancer whereas 5.8 were expected.[288] The one extra death in the study group could have been due to chance. The number of observed deaths would have needed to be higher for it to be laid at the door of EDB.

In another study an excess of lymph gland cancers (lymphoma) were reported in granary workers (in the US) exposed to a variety of chemicals including EDB. The presence of these other chemicals complicates the picture because it might be one of the other chemicals, and not EDB, that caused the lymphoma.[289]

However, it is not necessary to rely on the hazy evidence from human studies before taking action on EDB. This chemical has been tested on rats and mice in feeding, inhalation, and skin painting studies. In every test, no matter how the chemical was administered, EDB caused a number of different types of cancer in male and female animals of both species.[290] It was for this reason that the US Occupational Safety and Health Administration labelled EDB a potent animal carcinogen in 1983.[291] Because of these tests on animals, EDB must be considered as a chemical that could cause cancer in humans.

Fumigant
(v) chloropicrin

Entry into Body
By inhaling.
Symptoms and Consequences of Poisoning
Stinging eyes and mouth are common symptoms. The eyes may water, and coughing, giddiness, headache, feeling sick and vomiting may follow.[292]

In Japan people living close to a factory where tobacco leaves had been fumigated

Type and Example	Entry, Physical Signs, Symptoms and Consequences

experienced all of these symptoms, which lasted anywhere between a few hours and 3 days after exposure.[293] Some residents living 100 metres from the factory also had symptoms.[294]

Chloropicrin was also used as a chemical warfare gas for a brief period in the First World War and those exposed to it had diarrhoea, felt sick, and vomited, for weeks after exposure.[295]

Fumigant
(vi) phosphine, aluminium or magnesium phosphide

Entry into Body
By inhaling.
Symptoms and Consequences of Poisoning
The initial symptoms are feeling sick, stomach pain, vomiting and diarrhoea, followed by difficulty in walking; convulsions, unconsciousness and death occur within 24 hours.[296]

The sweet smell of phosphine gives little indication of how toxic the gas is. Phosphine is still widely used, particularly in warehouses. Many tragedies have occurred with this gas. Various reports refer to people dying in homes treated with phosphine and deaths amongst those living close to granaries and warehouses where phosphine (produced when water vapour reacts with aluminium or magnesium phosphide) had been used.[297-299] In one case 2 adults and a child living next to a granary in which the grain had been treated began showing symptoms the day treatment started. The husband was the first to start vomiting, followed by his wife, with their child falling ill the next day. The day after this both adults were found dead in their bedroom. Although the child was taken to hospital for treatment it was too late, and she died.[300]

Many deaths from phosphine poisoning have been reported in India.[301] The majority of fatalities – some 319 were recorded up to June 1988 – were suicides, although, as with other chemicals, some died through accidents, while others were deliberately poisoned by someone else.[302]

Fumigant
(vii) hydrogen cyanide

Entry into Body
By inhaling.
Symptoms and Consequences of Poisoning.
Symptoms of slight poisoning include a

Type and Example	*Entry, Physical Signs, Symptoms and Consequences*
	metallic taste, stinging, or soreness of nose and throat, dizziness, feeling sick, throbbing headache, tightness in chest, weakness in legs. Severe exposure will kill in minutes.[303]
	Cyanide is probably the poison of choice of most thriller writers. It is also a chemical used in many suicides. A number of pest control workers have died from cyanide poisoning including others who have remained in, or returned to, a fumigated area before it was thoroughly ventilated.[304]
	It is important to remember that a few deep breaths of air heavily contaminated with cyanide will cause someone to collapse and their breathing to stop. If the person is not removed from the area quickly and resuscitated, death will occur rapidly. Fatalities with cyanide have been reduced by workers operating in pairs so that there is always someone on hand to give help immediately.[305, 306]

6

Pesticide Residues in Food and Water

Introduction

Food, fresh and salt water are subject to extensive and often illegal contamination by a cocktail of toxic synthetic pesticide residues. As pesticides break down in the environment at different rates, the more persistent ones (which resist breakdown) can contaminate food and water for a long time. Even though we only consume these residues in small amounts, little is known about their long-term effects on our health.

Pesticide residues, both natural and synthetic, can be found in all the things we eat – fruit, vegetables, bread, meat, poultry, fish, and the processed foods made from them. Some of this contamination is legal, but does this mean it is safe? Much of it is illegal, with residues in excess of legal safety levels. Sometimes banned chemicals are used. Imported food can pose special problems as it is often sprayed with chemicals banned in Europe and the US. Most of this chapter deals with synthetic pesticides. It is important to remember, however, that plants contain natural pesticides and that we consume far more of these than we do of the synthetic pesticides sprayed onto crops.[1]

Pollution of drinking and surface water by pesticides and artificial fertilisers is another way in which these chemicals end up in the human food chain. Contamination of ground water is widespread in the UK, Europe and the US as a result of agricultural and industrial pollution. In Western Europe, agricultural policies over the last 40 years have led to many water pollution problems. The European Community (EC) Common Agricultural Policy (CAP) has aimed at self-sufficiency in basic commodities, working through a combination of price support, subsidies, tax relief and trade barriers. The result has been an increase in food production brought about by the use of artificial fertilisers and pesticides and it is these, along with large quantities of manure from intensive animal rearing which are now causing serious problems of water pollution.[2]

112

Pesticides are only one of the chemical contaminants to be found in food and water. There are also nitrate and phosphate residues from artificial fertilisers (and animal manure). Nitrates have been linked to the blue blood (methaemoglobinaemia) condition in babies and old people, and stomach cancer in the general population.[3] In rivers, lakes and on the margins of the sea, phosphates and nitrates cause an excess growth of algae (eutrophication) which depletes the water of oxygen, blocks out light, and kills off other aquatic life, including fish. Meat, poultry and farmed fish can also contain detectable levels of animal/veterinary vaccines, antibiotics, hormones, wormers (anti-helminthics), growth promoters and feed additives.[4] In addition to these unwanted residues, processed food can contain legally permitted food additives, colourings and preservatives, the long-term health effects of many of which are unknown.

Consumers are clearly worried about food safety. Results of a National Consumer Council survey showed that 61 per cent of the shoppers questioned said it was the Government's responsibility to provide advice and information on food safety.[5] Yet the UK government is not at all keen to see foods labelled 'treated with chemicals' and the pesticides used on them listed, as proposed in a new EC Directive.[6]

Food and water quality are now major public and political issues, especially with the 'greening of politics'. On a worldwide basis consumers are pressuring food manufacturers, water authorities and governments to maintain and improve the health and hygiene standards of food and water. This chapter looks at some of the issues involved as well as legal standards and enforcement.

Food

Pesticides are used primarily to increase crop yields and reduce post-harvest losses by killing pests. A further use of pesticides to improve the appearance of food is largely cosmetic. Consumers have become used to buying uniform, blemish-free produce, whether home-grown or imported. Cosmetic spraying does nothing to enhance the taste, or improve the nutritional value of the produce but can leave pesticide residues in the food. Many pesticides, especially systemic ones, are absorbed into the crop. These pesticides cannot be washed off fruit or vegetables. Peeling fruit and vegetable skins may help avoid some pesticide residues but will not remove these systemically absorbed chemicals. Peelings may also contain some of the valuable nutrients in the food.[7]

Use of persistent pesticides too close to crop harvest or sale is a particular problem which can lead to excessive residue levels. By law UK pesticide labels now have to specify the 'harvest interval' – the period of time that must apply between spraying and harvesting. The correct harvest interval can be difficult to judge for high-value, seasonal produce such as salad, vegetable and glasshouse crops, where the picking date is often determined by the price in the shops. Correct choice of pesticides is critical here in minimizing residue levels if the crop may have to be harvested sooner than intended. The newer synthetic pyrethroid insecticides generally have shorter harvest intervals than conventional, though often cheaper, organophosphorus insecticides, and could be used as alternatives to reduce the risk of residue contamination.

Illegal use of pesticides too close to harvest is hard to detect because there are too few enforcement inspectors and not enough sampling of foods. Even in the UK banned pesticides still appear in home-grown food. Food with illegal or excessive levels of pesticide residues is often imported into countries. In 1986, for example, North Americans purchased US$24.1 billion worth of imported food. The Food and Drug Administration (FDA), however, tests less than one per cent of all imported food shipments. Over a six-year period the FDA found that 6 per cent of its samples contained illegal pesticides residues.[8]

Britain also imports food and produce containing pesticide residues of chemicals banned or restricted in the UK. MAFF figures for 1986 showed that samples of cod and halibut liver oil and imported, processed poultry and pork all contained residues of highly persistent organochlorine group pesticides.[9] Tighter international controls, such as restrictions on the export of banned or restricted pesticides, would reduce the problem.

The crop protection industry argues that most crop protection products leave no detectable residues, since the chemicals break down long before the crop has been harvested. Even where traces of chemical can be detected in a food, the industry states that the presence of residues is not a health problem.[10] The industry also argues that many foods contain natural toxins and carcinogens but that, as with synthetic chemical residues, the amounts consumed are very small (1.5 grammes per day per person – natural pesticides; 0.1 mg per day per person – synthetic pesticides). They also claim that as new crop protection products work at a fraction of the dose rate of older chemicals, there is less likelihood of residues occurring.[11]

These claims are worth examining in more detail, and in the rest of this section we look at how residue standards are set, and then at what is found on crops and produce.

Setting Standards

Internationally, safety evaluation of pesticide residues in food is a continuous programme run jointly by the United Nations Food and Agriculture Organisation (FAO) and World Health Organisation (WHO), and is carried out on the basis of an annual Joint FAO/WHO Meeting on Pesticides Residues. Analytical and toxicological data are thoroughly reviewed and used for safety evaluation. To set standards, estimates are required of the amount of pesticide that can be consumed each day

Natural Pesticides and Food

Synthetic chemicals may not be the only hazards in plants. It has been claimed that there are many more natural pesticides in food crops. One prominent US research scientist, Bruce Ames, has gone so far as to claim that 99.99 per cent of dietary pesticides are natural. In other words, chemical residues from synthetic pesticides only represent 0.01 per cent of all the pesticides that we consume.[12]

Ames has a point. If natural pesticides are present in far greater concentrations than synthetic chemical residues should we be devoting as much attention to the latter when decisions are made about the safety of foods. The answer must be a qualified 'yes'. Simply because there are many natural pesticides in food it would be quite irresponsible to ignore the presence of the synthetic chemicals that we apply to crops to control pests. Ignoring additional chemicals simply because natural ones are already present would be quite wrong.

There is a more important message that Professor Ames is trying to get across in his recent series of publications on natural pesticides. Ames is throwing into question the whole basis of much of the animal testing that is performed to identify the chemicals that may cause cancer in humans.[13] Needless to say his attack has created a storm of protest, with many other cancer researchers claiming that his attack is misdirected, and his views oversimplified.[14] This is not the place to discuss Ames's specific attacks on cancer screening programmes. In fact, it is likely to be several years before the dust from this particular storm settles. Our concern, however, must be the question that he raised about the role of natural pesticides.

As Ames points out, of the many chemicals that have been tested for their carcinogenicity only 52 natural pesticides have been tested in animals, and of these some 27 cause cancer and are present in many common foods. Before we ban all the produce from the greengrocer, it is important to reflect on another argument that Ames is promoting, which is that eating fruit and vegetables is not bad for you. In fact, it would seem that they are actually good for you as epidemiological studies have shown that eating a diet rich in fruit and vegetables protects against cancer. Why this should be so is not clear. In years to come researchers will be investigating why fruit and vegetables have this protective effect.

Some believe that it is anti-carcinogenic factors in plants themselves that offer some protection. It is also claimed that there are detoxifying mechanisms in humans which have adapted over the years to deal with these problem chemicals in nature. Ames has suggested that humans and other animals have evolved defences to protect against natural toxins and that these defences are good enough to take care of synthetic chemicals as well. This is a rather sweeping generalisation. There are many synthetic chemicals that have known counterparts in nature. The polychlorinated biphenyls (PCBs) which have been very widely used in electrical equipment are one such example. We have no idea whether the defence mechanisms in humans are capable of protecting against PCBs even though these mechanisms may have evolved to deal with natural pesticides.

Ames has pointed out another potential problem with natural plant pesticides – cultivated food plants often contain far fewer natural toxins than their wild cousins. The wild potato *Solanum acaule* is the distant relative of cultivated strains of potato, yet it has a glycoalkaloid content three times higher than cultivated strains and is more toxic. The leaves of the wild cabbage *Brassica oleracea* (which is the progenitor of broccoli, cabbage and cauliflower) contains about twice as many of the natural pesticide glucosinolates as cultivated cabbage.[15]

Plant breeding to develop plants that are more insect-resistant may also give rise to some problems. In the US a major grower recently introduced a new variety of highly insect-resistant celery on the market.

The introduction was followed by a flurry of complaints from all over the country from people who complained that after handling the celery they developed rashes and burns when they were subsequently exposed to sunlight. On investigation it was found that the pest resistant celery contained 6,200 parts per billion of the carcinogenic psoralens instead of the 800 parts per billion present in normal celery. It is not clear whether other natural pesticides in celery were increased as well. The celery is still on the market. A second example is that of the new potato developed at considerable cost which had to be withdrawn because of its acute toxicity to humans. The reason for its withdrawal was the presence of very much higher levels of two natural toxins, chaconine and solanine, which both inhibit cholinesterase, thereby blocking nerve transmissions. Chaconine and solanine are present in normal potatoes at a level of about 75 parts per million, which is one-tenth of the concentration that would be toxic.[16]

As our knowledge about both natural and synthetic pesticides increases we will become better able to decide what is reasonably safe and what is not. Plant breeding programmes to develop insect-resistance in plants may have many pitfalls as the two examples above illustrate, but these will be problems that can be overcome. The search is on for safer food, but there is a wider agenda that is being addressed as well, and that is the reduction of the use of synthetic chemical pesticides which have caused so many environmental problems. It is the environmental concern that is driving the research for newer ways of protecting plants. We can expect hiccoughs along the way because we are attempting to accelerate a process that nature has taken millions of years to evolve.

---

without causing harm. Referred to as the Acceptable Daily Intake (ADI), this is defined as 'an estimate of a daily exposure to the human population (including sensitive sub-groups) that is likely to be without an appreciable risk of adverse affects'.[17]

ADIs are derived from long-term feeding studies with laboratory animals. Assessments are made of carcinogenic, mutagenic and teratogenic (birth defect) risks as well as possible neurotoxic (ability to kill nerve cells) and reproductive effects. The intake causing no toxicologically significant effect on animals is referred to as the 'no-effect level' (NOEL), and the NOEL is usually divided by a safety factor of 100 to give the ADI (see Table 6.1).

Table 6.1: Determining Maximum Pesticide Residue Levels in Food

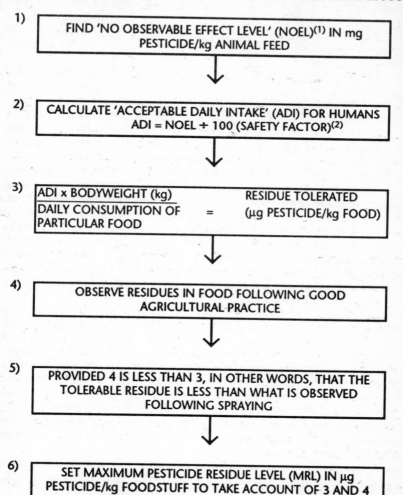

1) FIND 'NO OBSERVABLE EFFECT LEVEL' (NOEL)[1] IN mg PESTICIDE/kg ANIMAL FEED

2) CALCULATE 'ACCEPTABLE DAILY INTAKE' (ADI) FOR HUMANS
ADI = NOEL ÷ 100 (SAFETY FACTOR)[2]

3) $$\frac{\text{ADI} \times \text{BODYWEIGHT (kg)}}{\text{DAILY CONSUMPTION OF PARTICULAR FOOD}} = \begin{array}{l}\text{RESIDUE TOLERATED} \\ \text{(μg PESTICIDE/kg FOOD)}\end{array}$$

4) OBSERVE RESIDUES IN FOOD FOLLOWING GOOD AGRICULTURAL PRACTICE

5) PROVIDED 4 IS LESS THAN 3, IN OTHER WORDS, THAT THE TOLERABLE RESIDUE IS LESS THAN WHAT IS OBSERVED FOLLOWING SPRAYING

6) SET MAXIMUM PESTICIDE RESIDUE LEVEL (MRL) IN μg PESTICIDE/kg FOODSTUFF TO TAKE ACCOUNT OF 3 AND 4

Notes
[1] A NOEL, established by feeding studies in animals, is the highest concentration of pesticide in food that has no effect whatsoever on the animal.
[2] The safety factor of 100 takes account of a likely tenfold variation in susceptibility in animals and a further tenfold variation in humans.

The ADI is expressed in terms of milligrams of the residue ingested per kilogram of bodyweight per day (mg per kg per day). It is based on the average man having a typical weight of 60–70 kg (132–154 pounds).[18]

The ADIs serve as one of the bases for legislation in many countries to regulate pesticide residues in food. Many countries set a statutory limit, usually referred to as Maximum Residue Level (MRL) or residue 'tolerance', for residues in specific food commodities. In order to avoid serious inconsistencies in MRLs between countries, the Codex Alimentarius Commission (a United Nations body) has established the Codex Committee on Pesticide Residues at which governments discuss MRL's. The Codex Committee bases its work on the scientific approvals made by the Joint FAO/WHO Meeting on Pesticides Residues.

Individual governments translate internationally derived ADIs into their own MRLs. 'Total diet' and 'market basket' surveys are carried out in individual countries and involve the analysis of a standard set of prepared foods, or food groups, chosen to reflect the average dietary habits of the general public in a particular country, or of a specific sub-set of the population such as infants, ethnic groups or vegetarians. The results from these surveys are used to calculate an approximate dietary intake of the various pesticides and other chemical residues. These investigations require huge resources and detailed planning. In the US permissible pesticide residue levels, called tolerances, have been set by the US government department, the Environmental Protection Agency (EPA) for more than 300 different pesticides currently used on food.[19]

One of the main reasons for fixing residue limits is the risk of cancer in people eating contaminated food. In the US a scientific panel of the National Academy of Sciences reviewed the subject in 1987. The panel concluded that the greatest risk to consumers was from the use of fungicides, followed by the use of herbicides. From the evidence it had, the panel said that certain chemicals, if present in food, would cause the cancer rate to increase by more than one case in a million. It would be prudent, the panel argued, to try and phase out the use of certain chemicals. The two fungicides that presented the greatest cancer risk to consumers in the US are chlordimeform and permethrin. Chlordimeform, however, is not registered for use in the UK. As far as herbicides are concerned, linuron, because of its widespread use in the US, represented some 98 per cent of all the carcinogenic risk from this group of chemicals. The situation will be different in the UK owing to differences in usage of linuron, but it is sensible to review the use of this chemical in the light of its reported risk to consumers.[20]

Five other chemicals were listed by the National Academy of Sciences panel as having considerable carcinogenic potential: arachlor, chlorthalonil, asulam, cypermethrin and mancozeb.[21]

Residue Monitoring and Health Risks

Residues and Acute and Chronic Effects

Pesticide residues can cause immediate health problems. The existence of poisoning from carbamate and organophosphorus pesticide residues in the US and Israel is well established. In 1985 watermelons which had been illegally sprayed with the carbamate group insecticide aldicarb were linked to an outbreak of severe 'flu-like' illness among scores of people in the western US.[22] Some 1,000 people were ill within hours of eating the melons. Symptoms included profuse sweating, shaking, muscle cramps and twitching, nausea, vomiting, diarrhoea and blurred vision.[23] There is even greater concern over the long-term effects of pesticide residues on health. According to a study by the US Center for Disease Control, exposure to small amounts of the carbamate group insecticide aldicarb may alter the human immune system. Resistance to disease may be lowered as a result.[24]

On an individual basis pesticides may be linked to allergy. A scientific report at the Forum on Food and Health meeting in 1987 reported that:

> in 50–100 patients tested in double blind trials, reliable clinical symptoms occurred with ordinary but not organic food. 85% of these 'chemically sensitive' patients do have evidence of a genetic or acquired defect in, for example, sulphoxidation. These people are a very small percentage of the population, but are important clinically.

The Forum went on to report that: 'In the USA, secular trends have shown a reduction in sperm density which has been correlated with pesticide residues in semen.'[25]

Persistent pesticides such as DDT also contaminate mothers' milk and thus babies. In 1981 the entire milk supply on the Hawaiian island of Oahu was contaminated with the organochlorine insecticide heptachlor, a known animal carcinogen banned in the early 1970s in many countries. The heptachlor was used on Hawaii's pineapple crops.[26]

A number of pesticide residues, including DDT, were present in greater concentrations in the blood and placentas of Indian women who had spontaneous abortions compared with those who had full-term deliveries.[27, 28]

A US woman fieldworker who worked during the first three months of her pregnancy picking grapes previously sprayed with pesticides suspected of causing deformities, and whose son was born without arms and legs, received an US$800,000 dollar out-of-court settlement from the chemical companies involved.[29]

In one of the better-documented reports, five individuals who had consumed fresh fruit and vegetables in Israel were hospitalised with chronic intestinal pain, heartburn and diarrhoea caused by organophosphates. On investigation it was found that all the people affected had eaten fresh fruit and vegetables that had been recently sprayed with insecticide. When fruit and vegetables were removed from their diet all of the patients recovered and their cholinesterase values – which had been low on admission – returned to normal.[30]

The US EPA commissioned a report from the independent National Research Council on how to minimise the cancer risks of pesticide use.[31] The National Research Council compiled a list of the 15 foods likely to contain the highest levels of tumour-causing residues. They are, in decreasing order of toxicity: grapes, tomatoes, potatoes, orange, lettuce, apples, peaches, carrots, beans, soyabean, corn, wheat, beef, chicken and pork.

UK Residue Monitoring

The MAFF Working Party on Pesticides Residues (WPPR) reports to the Steering Group on Food Surveillance, and it has produced reports covering the periods 1977–81, 1982–85 and 1985–88. It monitors the following types of food:

Cereals and Cereal Products

Crude 'bucket and shovel' techniques are being used to apply large doses of insecticides to grain when it is stored and this can result in high residue levels. The WPPR Report for 1985–88 (the reports are published every four years) showed that wholemeal flour contained twice the level of residues in white flour. The worst levels were found in bran. The WPPR reported its concern at finding multiple and increasing residues in cereals and cereal products to MAFF's Advisory Committee on Pesticides, which as a result is reviewing its advice about the multiple treatment of grain with pesticides.[32] Pirimiphos-methyl, an organophosphorus insecticide, was detected in 28 per cent of samples of brown flour. Fifty-eight per cent of 223 samples of white bread contained organophosphorus residues. Residues of more than one organophosphorus pesticide were found in 9 per cent of samples in 1984–86, and 15 per cent of samples in 1987.[33]

The government-sponsored UK Home Grown Cereals Authority has echoed this concern.[34] In a 1988 report it stated that potentially toxic pesticides are:

> bonding with cereals making them impossible to detect, destroy or wash away. Once the pesticides bind chemically to the grain they cannot be easily detected by tests now in use. Over half of the pesticides left on cereals will escape detection because the recommended analytical procedures will not have extracted this portion.

Some pesticides used on grain will even survive high-temperature baking. Research by MAFF has shown that with some organophosphorus pesticides – commonly used to protect cereals grains during storage – up to 50 per cent of the pesticide present in the wholewheat flour survived the baking process. In certain cases the MRL was exceeded in bread baked from wholewheat flour obtained from whole grains for which the pesticide was below the MRL; the pesticides not only survived the baking process but were concentrated to exceed safety residue levels.[35]

Fruit, Fruit Products and Vegetables

MAFF routinely samples fruit and vegetables for pesticide residues as this type of produce is felt to be a high risk area for the consumer. The 1982–85 WPPR survey showed that most citrus fruit products contained residues, particularly of pesticides applied post-harvest.[36] In 1985 48 samples of citrus fruit drinks were analysed. Nearly half contained residues of thiabendazole and 2-phenylphenol at sufficiently high levels for the WPPR to seek the advice of the Department of Health.

Forty-three per cent of the 1,649 fruit and vegetables sampled between 1981 and 1984 contained detectable residues, with some 29 containing residues above the MRL.

Potatoes present more of a problem. Thirty-seven out of 67 samples contained residues of tecnazene (a sprout suppressant) at levels at, or in excess of, the EC MRL of 1 mg/kg. The WPPR Report notes that on tracing the origin and history of the samples, all but one had been treated in accordance with label recommendations. The fungicide, quintozene, was also detected in many potato samples. No survey was carried out for the 1985–88 Report in spite of the high risks associated with this product group.[37]

Imported fruit and vegetables can contain high residue levels. Excessive pesticide levels have, for example, been found in a range of fruit and vegetables grown in Cyprus, which exports the bulk of its produce to Britain. The Cypriot Health Minister said: 'the farmers use more chemicals than they should and they cut the crops before they should'.[38] Crops found by the government chemists to contain abnormal residues include cucumbers, tomatoes, green beans, lettuce, strawberries and potatoes. In 1989, Britain imported about 65,000 tons of potatoes from Cyprus. Precise figures have not been disclosed but according to the Head of the Government Laboratory 40 per cent of strawberry samples tested in 1989 showed pesticide residues above EC standards. The Cypriot government is trying to reduce the problem with farming education campaigns and tougher laws.

Animal Products

Only small surveys of specific animal products had been carried out in the UK before 1984, as a result the MAFF WPPR instituted a four-year survey of residues from animal products.[39] The Working Party survey results revealed that the most commonly detected organo-chlorine residues were pp'DDE, a metabolite of DDT, and gamma-HCH (lindane), a form of hexachlorocyclohexane. Residues were most frequently found in lambs, and the explanation is thought to be the use of DDT and HCH in sheep dips. Following the withdrawal of products containing these active ingredients in sheep dips from January

1985, levels in lamb were observed to fall but even in 1985–6, 10 per cent of samples still contained detectable HCH.

All samples of processed pork and poultry from China contained organochlorine residues, particularly alpha-HCH, a dangerous technical form of lindane never permitted for use in the UK. Levels exceeded the EC MRL in the majority of cases.

Residues of organochlorine pesticides were found in milk and milk products. 119 out of 176 samples of milk in 1985–86, and 109 out of 120 in 1987 contained such residues. Dieldrin and gamma-HCH were the most frequent. It is suspected that residues from animal feeding stuffs are finding their way into milk, and work is in progress to investigate this possibility.

Pentachlorophenol (PCP), used in wood preservatives, is now found in nearly half the samples of the food groups – offal, poultry, eggs and milk.

Wildlife: Game and Fish

According to the WPPR report, in a great number of cases eels in the UK contain 'unexpectedly high' levels of the organochlorine pesticides dieldrin, DDT metabolite, and HCH isomer.[40] The results suggest continued use of both dieldrin and DDT, even though all approvals for DDT use were cancelled in October 1984, and all approvals for dieldrin were revoked from the end of 1989. Advice was given by government departments and water companies to consumers relating to the consumption of eels. The problem is discussed further in Chapter 7.

Wood pigeons were also surveyed for organochlorine residues. As pigeons are a sedentary species, residues probably reflect local exposures. Of 122 samples, DDT metabolites were found in 42 per cent of which three were above the EC MRL. Gamma-HCH residues were found in 4 per cent of the samples. DDT levels appear to be decreasing, but the continuance of HCH levels is not adequately explained.[41]

Residues of a toxic organophosphorus pesticide, dichlorvos, have been found in samples of Scottish salmon on sale in supermarkets. The Scottish Salmon Growers' Association had said that such samples 'should not be allowed to be sold'.[42]

Wines

In February 1990 the US FDA, conducting routine tests on an Italian Asti Spumante wine, discovered traces of the pesticide procymidone. Procymidone is a fungicide widely used in almost every wine-producing country in the world, except the US. The FDA Survey resulted in the Asti Spumante shipment being banned.[43] Following this, the FDA tested over 200 more wines from 21 countries. Some Italian and French wines were found to contain 0.02 parts per million of procymidone which is not authorised for use in the US. In attempting to defuse the row over authorisation, the French Wine and Spirit Exporters Federation, which represents most exporters, assured US consumers that 'there is no risk to human health'.[44] France exported wine worth £322 million to the US in 1989, about 13 per cent of its exports.

Children: Infant Foods and Breast Milk

Children, especially under the age of 6, may be particularly at risk from pesticide residues, because they tend to consume more vegetables and fruit, or fruit products: these may contain residues. Babies may be particularly at risk from contamination of mothers' breast milk by pesticides, dioxins and other pollutants.

The Natural Resources Defense Council (NRDC), a US environmental pressure group, estimates that, relative to their weight, the typical pre-school child consumes 6 times as much fruit as the average adult, 7 times more grape products, 7 times more apples and apple sauce, and 18 times as much apple juice as an adult.[45] The controversy over the plant growth regulator, daminozide (sold as Alar) on apples has highlighted the risk to young children (see page 124).

In the UK the MAFF WPPR report (1982–85) stated that the detection in milk of relatively high levels of a persistent organochlorine insecticide, dieldrin, was 'unexpected'.[46] The government's Committee on Toxicity of Chemicals in Food, Consumer Products and the Environment, stated that: 'we are concerned at the levels detected and the frequency of occurrence of organochlorine residues in cow's milk, particularly as this may constitute a large percentage of the diet in infants and young children.' Scientists from the Department of Health's Committee on Toxicity warned, in an addendum to the WPPR report, that cow's milk was so often heavily contaminated by dieldrin that infants and young children could exceed the WHO's ADI levels, and that the long-term effects of this are unknown.[47]

Following the WPPR report, the UK government advised baby food manufacturers to reduce pesticide levels in rusk, milk and other baby foods. Tests showed that 11 out of 12 wholemeal baby rusks sampled contained pesticide residues.[48]

The WHO report, *Chemical Contaminants in Food: Global Situation and Trends*, makes the same point:

the reported levels of pesticides and other contaminants in human milk at certain times result in estimated intakes by the breast fed infant that exceed toxicologically acceptable intake levels ... whether such intakes are detrimental to the child's progress in terms of physical or possibly mental development is not known.[49]

In 1989 UK government scientists investigating chemical contamination of food and the environment announced they had discovered excessive levels of the toxic chemicals dioxins in women's breast milk. The findings followed tests of random items of 20 different foods available in UK shops in addition to breast milk. Some food samples contained dioxin levels 100 times the recommended 'safe' limits.

Daminozide (Alar) and Apples

Daminozide is a plant growth regulator (classed as a pesticide) used primarily on apples, peanuts and cherries to increase the red colour of the fruit and prolong shelf life. Daminozide contains the chemical, unsymmetrical dimethyl-hydrazine (UDMH), as a breakdown product and contaminant. Formation of UDMH is accelerated when apples containing daminozide are heated or processed; and residues of both daminozide and UDMH are generally found in apples, apple juice and apple sauce. Commercially, daminozide is sold under the product name Alar. In the UK official MAFF figures showed that Alar was used on some 100 million apples, about 7 per cent of the total crop.[50]

In the early 1980s the US Environmental Protection Agency (EPA) received a number of toxicity studies which indicated that UDMH caused tumours in mice. In January, 1987 the EPA's carcinogen assessment group classified daminozide as a probable human carcinogen.[51] In February 1989 the EPA announced its intention to ban Alar within 18 months. Alar has also been banned in Italy and New Zealand.[52]

Following the EPA's announcement, the UK Advisory Committee on Pesticides undertook a review of Alar. In May 1989 the UK government announced that Alar had been given a clean bill of health, based on a review of the evidence from the US.[53] Despite the Advisory Committee on Pesticides statement, the National Farmers' Union (NFU) states that: 'growers have however responded to consumer concern, and resulting commercial pressure. The NFU and other bodies representing apple growers have advised growers to conform with buyers' demands not to use Alar this season.'[54] Uniroyal, the company manufacturing Alar, has announced that it no longer wishes to market the pesticide for use on food.[55]

Dioxins in samples of cow's milk were 10 times the provisional draft guidelines used by the government's Toxicity of Chemicals in Food Committee.[56] A senior scientist commenting on the finding said: 'as a consumer I would say there is nothing to worry about but as a scientist one could never say anything is absolutely safe.' The government's food tests have been carried out in a specially-built, high-security unit at the MAFF's food science laboratory in Norwich. The risk of birth defects is considered so great that women scientists are not allowed to work in the laboratory.[57]

Cancer-causing Pesticides in Food

Maneb, Mancozeb and Zineb

In July 1989 the US EPA warned that three of the world's most widely used fungicides, sprayed for nearly 40 years on cereals, fruit and vegetables, might cause cancer. These three pesticide active ingredients, maneb, mancozeb and zineb, are sprayed before harvest on wheat, barley, hops, potatoes, lettuce, spinach, onions, leeks, tomatoes, apples, pears, blackcurrants and gooseberries. Mancozeb is also sold to amateur gardeners. The EPA suggested that continued use of these three fungicides might cause 125,000 additional cancer cases among the US population of 250 million.[58] Like daminozide, the fungicides become more dangerous when exposed to heat because their breakdown chemical, ethylenethiourea (ETU), causes thyroid cancer in animals. The EPA has ordered a ban on the use of these fungicides from mid-1991 onwards.

Maneb, mancozeb and zineb were on MAFF's list of more than 100 long-established pesticides which are suspect because they have not been tested to modern safety standards. As a result of the US evidence these three pesticides were subsequently reviewed by the UK Advisory Committee on Pesticides.[59] In January 1990 the government announced that the Advisory Committee had concluded that there was no risk to consumers from the use of these fungicides or from the metabolite, ETU. As a result pesticide products containing these products remain approved for use in the UK. Residue levels of these pesticides in spinach and lettuce are to be the subject of further investigations by the government.[60]

Captan, Chlorothalonil and Folpet

These three chemicals, all fungicides, are found in a wide range of pesticide products and are used on a variety of crops including fruit and vegetables. The NRDC has estimated that captan, chlorothalonil and folpet could cause cancer to between 48 and 400 of the current US pre-school children at some stage of their life.[61] Captan is currently being reviewed by the UK Advisory Committee on Pesticides.

Children and Carcinogenic Pesticide Residues

In February 1989 the NRDC published the results of a two-year study on the health risks of pesticides in food in the US. The NRDC states that children under the age of 6, more likely to consume fruit and fruit products than adults, are especially vulnerable to carcinogenic pesticides. It was the NRDC's opinion that some 5,500 to 6,200 pre-school children could develop cancer through their exposure to eight of the most commonly-used pesticides, which the NRDC considered to be carcinogenic. The NRDC study also considered possible neurotoxic effects on children from pesticide residues, and estimated the proportion could be as high as 2–3 per cent, or 2.2 million pre-schoolers aged 1–5.[62]

The NRDC study alleges that UDMH, a breakdown product of some pesticides, is the most potent of these carcinogens and it predicts extra cancer risks of 1 case in 1,100. This estimate is nearly 1,000 times that of the EPA, whose own guidelines state that a lifetime cancer risk of greater than 1 in a million is unacceptable.

The organophosphate methamidophos caused most concern. Average exposure to this pesticide was estimated to be more than 3 times the ADI set by the EPA in accordance with WHO guidelines. Unsafe levels were also recorded for the organophosphate pesticides parathion,

parathion-methyl, diazinon, azinphos-methyl, omethoate, monocrotophos and acephate.

Food Industry Control Over Residue Levels

Food retailers are concerned about pesticide residues and food quality because of the mounting consumer pressure for what is perceived as 'safer food'. The large supermarket chains can exercise a great deal of control over farmers and growers and can state which pesticides should be used and how often. Retailers also carry out their own testing programmes to check for microbiological and pesticide contamination of food. Much of the testing is in reaction to public concern about specific chemicals and not exploratory screening for all synthetic residues.

Large supermarket chains in the UK have resisted pressure to tell consumers about chemical contamination of food. Even though supermarkets carry out testing programmes to rule out contamination of food on their shelves, many still keep the results confidential. This secrecy flies in the face of the supermarkets' recent campaigns to present themselves as champions of the 'green consumer'. In particular, supermarkets have refused to publicise information on tests for residues on two pesticides, daminozide and tecnazene, which have given particular cause for concern (see pages 124 and below).

Tecnazene and Potatoes

Tecnazene is a fungicide and sprout-suppressant used on crops, especially potatoes. Residues of tecnazene are alleged to be higher in potatoes grown in Britain than in potatoes grown in other countries. Tecnazene may cause cancer.[63]

In early 1989 Friends of the Earth, working with Channel 4 Television, found that levels of tecnazene in samples of potatoes bought in London at three leading supermarket chains were double the international safety limits.[64] Friends of the Earth wrote to these three chains and to other leading supermarkets asking for details of their tests for tecnazene in potatoes. The replies were all unhelpful:

- *Safeway:* 'These results are for the company's internal use.'
- *Tesco:* 'The information cannot be disclosed for reasons of confidentiality.'
- *Waitrose (John Lewis):* 'We do not seek publicity for these measures.'
- *Sainsbury:* 'We consider that this information is confidential to ourselves and our suppliers.'
- *Budgen:* 'We see little reason to publicise data which would not be understandable by the majority of people.'

MAFF, in its consultation on the Residue Regulations originally proposed a MRL for tecnazene but then failed to set a limit.[65, 66] MAFF's Scientific Sub-Committee of the Advisory Committee on Pesticides (ACP) did, however, review the safety of tecnazene. As a result the ACP concluded that tecnazene in the effluent from potato-washing plants was a potential threat to fisheries, freshwater and marine organisms, and to drinking water if water was removed downstream of such a plant.

In 1989 the ACP considered new data from industry on tecnazene and subsequently set an ADI of 0.03 mg/kg body weight, 3 times higher than the 0.01 mg/kg limit set by the WHO in 1983. The ACP further recommended that a MRL of 5 mg/kg should be set for potatoes, but this was only a 'suggestion' – it was not a legal limit. In consequence the government and agricultural departments are advising users that:

- only one application to each crop is permitted
- the maximum application rate and 6-week interval between application and sale are legal requirements and must be observed
- tecnazene should be applied at a suitable point along the harvesting/grading line and not directly onto potatoes in the stack or in storage boxes[67]

The Marks and Spencer chain decided, independently, to adopt the WHO limit for tecnazene in potatoes, following the failure of the UK government to set a standard.

The Law and Enforcement of Regulations

European Community Law

The EC has introduced a Directive which aims to establish a set of mandatory MRLs for fruit, vegetables and certain other crops such as oilseeds and hops.[68] Two earlier Directives which already set mandatory MRLs for animal products and cereals (already incorporated into UK standards) will be reviewed as part of this process.

The EC will establish a priority list for crops and pesticide active ingredients where there is no current MRL. The UK government has proposed the pesticide active ingredients maleic hydrazide and pirimiphos-methyl, as its top priorities for inclusion in the EC list. Existing MRLs for specific pesticides would be systematically reviewed to see whether the levels should be altered.

UK Law

In 1988, in response to pressure from the public and EC, the UK government introduced legal controls over pesticide residues in food. The Pesticide (Maximum Residue Levels in Food) Regulations 1988, set legally enforceable MRLs for some 60 or so pesticides used in 'the most important fruit and vegetable components of the average national diet'.[69] Over 400 pesticide active ingredients are approved for use in the UK, and many of them are used on food crops; for many pesticides there is no MRL.

The new UK standards are based partly on international 'Codex' standards and EC limits but the criteria for setting MRLs are still far from clear. In some cases the regulations set MRLs which are higher than the government proposed in a consultative document in April 1988, and they fail to set limits for eight pesticide active ingredients included in the original 1988 proposals. The regulations set no MRLs for tecnazene on celery, lettuce and potatoes, even though it is widely recognised that the relevant (non-statutory) MRLs recommended by WHO are regularly exceeded in the UK.

The British Agrochemical Association (BAA), the pesticide manufacturers' trade association, in welcoming the introduction of legal MRLs, said that it:

> supports the continuing monitoring of pesticide residues on a routine basis and at a level which allows: (i) national agencies to prosecute those who disregard the approved recommendations for the safe and efficient use of a product; (ii) national and international agencies to assess possible long-term implications on public health. However, such agencies must have adequate resources to effectively achieve these aims without disrupting normal trading of perishable food.[70]

Both home-produced and imported food is constantly being monitored for residues by the MAFF WPPR: it analyses some 3,000 samples per year. The WPPR reports to the Advisory Committee on Pesticides and to the Steering Group on Food Surveillance, and these reports are published in the form of Food Surveillance Papers.[71] In fact, with the publication of the WPPR 1988–89 report, the government claims the UK is the first country to issue full pesticide monitoring results.[72] Out of nearly 4,000 samples analysed, 34 per cent contained residues below MRLs, and in 2 per cent of cases residues were found above MRLs.[73]

Enforcement

Enforcement of laws on residue levels is carried out in two ways: first by routine sampling and analysis of food by local authorities and second through farm/horticultural visits by health and safety inspectors checking on harvest intervals and maximum dose rates.

Sampling and Analysis

Local authority food sampling is carried out by Environmental Health Officers (EHOs) and/or Trading Standards Officers. Environmental Health Officers have control over food hygiene, whereas Trading Standards Officers are responsible for the composition, labelling and advertising. Port authorities at seaports have similar powers to Local authorities to conduct monitoring and take samples, which are analysed by the local Public Analyst. The Public Analyst is responsible for monitoring and analysing food content, as well as checking weights and measures.

Analysis of residues is a time-consuming and costly procedure. The monitoring carried out in the UK is widely recognised as inadequate.[74] No central government figures for monitoring are available. In 1982 EHOs tested 20,000 samples for bacteria and chemicals, but only 121 pesticide residue analyses were made.[75] The London Food Commission has stated that current testing budgets are so low that only one residue sample is tested for every 100 million food items purchased by UK consumers (assuming that each food item costs £1).[76] There are considerable analytical difficulties in measuring pesticide residues in food given the large number of approved pesticides; a lack of resources makes the problem worse. Residue analysis requires many different techniques, given the fact that the analyst often does not know what pesticides will be in what foods; there are currently 80 different approved pesticides which may be present in wheat, for example.[77]

Because of the difficulties of analysis, the low number of analyses, and legal limitations of the 1984 Food Act, legal action over pesticide contamination is rare. For all forms of food contamination there are only about 1,000 convictions each year under the Food Hygiene Regulations; the average fine is £250.[78] Prior to 1988, pesticide contamination of food could only be dealt with using the Food Act, 1984 (replaced by the Food Act, 1990) which requires proof either of a hazard to health, or that the sale of the food is substandard.[79] These are difficult issues to prove in court. There is a view that pesticide residues should not be considered as contaminants because they are the end result of a legitimate trade practice in food production.[80] From 1988 pesticide residues were governed by the 1988 MRL regulations; no prosecutions have yet taken place for failure to abide by these.

Inspection

The Agricultural Inspectorate of the HSE is responsible for checking that pesticides are being applied according to the conditions of approval, on farms, in glasshouses and on horticultural premises. Inspectors are required to check pesticide dose rates, approved tank mixes and harvest intervals. Crops breaching pesticide regulations were destroyed for the first time in 1988 following enforcement action by HSE inspectors. A lorry load of tomatoes and 1,800 lettuces bound for supermarket chains were destroyed before they left the farms in Kent because they were harvested too soon after being treated with pesticides, which is illegal under the Control of Pesticide Regulations 1986.[81]

In contrast with the UK parts of the US have tough standards and testing for residues.

California Dreaming – An American Comparison[82]

The California Department of Food and Agriculture (CDFA) is responsible for monitoring pesticide residues in food in California. CDFA produces an annual Pesticide Residue Report which reports on market place samples, target pesticides, produce destined for processing, and pre-harvest monitoring.

One of the major criticisms of pesticide residue monitoring in the UK must be the relatively small number of samples taken. The number of samples analysed in the latest UK WPPR report is 2–3,000, very few multi-residue tests have been done.[83] By comparison, the CDFA took over 13,400 samples in 1987, and in many cases each sample can be analysed for 100 different active pesticide ingredients.[84] In 1987 the US National Academy of Sciences released a report listing 53 chemicals as potential carcinogens, 35 of which are included in CDFA's pesticide residue programme.[85] Over 7,000 of the 1987 samples analysed were screened for these 35 suspected carcinogens. To our knowledge no such programme is planned for a similar list of potential cancer-causing pesticides in the UK.

The CDFA has set MRLs or 'established tolerances' for the majority of pesticide/commodity combinations. Previous reports of the WPPR have argued against statutory MRLs, stating that persuasion would encourage good practice and reduce residues. Although persuasion is always to be encouraged, it is clear that misuse of pesticides has occurred. MRLs are vital and we now have some in force. More are needed for commodity/pesticide combinations not already covered.

The Californian budget for the monitoring of pesticide residues is $4 million per annum. Britain spends just £1 million each year on its own programme for a population twice that of California. [86]

The CFDA publishes its findings annually, and in great detail. The UK WPPR publishes every four years, with only summaries of what it finds.

The Politics of Food Regulation

Once again the root of the problem with food quality regulation is that MAFF currently looks after the interests of both farmers and consumers.

The National Consumer Council, in a 1989 report called *The Future of Food Regulations*, said food safety should be the preserve of an independent agency, and not under MAFF.[87] Public confidence, the report argued, had been lost because of the 'secret and close' relationship between MAFF, farmers and the food industry. Other consumer bodies, including the Consumers' Association and the London Food Commission, have also called for a separate regulatory agency in the wake of food hygiene scandals, as has the chairman of the House of Commons Select Committee on Agriculture.[88] Trade unions, representing both agricultural and food production workers, have also called for a separate food agency.[89] The British Medical Association pesticide report has reached a similar conclusion.[90] Consumers in the

European Community Group have called for an integrated system of evaluation, approval and monitoring to be established throughout the EC.[91]

Not satisfied with its secret comings and goings, the UK government is also resisting an EC proposal that would force retailers to tell customers what pesticides have been used on fruit and vegetables in their shops. Recent moves in the EC to introduce labelling requirements for fruit and vegetables indicating when they have been 'treated with chemicals' have been opposed by the UK and the pesticide industry.

Water

Contamination of drinking and surface water by pesticides and other farm and industrial chemicals is a worldwide phenomenon. Drinking water in many countries where intensive agriculture is practised does not meet official, internationally recognised safety standards because of excessive contamination by nitrate and pesticide residues. Industrial weed control, poor disposal of chemicals by industry and the domestic use of pesticides contribute to the problem. Pesticides are not the only things to find their way into water. In EC countries some 700 synthetic substances, 300 of them pesticides, have been detected in water for human consumption. Many of the pesticides in drinking water, exceed official EC safety limits, known as the Maximum Admissible Concentrations (MACs). The pesticides found in excess levels most often include the herbicides atrazine, molinate and simazine.[92]

Most contamination with pesticides and fertilisers occurs at a variety of places, not from a single source, and this makes pollution control difficult. Contamination results from surface run-off from fields, verges, railway tracks and gardens into reservoirs and streams, or from the chemicals slowly leaching through the soil into the water table. The number of reported water pollution incidents in England and Wales rose by 16 per cent from 23,257 in 1987 to 26,926 in 1988 compared with 12,600 in 1981.[93] Prosecutions for water contamination are few and far between. In 1988, less than 1 per cent ended in prosecution, with the average fine being under £500. The maximum is £2,000 and three months imprisonment.[94]

There is growing public pressure for governments to clean up drinking and surface water and limit agricultural and industrial pollution. Acting on this concern, the EC is threatening to take the UK and Italian governments to the European Court to make them clean up their water and comply with the 1980 EC Drinking Water Directive.[95] However, there is still argument over the EC pesticide limits for water. Some governments and water supply companies say the standards are too strict and do not take account of the different properties of each pesticide. The EC is being pressed to review its water quality standards for pesticides by the UK and Italy,[96] as well as by industry. Meanwhile, the US EPA is proposing tough action against ground water pollution from pesticides, based on regular national Ground Water Supply Surveys.[97]

Agricultural Pollution

Agriculture, including horticulture and forestry, is the main source of drinking and surface water pollution in Europe and the US. In the UK intensive agricultural practices account for well over 50 per cent of water pollution incidents, involving chemical fertilisers and pesticides, and animal wastes.

Agricultural pollution of rivers is highly seasonal, the highest levels occurring in the autumn from leaching and land run-off following heavy rain. The problem is different in aquifers where some water may take up to 30 years to percolate from the soil to the water table, depending on the type of rock it has to pass through. Predicted peak contamination of pesticides and nitrates from current agricultural practices may occur in 5 to 20 years time, or even longer. Furthermore, long-term water storage in reservoirs can lead to a 50 per cent decrease in bacterial reduction of nitrates to nitrogen gas.[98]

In the UK official surveys have shown the widespread presence of pesticides in both surface and underground waters. The total herbicides atrazine and simazine were the most widely occurring along with selective herbicides, such as mecoprop, used for broad-leaf weed control in cereals and grassland. In 1988, the environmental pressure group Friends of the Earth (FoE) analysed the available records of UK water authorities and water companies for the period July 1985 to June 1987. FoE claimed that 298 water supplies in England and Wales were illegally contaminated with pesticides, the worst offences being in the Anglian, North West, Severn Trent, Thames, Wessex and Yorkshire water regions. The Anglian region, the most intensively farmed area in the UK, had the highest concentrations of pesticides in drinking water, where 11 pesticides were detected in quantities over the EC limit. In the Severn Trent region, 7 pesticides over EC limits were detected in the same period.[99]

A six-month survey of water quality coordinated by the Institution of Environmental Health Officers, based on 174 water samples from 25 London boroughs, found clear evidence of pesticide contamination. The worst cases involved the pesticides atrazine and simazine.[100]

Italy too is in breach of EC water standards. In the Italian provinces of Bergamo and Pavia in mid-1986, wells serving 200,000 people were found to be contaminated with the herbicides, atrazine and molinate. Alternative supplies had to be brought in by tanker. Breaches of the MACs were later found in other areas, generally involving atrazine, bentazone, molinate and simazine. Subsequently the Italian government passed decrees authorising the EC limits to be exceeded, setting limits just below the WHO guideline values. The decision was challenged in

Italy, and the European Court has been asked to rule on the legality of the decrees.[101]

In the US 60 different pesticides have been found in the drinking water of some 30 States as a result of farming.[102] Aldicarb, an acutely toxic pesticide, has been found in the underground aquifers of 50 Counties in 15 States. Farm wells in Wisconsin have been contaminated with aldicarb, and health officials have warned people not to drink tap water if aldicarb concentrations exceed 1 part per billion. In California more than 200 wells in 23 Counties were found to contain 57 different pesticides at various levels of contamination.[103]

Industrial Pollution

Industrial pesticide production is another major source of water pollution. Pesticides, along with other chemicals and waste products, are routinely discharged into rivers. Much of this discharge is legal, though the rules are being tightened. Illegal dumping goes on, however. In the UK, the chemical company ICI has admitted it has been illegally discharging excessive residues of the organochlorine insecticide lindane into the River Weaver from its production plant in Macclesfield, Cheshire. High levels of another banned organochlorine pesticide, dieldrin, are being discharged into the River Severn by the carpet industry at Kidderminster despite attempts by Her Majesty's Inspectorate of Pollution and National Rivers Authority to prevent it. Dieldrin use is banned in the UK but during the process of carpet manufacture it is being washed from fleeces from developing countries, where it is applied as a mothproofing agent.[104]

The German chemical company BASF has become embroiled in a dispute with Dutch water companies about discharges of the pesticide bentazone into the River Rhine from its Ludwigshafen site. Bentazone levels of up to 1.6 micrograms per litre have been found at abstraction points in the Netherlands, which is well in excess of EC standards. The water utilities insist that they should not be forced to install costly activated-carbon filters to reduce the contamination to 0.1 microgram per litre. On the other hand, BASF is referring to the WHO's proposed guideline value of 25 micrograms per litre for bentazone to argue that contamination does not present a health hazard. Even though BASF is developing a new treatment plant, the Dutch water companies, unhappy at the prospect of a two-year wait for the plant to be installed, may try to make the company pay for polluting its water source.[105]

Large-scale weed-killing programmes carried out by local councils and private firms, and extensive spraying of railway lines by British Rail is adding to water pollution by pesticides. Atrazine, a persistent herbicide used to control weeds on embankments, is one of the commonest water pollutants in the UK.[106] Its use for this purpose is banned in Switzerland where Ciba-Geigy, the Swiss manufacturers of atrazine, acknowledge that this weed control can easily be achieved with less persistent herbicides.[107]

The Netherlands government is currently taking legal action against Shell to make it clean up polluted waste, at an estimated cost of £50 million, from its pesticide factory near Rotterdam. Until 1990 the plant produced three chlorinated hydrocarbon 'drin' (aldrin, etc.) pesticides, related to DDT.[108]

Similarly, Shell in the US, after five years of court action, has agreed to pay half of the first US$500 million (£263 million) spent on cleaning up Colorado's Rocky Mountain Arsenal, where it produced 'drin' pesticides until 1974.[109]

Setting Standards

The setting of water standards has been and remains a controversial area and one where EC limits and WHO guideline values differ significantly.

As well as dealing with food contamination, the WHO has a programme on the development of Quality Guidelines for Drinking Water. This programme covers various categories of chemicals, including pesticides. The recommended values contained in the WHO 'Guidelines for Drinking Water Quality', issued in 1984, are based on toxicological and other support data available up to 1981. The WHO guideline values refer to lifelong consumption and in the course of their derivation the total intake from air, food and water of each substance has to be taken into account. The health concerns arise primarily from adverse effects after prolonged ingestion of chemicals which accumulate in the body and which have carcinogenic properties. WHO argues that acute health effects from pesticides in water are less likely although there are well documented incidents.[110]

In relation to pesticides and their metabolites, except those recognised as genotoxic carcinogens, WHO had adopted the principle that the guideline value should be based on the ADI approach. One per cent of ADI would be applied for substances for which exposure from food residues approaches the ADI. For other pesticides, especially a number of herbicides where residues were considered unlikely to be present in any significant amount in food at the time of consumption, a higher value of 10 per cent of the total ADI was allocated to drinking water.

The EC MACs are based on analytical criteria of detectability. These standards are much stricter than WHO guideline values, setting low levels of pesticide in drinking water which are often below the most frequently found concentrations.

A comparison of WHO and EC standards (see Table 6.2) shows that for the small number of pesticides and related compounds covered in the WHO 'Guidelines for Drinking Water Quality', the total concentration allowed on the basis of health criteria is 134 µg/litre compared to only 0.5 µg/litre for all pesticides and related products allowed by the EC Directive. The WHO safety figure is 268 times higher than the EC safety figure.

Such a discrepancy in official standards raises important questions about the basis of 'scientific objectivity' used in setting regulatory levels.

Table 6.2: Comparison of EC and WHO Standards on Drinking Water

EC Directive for Drinking Water Quality
All pesticides and related products (including PCBs)

Directive Level[111]

For any one substance	0.1 µg/l
For all substances together	0.5 µg/l

WHO Guidelines for Drinking Water Quality
Pesticides (only eight categories covered)

Guideline Value[112]

	(µg/l)
DDT	1
Aldrin and dieldrin	0.03
Chlordane	0.3
Hexachlorobenzene	0.01
Heptachlor and heptachlor oxide	0.1
Gamma-HCH (lindane)	3
Methoxychlor	30
2,4-D	100

Note: Total concentration suggested by WHO Guidelines (for only 8 pesticides is *268 times* the allowable concentration for *all* pesticides under the EC Directive.

Source: Tarkowski, S. *WHO's Role in the Safety Assessment of Pesticides*, WHO, Copenhagen, undated, p. 10.

Enforcement of Residue Standards in Water in the UK

The standards controlling pesticide residues in water are set by the European Commission, and responsibility for enforcement rests with the UK government. The 1980 EC (then known as European Economic Community) Directive on Drinking Water Quality, which came into force in 1985, sets two MACs for pesticides in water:

- for any single pesticide – 0.1 micrograms per litre
- for 'total' pesticides – 0.5 micrograms per litre

These standards were set at the then prevailing limits of detection for certain chlorinated (organochlorine) pesticides and not solely on toxicological criteria.

The UK has been slow to enforce the EC Directive and has its own set of advisory values, drawn up on Department of Health advice; these are much higher than the EC limits.[113] The UK has lacked an integrated approach to pollution control linking air, land and water pollution, though such a system is now promised in the Environment Protection Act, 1990.[114] Currently UK regulation of water quality is based on a twin system of control involving separate government organisations, operating under different Acts and Regulations:

- National Rivers Authority (NRA), created in 1989, is responsible for general water quality including pesticide and fertiliser contamination[115]
- Her Majesty's Inspectorate of Pollution (HMIP), part of the Department of the Environment, licenses and enforces industrial discharges to water of 'listed' dangerous substances, as well as having responsibility for air and land pollution.

The government is now proposing a national classification scheme for dangerous substances in ground water.[116]

Faced with the need to improve water quality, and linked to the privatisation of the water companies, a new UK statutory framework to regulate water quality has been established. The UK Water Act 1989, for the first time, makes it a *specific criminal offence* to supply water which is unfit for humans to drink.[117]

The Water Act establishes a new public body, the National Rivers Authority, to take over the responsibilities of water authorities in England and Wales in relation to water pollution, water resource management, flood defence, fisheries, recreation and navigation. Among other duties the NRA will have responsibility for water pollution control for inland, coastal and underground waters. It will regulate polluting discharges to water through a consent system and will be able to levy charges in respect of these consents. The NRA will also have powers to establish water protection zones and nitrate-sensitive areas in England and Wales. It plans to draw up a new code of practice on agricultural pollution. The NRA is to set up a public register of information on matters such as water quality objectives, consents to discharges and water samples. A new Drinking Water Inspectorate is to be established to enforce the Act and regulations made under it.

Various regulations will be made under the authority of the Water Act. The Water Supply (Water Quality) Regulations 1989 are now in force and under Regulation 29 of these, water companies (referred to as 'water undertakers' in the legislation) are required to prepare and maintain a record of *water supply zones*. Each water undertaker is required to develop a monitoring strategy for pesticides based on the likely risk of particular pesticides being present in the water source serving the zone. If a pesticide concentration in the water leaving a treatment works exceeds the prescribed concentration of 0.1 µg/l, the water undertaker should re-sample the water immediately and if this confirms the original finding then the relevant authorities, including the NRA, must be notified. Regulation 30(1) requires water undertakers to make available to the public, free of charge, the records of water quality they must maintain under Regulation 29.

The NRA and Water Act 1989 is an important step in the development of integrated pollution control in the UK but there is still confusion over the legislative framework. HMIP is being given new powers to control land, air and water pollution under the Environment Protection Act, 1990. How the role of the HMIP will integrate with that of the NRA remains to be seen.

The Politics of European Water Regulation and Enforcement

The EC Directive on Water Quality 80/778/EEC marked a turning point in measures to improve drinking water standards. Until the Directive was introduced the presence of pesticides in water was not a major issue. Little was known about the toxicological or epidemiological aspects of many pesticides and there were no reliable methods for measuring very low concentrations. The Directive highlighted very different views and approaches to water quality in countries in Europe and demonstrated widely differing enforcement actions. Many breaches of the EC Directive on Water Quality still occur and no country provides a shining example.[118] The fact that breaches of the EC Directive are fairly common has provoked a Europe-wide debate about pesticide regulation and water quality.

A large part of the debate focuses on the validity of the standards set by the EC. Unlike the MACs set for the other 50-odd parameters of water quality, those for pesticides were not derived solely from toxicological considerations but were set at the then prevailing limit of detection for certain chlorinated pesticides. The pressure from industry, water companies and some governments for a relaxation of the Directive and the adoption of different standards based on individual pesticide toxicities and their health effects has polarised the debate, with industry and government on one side and environmentalists on the other.

The nature and scale of pesticide contamination of water also varies between the EC Member States. Residues of pesticide have been found in ground and surface waters in France, Germany and the UK, which did not immediately enforce the Directive but preferred to proceed gradually on the basis of the study and verification of each substance. Other states such as Belgium, Luxembourg, Denmark and Spain do not appear to have encountered any particular problems. The position in individual countries with regard to implementation of the Directive is examined below:

Belgium

The national standard came into force in 1985. Most limit values are the same as those in the Directive but no guide levels are indicated.

France

Atrazine is the most commonly used herbicide, and small amounts of this pesticide and simazine, another persistent herbicide in the same chemical group, have been found in surface water and ground-water which is close to the surface. Traces of atrazine have been constantly detected in the waters of the Marne, the Oise and the Seine, with concentrations increasing in the early spring when spraying takes place. In the southern parts of the country, of 100 wells drawing water from underground sources near the surface and from springs, 15 per cent had atrazine concentrations of between 0.05 and 0.1 µg/l, and 1 per cent of these had higher concentrations, still under 0.5 µg/l. In areas where spraying is routine, only deep wells seem to be adequately protected from contamination by pesticides.

Legislation, which incorporates all the limits of the EC Water Directive is in force. The view of the French water services is that, given the large number of pesticides in use, it is difficult to fix a MAC for every substance and to carry out daily analytical control. They argue that as substances have different degrees of toxicity or may not be toxic at all at the same concentration, what they perceive as unduly stringent limits may result in them incurring unnecessarily heavy costs. The distribution services believe, for the time being, that the only solution to the problem is one of derogations (exemptions) from the EC Directive.

Italy

Pesticide contamination of water supplies has also been a problem in Italy. A systematic survey of drinking water revealed the presence, in certain provinces, of concentrations of the widely used herbicides atrazine and molinate which were higher than the official limits. The contaminated wells were closed and emergency systems had to be used to ensure the supply of water to areas whose normal supply had been temporarily cut off. In 1986 in the province of Bergamo and Pavia this water contamination affected more than 200,000 people. Subsequently the whole of the Po plain was included in this water emergency. Atrazine is used on maize crops and the water contamination could be closely matched with intensive maize-growing areas.[119] The Italian government Decree on the quality of drinking water endorsing the MAC values of the EC Directive came into force in 1986 but there have been major problems with its enforcement. Italian water suppliers believe that the enforcement of the limits in the Directive should be postponed in order to enable scientific determination of individual limits for the more widely used substances. In this respect, the positions of the Italian and UK governments and water suppliers are very similar.

The Netherlands

Maize and potato crops are the main areas where pesticides are used, and atrazine, bentazone and molinate have been detected in ground and surface waters nearby.

The Netherlands has enacted the EC Directive. Dutch water suppliers feel that it is unacceptable that the Directive makes no distinction between concentrations of individual pesticides. Practically, however, they fear that if legal limits were to be higher than those of the Directive pollution would increase, or at least it would prove much more difficult to reduce present levels of pollution. As a result, water supply companies decided not to oppose the enforcement of the Directive's MAC values.[120]

Germany

Prior to reunification in October 1990, in what was formerly West Germany, some 40 pesticide active ingredients have been found in water, especially the herbicide atrazine, which was detected in all intensive agricultural areas. The use of atrazine is now banned in protection zones for drinking water catchment areas.[121] Regulations on drinking water have been in force since 1985 and the EC Directive was endorsed by a decree in October 1986. The quality limits laid down are the same as in the Directive and also include the main breakdown products of the substances regulated. The government is planning further measures to ensure that pesticides do not pollute groundwater.[122]

German water suppliers have been broadly in favour of proper controls. Their view is that imposing strict limits will stimulate industry to produce compounds which are more effective when used in smaller quantities. They have asked the government, in particular, to take measures to further control the production, sale and use of pesticides, to encourage industries to produce substances which are more biodegradable and less harmful, and to protect water sources against pollution caused by agriculture and industry.[123]

In what was formerly East Germany, there is no data available on contamination of water by pesticides, though it is acknowledged that there is widespread environmental damage from agricultural and industrial pollution.

United Kingdom

Pesticides and in particular herbicides are regularly found in ground and surface waters and the UK government has been in constant breach of the EC Water Directive in many parts of the country.

In 1989 a Water Act was passed which set the same MAC levels as the Directive but these standards remain to be enforced and the situation will remain fluid while there is pressure on the EC to revise its own limits. In the meantime, advisory standard values for pesticides have been determined by the Department of Health and these values range between 0.1 µg/l and 400 µg/l depending on the toxicity of the pesticides. While monitoring for pesticides continues, no limitations or restrictions are applied to the use of water sources containing concentrations which do not exceed the limits laid down by the Department of Health. According to the water distribution services, the limits in the Directive should be reconsidered on the basis of individual toxicity and, for certain substances, should be made more stringent if necessary.[124]

Spain

The most commonly used pesticides are atrazine, lindane and molinate which are used for citrus fruits, grapes, olives and rice. A 1982 law endorsed the limits indicated in the EC Directive. Water distribution services feel that systematic checks for the presence of pesticides in water are essential but consider that different concentration levels should be adopted for individual substances on the basis of their toxicity.[125]

Comparison of European Water Standards

The Dutch and German water utilities' approach to the pesticide problem is unqualified acceptance – at least in public – of EC standards. At the 1988 European Institute for Water Conference Dutch companies stated that 'pesticides should not be present in drinking water', and that compliance with EC standards should be achieved by tackling the problem at source. Their German counterparts adopted the same position,[126] adding that: 'the waterworks are often presented as the guilty party for problems they have neither created nor bear responsibility for; indeed, problems they cannot solve.'[127]

In response to these concerns, the then West German government issued a new groundwater policy in late 1988. The policy stated that the essential thing was for ground water to be maintained in as natural a condition as possible. Quality standards for ground water which define acceptable pollution loads are not admissible and harmful chemical applications must be prevented wherever possible.[128] The conflict between this philosophy and the UK's approach could hardly be more fundamental.

The UK has failed to meet EC standards for drinking and bathing water purity. The EC Commission is questioning the right of the UK government to grant exemptions to water authorities from current EC pollution limits – limits which carry the force of European law. The government granted these exemptions as part of its plan to privatise the water industry, diluting the polluter-pays principle to make water a more attractive proposition to investors. In the run-up to the privatisation of the water industry in 1989, the government legalised thousands of spillages of raw sewage into rivers and streams and took powers to force pollution inspectors to authorise currently illegal discharges from sewers.[129] Since then the UK government has vetoed plans to force privatised water companies to comply with EC quality standards by 1993, substituting a target of 1995 instead. In fact, the government is arguing that the polluter-pays principle should not apply to water and that the cost of cleaning up water supplies should be borne by the customer.[130] At the heart of the debate over European water quality is the question of the cost of cleaning up water supplies and who will pay for it – the polluter or the consumer. In the UK it has been estimated that a huge investment of £10,000 million for water purification and sewage works would be needed to meet EC standards.[131]

Thames Water, for example, has said it would cost hundreds of millions of pounds to install filter beds to meet the EC standards. Severn Trent Water Authority maintains that filters would need changing so often that plants would be impossible to operate. It has been estimated that the cost of removing nitrates to satisfy EC limits would be over £70 million.

Other Solutions

Water Protection Zones

The best way to protect water supplies from contamination by agricultural pesticides and chemicals is to restrict their use in certain agricultural areas, and establish water-protection zones. These have been tried in West Germany, Italy and Switzerland and have proved very effective. The UK government has declared its intention to set up similar zones, but farmers want compensation. It is on the question of compensation that UK plans to prevent agricultural water pollution have so far floundered. The 1989 Water Act, however, now gives the new National River Authority powers to set up water protection zones and nitrate-sensitive areas in England and Wales.[132]

Removal at Water Treatment Works

Simple treatment processes such as filtration generally remove rather small amounts of pesticide. More sophisticated, combined processes, such as granular activated carbon, ozone and perhaps ultra-violet light (UV)/hydrogen peroxide, can achieve greater pesticide removal.[133] Their high cost is likely to limit application and they also require careful monitoring and skilled management.

Conclusion

Pesticide pollution of food and water remains a topical and controversial subject, especially in Europe and the US. Pesticide pollution is, however, closely bound up with the type of agriculture practised. The more intense the agriculture the higher and more widespread the levels of pollution. Agricultural reform, as well as stricter control of industrial discharges, means this issue is linked to the need for pesticide and fertiliser reduction strategies and the development of alternatives, as discussed in Chapter 12.

7

Pesticides, Wildlife and the Environment

Introduction

Pesticides directly affect the health of other living organisms, small or large, much as they affect human health. They also have major indirect or 'ecological' effects and alter the diversity, abundance and distribution of species. The wildlife that is affected includes plants and trees, vertebrate species (with a skeleton) such as fish, reptiles, birds and mammals, and invertebrate species ranging from arthropods (insects and their close relatives), soil and aquatic organisms to micro-organisms such as fungi, bacteria and viruses.

Both the direct and indirect effects of pesticides on wildlife are closely bound up with the type of agriculture (including horticulture and forestry) practised. The spread and intensification of agriculture, involving increased use of pesticides and artificial fertilisers has had a major impact on the abundance and diversity of wildlife species. This is especially the case where broad spectrum, non-selective pesticides are used which kill beneficial and non-harmful animals as well as pest species. A related problem is the illegal use of pesticides as poison baits which kills both wildlife and domestic pets and animals. In the UK, many protected rare hunting birds such as the red kite are still being killed by unlawful pesticide baits.[1]

Direct effects of pesticides on wildlife include:

- direct poisoning
- 'food chain' or secondary poisoning
- chronic or long-term effects on health

Indirect or *ecological effects* include:

- reduction and disappearance of species
- loss of food and habitat sources
- creation of new or 'secondary' pests
- pesticide resistance

142

Direct Effects

Direct Poisoning

Many pesticides are directly poisonous to wildlife. There are many well documented cases of fish, bird and mammal 'kills'. Smaller organisms such as earthworms and aquatic creatures are also directly killed, though as the effects here are less spectacular or evident, the subject is less well documented.

Direct or 'acute' poisoning of creatures involving pesticides which are persistent, as well as toxic, can also lead to food chain, or secondary poisoning. This type of poisoning occurs as the pesticide-contaminated corpses are eaten by scavengers and animals higher up the food chain, and this phenomenon is discussed at length later in this chapter.

Herbicides

Herbicides obviously have major effects on plants but are generally not as toxic to vertebrate and invertebrate species as other categories of pesticide. Herbicides can, however, be transported in the run-off, or drainage water, from agricultural land and end up in surface waters. Studies have shown that losses of up to 5 per cent of the herbicide can be expected from fields of moderate slope. Losses may be even higher if say, heavy rain occurs within two weeks of application.[2]

Vertebrates – direct poisoning is limited to a few herbicides. Examples include:

- Paraquat – a widely used total, contact herbicide. It can be fatal to rodents and small mammals, especially hares, which pick up the chemical on their fur and skin when passing through sprayed vegetation.[3]
- DNOC – an early generation herbicide, responsible for adult partridge deaths.[4]

Invertebrates – herbicides are directly harmful to aquatic invertebrates as well some species of arthropods. Examples include:

- Atrazine – widely used soil-acting herbicide for agricultural and industrial weed control – has reduced populations of water-borne insect larvae. The emergence of chironomid larvae is significantly reduced by atrazine concentrations as low as 20 μg per litre.[5]
- 2,4-D, DNOC and Dinoseb – three herbicides which are toxic to insects including non-harmful species. Dinoseb and DNOC are now banned in the UK.[6]

Fungicides

Fungicides are generally more toxic to wildlife than herbicides but not as harmful as insecticides. However, some fungicides have unexpectedly broad spectrum effects on wildlife, especially invertebrate species including marine organisms. Fungicidal seed dressings can be particularly hazardous to wildlife especially when waste seed is carelessly dumped in the open or in water courses. The following are some examples:

- *Organomercury fungicides (OM)* – widely used as seed dressings, are highly toxic to fish and aquatic organisms but generally less so to land organisms. OM compounds have differing toxicities. Phenyl Mercury Acetate (PMA), the main OM seed dressing for cereal seed, is highly toxic to aquatic life but much less so to birds and mammals. Methyl Mercury is very toxic to wildlife as mercury can be converted into the methyl form by bacterial activity within the gut of ruminants, at the bottom of a lake, or in a swamp.[7]
- *benomyl, thiophanate-methyl* and *triadimefon* – three widely used cereal fungicides in the UK can cause increased adult or larval mortality rates of a chrysomelid beetle.[8]
- *MBC fungicides* – a widely used group of fungicides in agriculture, horticulture and the garden, have been shown to kill earthworms and other soil organisms. Birds feeding off earthworms killed by fungicides may themselves be poisoned via a 'food chain' effect. Benomyl can reduce earthworm populations by 85–90 per cent from a single application, and is even applied for this express purpose.[9] Repeated application can cause local extinction of some earthworm species such as *Lumbricus terrestris*.[10]
- *dithiocarbamate fungicides* – cell-killing (cytotoxic) chemicals are harmful to a wide range of aquatic vertebrates and invertebrates.[11]

Marine organisms are also being directly poisoned by pesticides as well as suffering sub-lethal chronic effects. Fungicides contained in anti-fouling paints are thought to be the main problem.

In 1987, following a review by the Advisory Committee on Pesticides (ACP), the UK government banned the use of anti-fouling paints containing tributyltin oxide (TBTO) on boats less than 25 metres long.[12] Use of TBTO-based paints on larger vessels is, however, still damaging marine organisms. The Marine Conservation Society claims that continuing damage to dog whelk and oyster populations in Sullom Voe, Shetland and Milford Haven, Pembrokeshire is probably attributable to TBTO leaching from the hulls of tankers and other large vessels. In contrast, oyster populations in inshore areas where large vessels are absent have begin to recover since the ban on the use of TBTO on small boats.[13]

In 1989 a UK House of Lords Select Committee Report recommended an EC-wide ban on all products containing TBTO, whereas the EC is only proposing a TBTO ban on boats up to 25 metres in length. Less harmful anti-fouling paints, mainly copper-based biocides, are available and their commercialisation should be encouraged. Although the Paintmaker's Association claimed that these are likely to be twice as expensive as formulations based on TBTO, the Select Committee felt that this additional expense was necessary for the protection of the marine environment.[14]

Friends of the Earth (FoE) has, however, expressed concern over these new copper-based, anti-fouling paints. FoE claim they still release a lot of harmful copper into the marine environment.[15]

Insecticides

Insecticides are by far the most toxic category of pesticides. It is the total, non-selective nature of many insecticides that causes problems, with non-harmful and beneficial species such as bees being killed. Their use has resulted in many wildlife poisoning incidents and consequent banning, restricting or reformulating of the chemicals involved. Examples, examined in terms of the main chemical groups of insecticides, include:

Organochlorine insecticides – these insecticides are directly toxic to many forms of wildlife. They are also important in 'food chain' or secondary poisoning as well, due to their storage and persistence in body fat and tissue of prey animals, passing from prey to predator and accumulating at each step.[16]

- *Dieldrin, aldrin, endrin* and *heptachlor* caused major wildlife populations losses, vertebrate and invertebrate, in the 1950s in cereal-growing regions of the UK. Their use nearly wiped out many species of hunting bird such as sparrowhawks and peregrines. Aldrin, dieldrin and heptachlor seed-dressings were responsible for the deaths of sparrowhawks, kestrels, hen harriers and thousands of seed-eating birds. On one estate in Lincolnshire these seed dressings killed individuals of 18 species, including 5,700 wood pigeons, over 100 stock doves, 90 pheasants and 14 long-eared owls.[17] Secondary poisoning of mammals occurred when they fed on the corpses of poisoned birds. One study attributes the death of some 1,300 foxes in eight English counties during the winter of 1959–60 to the consumption of wood pigeons killed by dieldrin-dressed seed.[18] During the same period, farm dogs and cats, badgers and carnivorous birds were also found dead, and there is little doubt that they had also eaten birds killed by seed dressings.[19]
- *Dieldrin,* in particular, has been responsible for poisoning a wide variety of creatures. Dieldrin poisoning has been directly responsible for the deaths of some grazing mammals, such as horses, when sprayed onto fields in which they feed.[20] Wood

mice feeding on dieldrin-dressed wheat showed a rapid increase in body loads of this pesticide, with some dying immediately, while others showed sub-lethal effects making them more vulnerable to predators such as kestrels, which were in turn at risk from secondary, or food chain poisoning.[21] Dieldrin was also responsible for the deaths of otters and other aquatic vertebrates.[22] In 1988 a UK national freshwater fish survey revealed residues of dieldrin and DDT in eels. It showed that illegal use of DDT was continuing in some areas, and also hastened government moves to ban dieldrin.[23]

• *Endrin* is the most toxic of the chlorinated hydrocarbons, being 15 times as poisonous as DDT to mammals, thirty times as poisonous to fish, and about 300 times as poisonous to some birds. Its use killed enormous numbers of fish and fatally poisoned cattle that wandered into treated areas.[24]

Organophosphorus and carbamate insecticides – following problems with the organochlorine compounds, the main group replacing them was the organophosphorus group of pesticides. Carbamate pesticides, which have a similar toxicity mechanism to organophosphorus compounds, were also introduced as replacements for organochlorines. The problem with both organophosphorus and carbamate group pesticides is their acute or immediate toxicity, which is often more than the organochlorine compounds they replaced. Birds and bees are particularly susceptible to organophosphorus and carbamate poisoning.

Table 7.1: Persistence of Organophosphorus Insecticides

Pesticide active ingredient	Half-life in days
phorate	2
demeton-s-methyl	26
dimethoate	122
parathion	180

Source: Menzie, 'Fate of pesticides in the environment'.[25]

Organophosphorus pesticides are a large group of chemicals and their persistence is also variable. Some examples of half-lives (a measure of their persistence in the environment) are given in Table 7.1.

In the mid-1970s the use of the organophosphate carbophenothion as a cereal seed dressing in Britain led to the deaths of large numbers of greylag and pink-footed geese and whooper swans, which ate treated seed sown in the fields in which they were over-wintering. Carbophenothion's use as a seed dressing was voluntarily withdrawn in Scotland and restricted in the rest of the country.[26]

Three of the most widely used carbamate insecticides in the UK are aldicarb, carbaryl and carbofuran. All three are toxic to earthworms at normal application rates, and 60–80 per cent reduction in earthworm populations have been recorded following applications.[27] Aldicarb, first marketed in the 1970s as a liquid formulation, was also responsible for the deaths of many bird species (gulls, lapwings and stone curlews), especially in sugar beet areas where it was extensively used for eelworm control. Since its reformulation as a granule there have been fewer reports of direct wildlife poisoning with this chemical.[28]

Insecticides tend to be non-selective in the range of insects they kill, often controlling friend as well as foe. Even the selective carbamate insecticide pirimicarb which, it is claimed, leaves beneficial ladybird larvae unharmed while controlling aphid pests (greenfly, blackfly etc.), has been shown to reduce by 50 per cent populations of non-target insect species on which partridge chicks feed.[29]

Pesticides and Potentiation

According to scientists from the University of Reading and the Institute of Terrestrial Ecology, Monks Wood, cocktails of pesticides may be killing wild and game birds in Britain even though the levels of individual pesticides are well below the approved safety limits. Safety limits are set for each pesticide in isolation. On farmland where birds are exposed to several pesticides, the initial effects of one pesticide may make an additional pesticide more poisonous to the bird (potentiation).[30]

Various researchers have all studied a specific class of commonly used cereal fungicides to see how they altered a bird's tolerance to organophosphorus insecticides. A potentiation effect was expected because certain organophosphorus compounds, which are applied in an inactive form, become harmful to birds if they are activated. A group of enzymes, cytochrome P450s, present naturally in all animals, are responsible for activating the organophosphorus insecticides.[31]

In partridges, the activity of P450 enzymes increased greatly when birds were given the fungicide prochloraz at doses of 180 milligrams per kilogram (mg/kg). The partridges were then injected with an organophosphorus insecticide, malathion, at very low doses ranging between 5.6 and 11 mg/kg. These doses proved lethal to the partridges. In control birds which were not given an initial dose of prochloraz the malathion had no effect. When the researchers gave the same doses of malathion orally the birds did not die. However, the activity of cholinesterase, also present in birds' blood, decreased; the decrease is a sure sign of organophosphorus poisoning.[32]

Synthetic Pyrethroid Insecticides

The newer synthetic pyrethroid insecticides which kill insects rapidly are also highly toxic to aquatic vertebrates and invertebrates. Pyrethroids' have generally low acute mammalian and bird toxicity, although deltamethrin is an exception. Because of their relative novelty as commercial insecticides, there is limited evidence about pyrethroids' effects on wildlife. Bees and beneficial insects have been inadvertently controlled by pyrethroids and they can be highly toxic to fish.[33] Pyrethroids' main effect is likely to be an indirect one of reducing insect populations on which mammals and birds feed. On the plus side, wood preservatives containing newer, synthetic pyrethroid-type insecticides are safer to bats than traditional woodworm timber treatments based on gamma-HCH (lindane).[34]

Many pesticides are toxic to fish and major kills have been recorded involving a variety of compounds. Fish, shrimps and other aquatic organisms absorb pesticides from water or minute particles and concentrate the chemicals, increasing the concentration by 100 to a million times the original level.[35]

Even small quantities of pesticide released in water, and especially insecticides, can result in large kills of fish and other water animals. In 1983 the UK Royal Society for the Protection of Birds (RSPB) recorded major fish kills traced to a disused, dieldrin-contaminated sheep dip.[36] The carbamate insecticide aldicarb has been shown to kill off water fleas.[37] The fire at the Sandoz pesticide plant, Switzerland in 1987, when thousands of litres of pesticides poured into the River Rhine, demonstrated their lethality in water. Hundreds of miles of the river were seriously polluted and fish and aquatic organisms were killed in their millions.[38]

An earlier incident when the organochlorine pesticide endosulfan escaped into the River Rhine also led to a large number of fish deaths.[39] Dichlorvos, an organophosphorus insecticide legally used in fish farming, has caused damage to other marine organisms, such as lobster and crab larvae, and mussels and oysters.[40]

Food Chain Poisoning, or Delayed or Secondary Poisoning

The term 'food chain' refers to the natural process where larger creatures (including humans) feed off smaller creatures and organisms who in turn eat smaller prey.

Food-chain poisoning occurs when creatures 'higher up' the food chain such as birds of prey and foxes feed on pesticide-contaminated creatures or corpses of smaller birds, fish, mice, rodents or insects. In so doing, these larger creatures accumulate pesticide residues in their own body tissues and organs, and particularly in body fat. If sufficient pesticide-contaminated animals are eaten, the predator can accumulate a lethal dose and die. If the amount consumed is not enough to kill, the pesticide may be there in sufficient amounts to lead to poor reproduction and fertility, increase the likelihood of disease and reduce the animal's ability to survive.[41]

Insecticides

Insecticides provide the clearest examples of food chain poisoning, especially the persistent, organochlorine group of insecticides such as dieldrin and DDT, which are not only toxic, but chemically extremely stable and are stored in body fat.

Dieldrin, a highly toxic and persistent organochlorine insecticide, was widely used as a cereal seed dressing, and caused the widespread

Bees and Insecticides

These beneficial, crop-pollinating and honey-producing insects are particularly susceptible to pesticides. There are three main problem areas:

(1) Contact insecticides kill bees directly and up to 48 hours later if they also have short-term persistence. Any chemical taint may also lead to worker bees being rejected on returning to the hive. Organophosphorus and carbamate insecticides seem to be more hazardous to bees than the organochlorines they replaced.[42] Research has shown that some synthetic pyrethroid insecticides have a repellent effect on bees, discouraging them from foraging in treated crops so they avoid being killed.

(2) Systemic insecticides which persist inside the plant can be as fatal as contact materials. Some pesticide is transferred to the nectar and the bees are killed off as they feed on it, often at pesticide concentrations as low as 1 part in 5 million.[43]

(3) Micro-encapsulated formulations of insecticides, where the spray liquid is made up of millions of tiny resin-coated droplets which release the insecticides more slowly, giving a longer period of pest control, can kill bees. These capsules can attach to hairs on bees and are taken back to the hive where the pesticide is released with disastrous results. In the US micro-encapsulated formulations now have a latex 'sticker' added to fix the capsules more firmly to the plant, reducing the risk of bees accidentally picking them up.[44]

---

poisoning of pheasants, pigeons, birds, and small mammals which ate the seed. Birds of prey and rodent predators were in turn poisoned when they ate the pesticide-contaminated corpses of pigeons and mice.

In the 1960s, especially in the cereal-growing regions of the UK, birds of prey like peregrine falcons, sparrowhawks and owls, as well as foxes and badgers were reduced to dangerously low levels by dieldrin poisoning through the food chain.[45] Dieldrin was subsequently withdrawn as a seed dressing and, along with other organochlorine insecticides such as DDT, aldrin, endrin and heptachlor, is now completely banned in the UK. However, these insecticides continue to be used illegally and dieldrin was found as recently as 1986 in the corpses of herons, otters, and in live eels.[46]

Secondary poisoning can occur with organophosphates and carbamates, although it is less prevalent. The organophosphate famphur, used as a pour-on treatment for warble fly on cattle, has killed a wide variety of bird species which ate the dead, pesticide-contaminated flies. In the UK, crow and pigeon family species have been killed, while in the US species such as black-billed magpies have been inadvertently killed.[47]

Illegal Use of Pesticides

Illegal use of pesticides as poison baits is directly responsible for deaths of many animals, including wild birds and mammals, farm livestock, pets and honeybees. Poison is laid in carcasses and in eggs. The main pesticides of abuse are mevinphos, alphachloralose, and strychnine.[48]

Under the Wildlife and Countryside Act 1981 it is an offence to poison wild animals. Proper use of pesticide is regulated through the Control of Pesticide Regulations 1986. In 1990 the Ministry of Agriculture, Fisheries and Food (MAFF) launched a campaign to reduce the illegal poisoning of animals.[49]

MAFF claims that the majority of 'wildlife incidents' result from the illegal use of pesticides, principally as poison baits, and that when pesticides are used as approved there are few problems.[50] While acknowledging government efforts to stamp out this type of illegal pesticide use, the MAFF claim needs some qualification. As with human poisoning, it is only the incidents involving acute toxicity that are reported, and again there is likely to be under-reporting of incidents. Information and data is lacking on the chronic health effects and indirect effects on wildlife.

'Wildlife incidents' are reviewed by the Environmental Panel of the ACP, and its latest report on pesticides in the environment shows that 641 wildlife incidents were investigated in England, Scotland and Wales during 1988:[51]

- 34 per cent of cases showed evidence that pesticides were involved
- 62 per cent involved some deliberate abuse of pesticides
- 19 per cent involved misuse

Official UK government figures show that between 1975 and 1988, 427 bird deaths – including 152 birds of prey – have been confirmed as pesticide poisoning incidents. The list includes at least 26 golden eagles, a protected species. The true figure for bird deaths from illegal use of pesticides is believed to be many times higher than official figures. The RSPB says that 282 birds of prey, 17 red kites and 29 golden eagles died of poisoning between 1971 and 1987.[52]

A variety of pesticides seem to be involved in the illegal killing of wildlife. Some baits, such as the organophosphorus compound phosdrin, are lethal to humans as well as animals. In 1983 a gamekeeper on a Scottish estate in Perth and Kinross died after accidentally drinking a small quantity of phosdrin diluted in water. In a report on the death the local sheriff recorded that Phosdrin was: 'widely misused by gamekeepers as a method of poisoning vermin'.[53] Official figures show that 100 dogs and 62 cats have been killed by poison bait since 1975. Strychnine is one of the poisons used. This cruel method of poisoning causes convulsions so violent that mammals have been known to break their own spines.[54] Another pesticide, alphachloralose, used under licence for killing pigeons, has been implicated in golden eagle deaths in Scotland. In 1988 a male eagle was discovered dead in its eyrie in a remote national reserve in Scotland. The subsequent discovery of a mountain hare poisoned by alphachloralose suggested that the eagle died after feeding on another hare which had been deliberately poisoned and left as bait.[55]

Thousands of birds, including at least two rare white-tailed sea eagles, are feared to have been killed by the illegal use of poison bait in Scotland. The white-tailed sea eagle was extinct in Britain and has only been reintroduced in the last few years as a result of a project by the Nature Conservancy Council, estimated to have cost £750,000. Red kites have also fallen victim. Two of the 11 red kites released in England and Scotland as part of a reintroduction programme have now been found dead. One of the kites had been poisoned with an organochlorine pesticide, endrin, banned since the early 1980s in Britain. In 1990, MAFF prosecuted a number of gamekeepers and landowners in connection with the banned pesticide endrin and these red kite deaths, as part of a wider campaign against illegal pesticide poisoning.[56,57]

In New Zealand parathion and fensulfothion, two organophosphate insecticides used in pastures to control soil-borne pests, killed hundreds of birds in two areas; the birds ate contaminated earthworms.[58] Similar effects have been recorded with carbamates.[59]

Long-term, Chronic Health Effects

Pesticides also control wildlife species through sub-lethal, or chronic, effects on reproduction and the general health of animals. Many of these chronic effects occur through the food chain.

Organochlorine Insecticides and Birds

The effects of DDT are generally sub-lethal, or chronic, as this pesticide and its metabolite DDE interfere with reproduction in both birds and mammals, primarily by inhibiting calcium metabolism needed for egg-shells and bones. Predatory birds, especially fish-eating species, accumulated sub-lethal doses of DDT or DDE in their body fat and organs. Fertility and reproduction was affected. Many predatory birds laid eggs which had thin shells. These broke under the nesting bird, preventing hatching. The decline in the populations of peregrines, sparrow hawks, merlins, hobbies, golden eagles, ospreys, grebes and pelicans in Europe and the US was linked to egg-shell damage caused by DDE.[60] Some of the more abrupt population crashes coincided with the introduction of the cyclodiene group of organochlorine pesticides such as dieldrin, aldrin and endrin. Where organochlorine use has been banned or restricted, populations have recovered, although recovery has often been slow. Reduced use of DDT, aldrin and dieldrin in Britain has led to a decline in the residue levels of these chemicals in the bodies and eggs of affected species, breeding success has improved, and populations have recovered. However, sparrowhawks and peregrines have still not fully recolonised parts of south-east England, where organochlorine use was heavy.[61]

Bird migration highlights the problem of the need for the international control of pesticides to protect wildlife. For example peregrine falcons, protected in their breeding sites in Canada, are being killed by organochlorine pesticides in over-wintering grounds in Mexico.[62]

DDE and the Western Grebe

As well as being a breakdown product of DDT, DDE is an insecticide in its own right. In Clear Lake, California, a popular tourist resort, DDE was applied directly to the lake in concentrations designed to control gnat larvae but leave fish and other aquatic organisms unharmed. However, fish in the lake accumulated pesticides, which in turn poisoned fish-eating birds. Some birds died as a result of this poisoning, but the main effect was on fertility and reproduction. In 1949 there were some 1,000 breeding pairs of western grebe. By 1962, following repeated DDE treatment of the lake, only one young grebe chick was observed. The concentration of pesticide in body tissue of western grebes was estimated to be 6,000-8,000 times the original application rate of DDE. Once treatment stopped the bird population began to recover.[63]

Indirect/Ecological Effects

The effects of pesticides on all forms of wildlife, from the larger mammals down to the tiniest micro-organisms, is inextricably bound up with the type of agriculture used. Changes in the way crops and plants are grown and animals reared have important ecological effects on the diversity and abundance of wildlife species. Modern, intensive agriculture has had a major impact on wildlife. The development of pesticides and their application in increasing quantities as farming intensified has led to changes in the abundance and diversity of wild species of plants, insects and animals.[64]

The use of artificial fertilisers and pesticides, linked to mechanisation and crop breeding, has made it possible for conventional farms to specialise in a few crops – and in some cases a single crop – on the same land, year in and year out. Monoculture has become the rule in many areas, with a consequent decline in the rotation of crops in mixed-pasture arable farming. Economic pressure has encouraged the trend towards more specialisation and continuous monoculture.

In Europe rising land prices, combined with high arable crop subsidies from the EC, have been the prime movers, and resulted in marginal land, not particularly suited to arable cropping, being ploughed up, and in hedgerows being removed to increase field acreages.[65] Farming has become very industrialised. No sooner is one crop harvested than the next is ready for planting. Cropping has also changed: the emphasis is now on higher-yielding winter-sown cereals and crops. This type of agriculture is only possible because of the use of pesticides and chemical fertilisers. Livestock, horticultural and forestry production has been similarly intensified. This trend to monoculture, combined with the increased use of pesticides and artificial fertilisers, has had a devastating effect on wildlife. The effects have been greatest where broad-spectrum, non-selective pesticides have been used: these kill friend and foe alike.[66]

The main consequences have been the reduction and disappearance of species; the emergence of new, or 'secondary' pests; the removal of food sources and habitat, and the growth of pesticide resistance.

Reduction and Disappearance of Species

Intensive pesticide use can lead to the reduction, and ultimately disappearance, of some species of wildlife. Often it is the more sensitive species of plants or insects, such as butterflies, which are aesthetically pleasing in their own right, that are most harmed. Species reduction and disappearance has important knock-on effects in terms of removal of food and habitat sources for other wildlife and the creation of new, 'secondary' pests.

Herbicides

Herbicides give the best illustration of this type of effect. Herbicide use, especially spraying round field margins and into hedge bottoms, has caused major changes in the diversity and distribution of 'weeds', hedgerow and verge plants. Field margin spraying can also be counter-productive, allowing what were previously hedgerow 'weeds' to become field weeds which compete with crops.

Herbicides and Hedgerows

In the UK hedgerows provide the main habitat for some 140 species of fern and flowering plants (approximately 10 per cent of UK flora). Either deliberate spraying into hedge bottoms or spray drift has led to a decline in many wild plants, more sensitive species disappearing first, with even the shrubbier species such as wild rose and hawthorn being affected. Herbicides have played a major role in virtually eliminating once-common crop weed species such as corn cockle and cornflower.

Insecticides

Pine Trees and Spray Drift
Swedish scientists studying pesticide residue levels in pine tree needles have stated that air contaminated with the banned organochlorine insecticide DDT drifted into Western Europe from East Germany in 1984. The scientists collected pine needles of different ages from across Europe and analysed them for pollutants that contaminated the needles as they grew. Needles which grew in 1983 and 1984 had high levels of fresh DDT, banned in Western Europe since the 1970s. The

highest concentrations appeared closest to East Germany. Pine needles of all ages also had high levels of lindane, coming from a constant source in the south of Europe, most probably southern France.[67]

Emergence of New or 'Secondary' Pests

The reduction and disappearance of some species of wildlife due to pesticides, along with the elimination of the competition they offered to other species, creates natural vacuums or 'ecological gaps'. Other creatures previously held in check by the dominant species may then move in to fill these gaps or niches.

Some of these 'colonisers' may be harmful. Before pesticides were used they were held in check by other species and were not major pests. As they are not susceptible, or less so, to the pesticides in current use, these secondary 'pests' can become major problems, requiring increased doses of conventional pesticides, or even new chemicals to control them.

Herbicides

Where herbicides kill off, or severely check, dominant, target weed species, minor weeds move in to fill the vacuum and become major weeds in their own right.

Field pansy has become a more common and serious weed in cereal crops as it is less susceptible to many current herbicides. Sterile brome, once a hedgerow bottom weed, has achieved field grass weed 'pest' status in many cereal crops as a result of a combination of minimal field cultivation techniques (that is, no ploughing) and the use of herbicides to control more dominant grass weeds.[68]

Insecticides

Insecticides, by reducing or removing 'natural controls' through the elimination of beneficial species, may create new pests. The Game Conservancy Trust has argued that outbreaks of aphids are now more frequent, and are caused by the insecticides used to control predator species such as earwigs, rove and ground beetles, and ladybird larvae.[69]

Cotton Whitefly in Sudan

Cotton whitefly is an example of an occasional, secondary pest of cotton in Sudan which became a major pest when intensive insecticide spraying killed off the natural parasites which controlled it.

DDT and other broad-spectrum insecticides were used in Sudan to control a major cotton pest, the cotton jasid. As cotton jasid numbers fell, cotton bollworm populations rose until they, in turn became a major pest in the late 1960s. To control the bollworm the initial season spraying was increased to as many as eight seasonal sprays with organochlorine and organophosphate insecticides. This increased spraying killed off the natural enemies of the cotton whitefly, small parasitic wasp species. Cotton whitefly is itself difficult to kill with conventional insecticides, as the immature whitefly 'nymphs' congregate on the lower sides of the leaves away from the spray. The nymphs also have a waxy protective coating which is difficult to penetrate.[70]

So, with natural, parasitic control reduced and insecticides having little direct effect, the whitefly developed from a secondary pest into a primary one. The situation gradually worsened and by the 1980–81 cotton growing season uncontrolled whitefly outbreaks were recorded. It is estimated that between 1975 and 1981, cotton whitefly caused cotton production to fall by 40 per cent and insecticide spraying costs to increase by 600 per cent.[71]

Removal of Food Sources and Habitat

By killing 'pests', non-harmful, and even beneficial species of wildlife, pesticides remove the habitats and food sources of a wide range of other animals, even where pesticide use is selective and efficiently targeted. For many pesticides, the main hazard to wildlife may be the indirect one of removing the food sources and habitat on which other creatures depend.

Herbicides

Herbicides control 'weed' species which provide a great variety of habitats and food sources for a wide range of animals and insects. Herbicide use in cereal crops has been most intensively studied, but high levels of weed control in other crops such as brassicas (cabbages, cauliflowers etc.) have been shown to reduce the diversity and abundance of non-target insects.

Game Birds and Pesticides

The UK Cereals and Gamebirds Project, carried out by the Game Conservancy Trust, has linked the management of cereal fields and intensification of grain production – especially the increased use of pesticides – with the long-term decline of the wild grey partridge.[72] Since the 1950s populations of wild grey partridges have been reduced to less than 20 per cent of their pre-war densities.[73] A major factor in this decline is thought to be the indirect effects of pesticides in removing the plant-eating insects on which partridge chicks depend for survival, especially in their first weeks of life. These non-target insects include various species of beetles, plant bugs, sawfly, and caterpillar larvae.[74]

The Game Conservancy Trust has concluded that the indirect effects of herbicides have probably been the most important factor in increased chick mortality: 'because of their ability to remove cereal field weeds that are the host plants of many of these plant-eating, chick-food insects'. The Conservancy goes on to note that:

the indirect effects of pesticides, acting on invertebrate food items in the diet of young chicks, have however become far more of a problem than direct toxicity ever was ... Fortunately, we have found a way of dealing with the problem, providing only that the political will is there to encourage farmers to use conservation headlands (see below).[75]

Fungicides

Fungicides can have unexpectedly broad-spectrum effects on wildlife, especially as some 'fungicidal' compounds have insecticidal properties as well. Pyrazophos, an organophosphate 'fungicide', controls aphids as well as leaf diseases in cereals, and its insecticidal activity is used as a major selling point by the manufacturer. The Game Conservancy Project also highlighted the use of pyrazophos as a major reason for the reduction in non-target insects on cereal crops on which grey partridge chicks feed. The manufacturer involved labelled evidence produced by the Game Conservancy as a 'poor piece of work'.[76]

Insecticides

The total, non-selective nature of many insecticides means that they not only control pests, but beneficial and non-harmful species as well.

Insecticide Selectivity and Conservation Headlands

As described above the Game Conservancy Research Project has highlighted the indirect effects of insecticides and insecticidal fungicides on grey partridge numbers, through killing of non-harmful insects on which the chicks depend for food.

The Conservancy has shown that the introduction of 'conservation headlands' – six-metre-wide strips on the edge of fields where herbicide and insecticide use is selective – results in increased numbers of broad-leaved weed species and increased densities of chick-food insects.[77]

> Monitoring of spring pair densities of grey partridges showed that breeding density over the entire study farm increased from 3.7 pairs per square kilometre in 1979 to 11.7 pairs per square kilometre in 1986. No similar increases were observed on adjacent farms where pesticide use was unchanged.[78]

The Game Conservancy is also trying to quantify the impact of conservation headlands on 'butterflies, rare arable weeds, small mammals, songbirds and predatory insect species.'[79]

The Boxworth Pesticide Project

The 'Boxworth' project, a seven-year, government-funded research programme to investigate the environmental effects of sustained use of pesticides in intensive cereal production, showed that intensive, 'calendar' pesticide use not only kills wildlife but also wastes farmers' money. By contrast, carefully supervised pesticide applications were environmentally safer and cheaper than high chemical input 'full insurance' applications designed to maximize yields.[80]

The project was carried out from 1981 to 1988 at the Ministry of Agriculture's Boxworth Experimental Husbandry Farm in Cambridgeshire. During this period areas of the farm received three contrasting pesticide regimes:

- *'Full Insurance'*: a pre-planned preventative application (designed to kill everything) programme of fungicides, herbicides and insecticides;
- *'Supervised'*: treatments applied only if diseases, pests or weeds exceeded thresholds of economic damage;
- *'Integrated'*: as 'supervised', but with additional modification of crop husbandry to further reduce pesticide use (for example, choice of disease-resistant varieties).

Economic and management aspects of the three regimes were closely monitored, and scientific studies were carried out on the effects on animals and plants.

The Boxworth Project demonstrated that the 'full insurance' regime reduced populations of several insect species markedly. The most severely affected insects were beneficial parasites and predators, like *Bembidion obtusum*, a beetle which over-winters on the soil surface within the crop.[81]

The project further demonstrated that the 'supervised' approach was more cost-effective over a five year period than routine, preventative spraying. While the 'full insurance' approach produced slightly higher yields it was not enough to justify the extra money spent on chemicals.[82]

Pesticide Resistance

Growing numbers of plant diseases, animals, insects and weeds are now resistant to pesticides. Pesticide resistance is a worldwide problem. Resistance develops most rapidly and severely where the chemicals are applied frequently, at high dose rates and in tank mixes, with up to three or more different pesticide products being used in the same spray or treatment. Resistance to one pesticide can mean a 'pest' species becomes resistant to other pesticides in the same chemical group.

As well as threatening the commercial usefulness of a product, pesticide resistance also has an indirect effect on wildlife. Resistant populations soon become the dominant ones and can compete with other species reducing even further the diversity and abundance of other wildlife.

Herbicides

Herbicides are the category of pesticides where resistance has developed most slowly. The exact scale of herbicide resistance is not clearly known, but by 1984, 49 weed species in 13 different countries (mainly in the US, Canada and Western Europe) had, under repeated field herbicide applications, evolved populations resistant in varying degrees to a total of 11 herbicides.[83] By 1988, known resistance had risen to 59 weed species in 20 countries, resistant to 12 herbicides or groups of herbicides.[84] For example, in Warwickshire, UK, mayweed has been found which can tolerate the herbicide simazine at 1,000 times the dose lethal to susceptible types.[85] Herbicide resistance is a problem likely to become apparent within the next few years in an increasingly large number of crops where repeated herbicide use is a more recent phenomenon. Resistant weed species can multiply rapidly and become the dominant plants in an area, competing with other species, including non-harmful ones, to gradually degrade plant life.

Fungicides

Fungicides are compounds where resistance problems can develop very quickly, and can lead to severe restrictions on a chemical's use.

Within two years of the new systemic fungicide metalaxyl (commercial name Fubol), being introduced in the UK, major resistance problems were encountered with resistant strains of potato blight. Metalaxyl-resistant strains of lettuce mildew, which were 10,000 times more resistant than susceptible strains, led to further restrictions on the chemical's use in glasshouse crops.[86]

By the early 1980s the intensive use of the fungicide carbendazim led to the emergence of resistant strains of eyespot – a major fungal disease of cereals – in all areas of intensive cereal production in the UK.[87]

Insecticides

Insecticide resistance is a well established phenomenon in areas of intensive pesticide use such as hop and fruit growing, horticulture, intensive glasshouse production and in controlling insect vectors of disease. By 1980 resistance was established in 168 insect pests of human and animal health, and 260 pests of crops, forests and stored products, and the figure was predicted to rise.[88] By 1985 some 450 species worldwide were resistant to pesticides and this figure is also projected to rise.[89]

The hop-damson aphid is a major and much-sprayed pest of hops in the UK, and is resistant to many organophosphate, carbamate and organochlorine aphicides. The possibility of synthetic pyrethroid insecticides becoming redundant is considerable given that aphids have already developed a low level of resistance to some of them, and a number of other hop pests show significant resistance.[90]

Insects can develop astonishingly high levels of resistance, and some species seem to have incorporated toxic insecticides into their lifestyle. Along the Ituxi River in Brazil, the male of a bee species, *Enfriesia purpurata*, not only finds the odour of the organochlorine insecticide DDT attractive but is so stimulated by it that it brushes the insecticide from the walls of houses routinely sprayed with DDT to control the malarial mosquito. Bees collected from houses had levels of DDT residues as high as 42,000 micrograms/gram without apparent ill effects. Normally DDT residues of 6 micrograms per bee are fatal.[91]

Conclusion

Problems created for wildlife by the toxicity and overpersistence of many pesticides have led to banning, restrictions on use, and the introduction of less persistent groups of pesticides. The effects of pesticides on wildlife have, in fact, illustrated the difficulties of trying to phase out highly toxic or persistent materials, as many of the substitutes have proved to have similar properties. For example, the organophosphorus and carbamate insecticides which were designed to replace the persistent organochlorines (like DDT), were less persistent but more toxic. Similarly, synthetic pyrethroid insecticides, while generally recognised as being safer to humans, can be highly toxic to wildlife. Furthermore, the indirect effect of pesticides on wildlife in removing food sources and habitat, which is much less researched and documented, have probably been underestimated. Enough work has been done to tell us that all categories of pesticides, including herbicides and fungicides, can adversely affect both the numbers and diversity of wildlife species.

8

Pesticides in the Home

Over 100 different types of chemicals are available over the counter and can be sold to people of any age. Domestic applications of pesticides can be dangerous because most amateur users know little of the potential dangers. In this chapter we look at the problems – and suggest ways of reducing the risks – in the most common domestic uses of pesticides: garden pest and disease control; household use, particularly timber treatment; pest infestations affecting children and pets, and rodent control.

Pesticides in the Garden

Every year Britain's amateur gardeners spend about £15 million on pesticides. There are well over 500 products on the market, made up from more than a hundred different chemical ingredients, mixed in a variety of formulations.

Neither the government nor the agrochemical industry publish the figures for the quantity of pesticides involved. A few years ago the organic farming body, the Soil Association, made a rough estimate from sales figures, and calculated that about a kilo of active pesticide ingredient is used, on average, on every acre of British back garden every year.[1]

Spraying pesticides in back gardens poses a number of particular health and safety problems not normally encountered by farmers or professional users. These include:

- use of chemicals by untrained people
- use of chemicals close to houses, on vegetables gardens and in places where children are likely to be playing
- storage of chemicals where children have easy access
- damage to wildlife which may be specifically attracted to the garden by bird tables, nest boxes etc.
- risk to pets.

Are Garden Chemicals Dangerous?

The UK government's view is that garden chemicals pose no appreciable risk to human health or to the environment. It argues that this is because the most hazardous chemicals are not approved for domestic use and for those that are the concentrations of the chemicals are so low that there are no significant health risks. Government spokespeople, and representatives from the industry, have repeatedly stressed that there is 'absolutely no danger' from correctly used garden chemicals.[2]

Although some of the most hazardous chemicals are, indeed, not released for general sale, a survey by the Soil Association carried out in 1985 found a substantial number of hazardous chemicals for sale over the counter in chemists and garden centres.[3] The survey used toxicity data from Europe and the US, and concentrated on pesticides sold in Britain in garden centres and by large retail groups. A brief review of trade literature in 1990 showed that, with the exception of a few chemicals withdrawn by the government (for example, ioxynil), little had changed in five years.

The Soil Association Survey of Garden Pesticides

The Soil Association found the following hazardous pesticides for sale in garden shops and centres:

- 38 chemicals irritating to the eyes, skin or respiratory tract. These include *benomyl* and *carbendazim,* two systemic fungicides used on a wide variety of fruit, vegetables and ornamental flowers; and *glyphosate,* a translocated, foliar acting herbicide used against weeds.
- 25 chemicals which are known or suspected carcinogens. These included *captan,* a protective fungicide used on fruit and ornamentals; *trichlorfon,* a contact and ingested organophosphate insecticide used against caterpillars, cutworms, earwigs etc., and *dicofol,* a systemic acaricide used on various horticultural crops.
- 29 chemicals which are known or suspected mutagens or teratogens. These include *malathion,* a broad-spectrum, contact organophosphate insecticide and acaricide (a pesticide used against spiders and mites) employed, among other things, against aphids, red spider mites, pollen beetles and codling moth, and *MCPA,* a translocated, post-emergent herbicide used on lawns.
- A number of chemicals which are acutely poisonous. The best-known of these is *paraquat,* which has no known antidote if swallowed, but the commonly used *nicotine* is also intensely toxic, both if swallowed or absorbed through the skin.

The survey looked at *chemicals* rather than specific *products* for sale in shops. Each chemical is likely to be sold under a range of different trade names. This means that there are literally hundreds of hazardous pesticide products being sold for use by householders in their own gardens.

Garden pesticides are usually supplied in very small amounts compared with those used on farms or for municipal pest control. However, this advantage is at least partially offset by the lack of training for people using garden pesticides and because the pesticides are used where people live.

There is no proof that low doses of pesticides are harmless, particularly if the dose is repeated many times. As explained earlier, testing of chronic pesticide toxicity is still in its infancy. Other countries do not share the UK government's optimism about the safety of garden chemicals; several have banned chemicals specifically from home garden use while allowing them to be applied by professionals. For example, Sweden banned the use of paraquat for domestic use in 1982 because of its toxicity.[4]

Britain in effect follows the same principle by only passing certain chemicals for home garden use, but does not make this policy explicit. In 1987, weedkillers containing ioxynil and bromoxynil were banned from home garden use after research suggested links with birth defects. However, products containing these chemicals can still be found in many garden shops despite the ban.[5]

Special Problems of Domestic Use

Untrained Users

There is a high degree of ignorance about the real and potential risks of using pesticides in the garden. Advertising is partly to blame. The Soil Association survey found that the advertising stressed the ease of use of pesticides, and minimized the risks.

Although all advertisements are obliged to refer users to the safety instructions on the bottle or packet, the survey found a number which showed photographs or drawings of people using pesticides in a dangerous way! For example, one photograph showed a women holding a spray nozzle at the same height as her face and spraying tree foliage, thus risking spray blowing into her face. These problems are much more acute in many developing countries (see Chapter 9).

Greater Community Impact

Unlike most agricultural situations, garden pesticides are likely to be sprayed very near people. Particular hazards include:

- spraying near infants or young children, who are likely to put plant material or other objects directly into their mouths or lick their fingers after touching sprayed foliage;
- spraying a lawn where children are likely to play and brush against the chemicals with bare arms and legs;

- spraying near domestic pets, or leaving slug pellets where they can be eaten;
- spraying next to open windows, or in windy conditions where drift can enter the house;
- spraying when clothes are drying out of doors.

Although the home user is applying less pesticide, the chemical is much more likely to reach places where it can cause harm to people.

Possible Access for Children

In the past, there have been some tragic accidents where pesticides stored in old drink bottles have been drunk by children, who have mistaken the contents for soft drinks (some look quite like cola for example). In Britain, as a result of the 1985 Food and Environment Protection Act, there are now legal requirements to store pesticides in their correct containers, complete with warning label. However, there is no obligation to keep pesticides in childproof containers, in the way that most medicines are now controlled. Nor do most pesticide advertisements or labels stress that pesticides should be kept in locked cupboards and out of reach of young children.

Risk to Pets and Wildlife

Beneficial garden insects, amphibians, birds and mammals can all be killed by misuse (or even just by routine use) of garden chemicals. Spraying against insect pests will often kill butterflies, moths and bees.

Slug pellets are a particular problem. Urban hedgehog populations have been badly affected by slug pellets, or because they eat slugs which have themselves been poisoned. A similar fate will befall many small birds. It is rare to see a corpse, because once the animal begins to feel unwell it will try to find somewhere quiet to hide away. But the decline in hedgehogs, frogs, toads and many species of urban butterflies is accepted by most ecologists as being at least partially due to the widespread use of pesticides in cities.

Reducing Risks From Garden Spraying

Garden pesticide users could reduce risks to themselves and others by a number of simple steps:

- Wear long trousers, a long-sleeved shirt or jacket, and a mask when spraying pesticides, avoid spray drifting into your face, and never spray in windy conditions or when other people are nearby.
- Always wear gloves when handling or using pesticides, and keep one pair of gloves solely for pesticide use.
- Shut windows, and remove washing, deckchairs, children's toys

and, of course, food and drink from the garden before starting spraying.

- Wash your hands thoroughly after spraying or handling pesticides.
- Always store pesticides in the correct containers, and in a locked place where children cannot gain access.
- Spray fruit and vegetables in the early morning or late evening, when there are fewer flying insects about.
- Hide slug pellets in places where birds and hedgehogs cannot reach them, and clear away any dead slugs you see (as these will still contain poison).
- *Always read the label on pesticide containers before use.*

It is worth noting that pesticides are not always effective. A survey published in *Gardening Which?* in 1988 found that only two of the products – out of a wide range tested – kept a garden path free of weeds as advertised. The rest were, at best, temporarily effective and some did not clear all weed species.[6]

If you are worried about pesticides, you could consider cutting them out altogether. Thousands of gardeners already practice *organic growing methods*, where pests are controlled by biological and cultural techniques rather than chemicals. These include encouragement of predators of pests, such as ladybirds to control aphids; using barriers on fruit trees to stop insects from climbing up the trunk; practising crop rotation to stop crop specific pests from building up in the soil, and companion planting of different species to confuse pests which hunt by scent.

Organic gardeners also use a few relatively non-toxic and non-residual plant-based insecticides to control occasional pest outbreaks in an emergency, although some plant-based pesticides, such as nicotine, remain highly toxic and hazardous. Further information on organic gardening is available from the Henry Doubleday Research Association and the Soil Association, both of which are listed under 'Contacts'.

Pesticides in the Home

Pesticides are increasingly used in the home, in aerosol fly sprays, on house plants, and in timber treatment. There are risks in using these products.

Pesticides in Aerosols

Current concern about aerosols has concentrated on their role in depleting the ozone layer through the release of chlorofluorocarbons (CFCs). However, aerosols also pose a more immediate health hazard through pollution inside buildings. Pesticides from aerosols are released

Wood preservatives

Wood preservatives are pesticides containing insecticides and/or fungicides. All wood preservatives (pesticide products) must be approved under The Control of Pesticides Regulations 1986 before they can be advertised, sold, supplied, stored or used.

Formulations available are based on a number of active ingredients and fall into seven main categories:

(1) *Organic solutions:* normally solutions of an insecticide or fungicide in an organic solvent, but dual insecticide/fungicide formulations are marketed. Kerosene-based pesticide products are thought to give the most rapid results, as the organic solvent is believed to penetrate the timber ensuring high initial larva kill with persistence of the pesticide. However, solvent flammability, toxicity, transport, storage and handling of large amounts of preparations are practical disadvantages. Pesticide products such as these are commonly applied by spray in confined areas like roof spaces.

(2) *Aqueous preparations:* usually organic solutions as described above but sold as a concentrate to be dispersed in water on site. Some aqueous (water-based) solutions are also available. These formulations are frequently applied by spray in confined spaces but are not noted for deep penetration into wood. In concentrated form flammability and toxicity are disadvantages. When dispersed in water there are dangers in their use near electrical equipment, circuits and wiring.

(3) *Pastes:* a special form of emulsion which are applied as a thick paste to the surface of timber where deep penetration is required, and/or where it is necessary to keep a concentration of a pesticide product in a precise place. They are normally applied by trowel, caulking gun or brush and are both insecticidal and fungicidal in action.

(4) *Insecticidal smokes:* specialist treatments designed for the control of certain insects, for example, death watch beetle, where there is a need to kill emerging adults. They are pyrotechnic mixtures of insecticides which produce, via a 'smoke', a deposit which covers the surfaces of exposed timbers. Annual treatments are normally required.

(5) *Solid plugs:* fused rods of soluble boron compounds which are inserted into holes pre-drilled in the wood, dissolved by moisture and diffuse through the wood. They have a fungicidal action and are normally used in window frames and similar products.

(6) *Creosote products:* pesticide products produced from the distillation of tar. The active ingredients of these products are a complex mixture of phenols, cresols and pyrroles. They can be applied industrially by impregnation or by brush and are sprayed externally, for

example, on fences and outside woodwork. They are not approved for use indoors.

(7) *Surface coatings:* protective paints based on organic solvents (white spirit). They are designed to give some protection to the wood from the elements and give it an attractive appearance. They are applied by brush to outside woodwork. They require approval under the Control of Pesticides Regulations 1986 only if they contain active ingredients which penetrate wood.

Source: *In situ timber treatment using timber preservatives.* HSE Guidance Note GS46, HMSO, London, 1989.

in a fine spray which hangs in the air and can be inhaled, or contaminate skin and eyes. Products are marketed as highly convenient and clean, and for use indoors. Often, advertisements do not stress the care needed when spraying near food or young children.

In the past a number of toxic chemicals were allowed for use within the home, including DDT. Today, although aerosol pesticides are supposed to be safe, toxic, or potentially toxic, pesticides are still sold for use within the home. A survey carried out by Earth Resources Research in 1988 found a number of pesticides used in aerosols which have suspected chronic health effects, including piperonyl butoxide – a suspected carcinogen.[7]

Timber Treatment

Over the last few years a large-scale timber treatment business has grown up in Britain. Firms offer to treat existing building timber against a range of pests, including woodworm, various forms of fungi (especially dry-rot), death watch beetle and others. Many firms offer tempting terms; promises of a guarantee of 30 years are by no means uncommon, although given the fluctuating nature of the business they probably do not amount to much!

There are a number of serious problems with this approach to timber treatment. The pesticides used are, by their nature, extremely persistent, and many are very toxic. They are dangerous to people and have had catastrophic effects on populations of bats in Britain, which have relied on nesting in eaves of roofs for hundreds of years.[8] There are also serious doubts about the effectiveness and necessity of most of the treatments.

Toxicity to Humans

Some of the most hazardous pesticides known are routinely used for timber treatment. They can be absorbed into the body through breathing contaminated air, especially in the initial period after treatment, although sometimes for a very long time afterwards as well. Pesticides can also contaminate through ingestion after touching treated timber, or straight through the skin.

Examples of the most toxic pesticides used in timber treatment are:

Arsenic: used in some pre-treatments as an insecticide and fungicide, usually with copper and chrome. It is, of course, a deadly poison, and poisoning has occurred through handling wet timber. Splinters tend to fester painfully.

Creosote: has been commonly used in timber treatment as a fungicide and insecticide, although most often used for fencing and garden furniture. Despite still being a widely used DIY product in Britain, it has been banned for all but professional use in the US. It is, again, highly poisonous, causing skin and eye irritation with the possibility of permanent damage to the cornea of the eye. Acute bronchitis can be brought on in people caught in the spray mist, along with nausea and headache. It is carcinogenic.

Lindane or *gamma-HCH:* which has already been mentioned on several occasions in this book, is a highly poisonous insecticide. Lindane has been implicated in illness and death in several people after their homes have been treated against woodworm. Lindane is highly poisonous to wildlife. Use of lindane in roof spaces has killed many colonies of bats, and has been one of the main reasons for their decline in Britain over the last few decades. Lindane can remain at lethal concentrations to bats for more than 20 years after treatment. It is banned or severely restricted in many countries, including Japan and the US but not in the UK.

Pentachlorophenol or *PCP:* is a fungicide used in pre-treatment or remedial treatment, and is available in many DIY formulations, sold over the counter. PCP is a highly toxic poison and extremely persistent. It has also been banned in many countries, and in the US it is confined to professional outdoor use.

Tributyltin oxide or *TBTO:* is another highly poisonous fungicide, which is persistent and can be absorbed through the skin. TBTO was banned for use as anti-fouling paint on boats in Britain in 1987 because of the effects on marine life, including commercial mussel beds; populations of mussels have increased following the ban. TBTO is used in pre-treatment and remedial use on timber and is widely available from DIY shops and timber merchants.

Permethrin: is an insecticide which is generally considered to be far safer to bats and other wildlife and is recommended as a safe alternative by the Nature Conservancy Council in their advice leaflets about protecting bats.[9] However, permethrin has been the subject of acrimonious dispute in the US, and some senior researchers at the Environmental Protection Agency believe that it may be a carcinogen. The carcinogenic risk from DIY use, however, is likely to be small (see Chapter 5).

In January 1989, the London Hazards Centre published a report, *Toxic Treatments,* examining a number of case studies of people affected by poisoning as a result of domestic timber treatment.[10]

One of the cases highlighted by the study was that of Eric Riley, who died in January 1988, almost exactly a year after professional timber specialists treated his house with pentachlorophenol and lindane. Eric and his wife Ann felt ill for the year following treatment and, in April Eric had what appeared to be an epileptic fit, during which he was unconscious for about 40 minutes. Although he had no history of epilepsy, it did not occur to either the Rileys or the hospital that he could have been poisoned, so blood samples were not taken. Both of the Rileys continued to feel unwell, unnaturally tired and dizzy, with cramps and nausea. It was not until October that reports of similar incidents in a national newspaper suggested to them that they might be affected by the chemicals used in the timber treatment. Belatedly, they arranged for Eric's blood to be tested at the Poisons Unit of Guys Hospital, London. Before the results were available, Eric had a second fit in the bath and drowned. The Coroner recorded an open verdict.

Later in the year over 100 UK MPs signed an Early Day Motion demanding the banning of lindane and pentachlorophenol (PCP), although the initiative was rejected by the government.

Timber treatment by powerful insecticides is not only hazardous but, in most cases and in most parts of the country, unnecessary. Buildings maintained in good condition, with a damp-proof course to prevent rising damp, and at an adequate temperature, will not usually develop problems of insecticidal or fungal attack. Pressures towards chemical treatment are spurred on by the large profits being made by firms specialising in remedial timber treatment in Britain, and by the results of poor building construction, rather than any inherent need for heavy use of pesticides.

Pest Infestation in Children and Pets

Pesticides and Head and Body Lice

Various species of body and head lice can infest humans, and potentially anyone is at risk. 'No amount of personal hygiene and cleanliness will prevent temporary lousiness if there is association with unclean and careless companions.'[11] Human head lice, *Pediculus humanus capitis*, are biting insects which spend their entire life cycle on the human head, sucking blood. Lice eggs are glued firmly to the hair and the insects develop through various stages to the mature adult. The effects of bites include redness, itching, irritation, and inflammation. The effect of louse bites varies greatly with individuals, and with the degree of sensitivity to them.[12]

Chemical treatments based on pesticides are widely used for head lice control, and given that the treatments are applied directly to the head region, safety is a problem, especially for children and babies. Toxicity to the insect must be balanced against toxicity to the human, especially if persistent pesticides which bind to the hair to give longer protection are used. To be successful, pesticides must penetrate the outer layer (cuticle) of the insect and the shell of the egg, but preferably fail to penetrate the skin of the human head: something difficult to achieve in practice.

A variety of pesticides are used for body and head lice control. DDT was the first modern pesticide used for this purpose, originally with spectacular results. In World War 2, for example, most US soldiers and sailors on active duty were dusted with a powder containing 10 per cent DDT and were free of lice for the first time in any war.[13] The dust was applied by blowguns to the head, skin and clothing without undressing. DDT was subsequently used worldwide for lice control. Lice resistant to DDT soon developed in many parts of the world, particularly in the Middle East, Korea and Japan. The problems of insect resistance, combined with health concerns on DDT's overpersistence in the body, have led to restrictions on its use in some countries; for example, DDT is no longer authorised for lice control in the UK.

Currently in Britain five insecticides are licensed (under the Medicines Act) for use in medicinal preparations to control head lice – lindane, carbaryl, malathion, permethrin, and phenothrin. They are variously formulated as creams, lotions, powders, or shampoos.[14]

As well as concerns on the health effects of some of these materials, insect resistance is also a growing problem. Malathion, an organophosphorus (OP) insecticide, is generally recognised as the most effective of the current chemicals. It is feared, however, that over-use could lead to the development of malathion-resistant strains of head lice;[15] and that the most effective weapon in the chemical armoury may be lost. The other pesticides are either less effective or have their own special problems. The carbamate insecticide carbaryl gives variable control of head lice,[16] and there is the problem of the development of broad-spectrum carbamate resistance to this chemical. Lindane resistance has also been documented, and as it is the most toxic of the five insecticides, there are doubts over its long-term safety.[17] The other two insecticides, permethrin and phenothrin, were only licensed for head lice control in the UK in 1990, so it is too early to assess their efficacy.

In Britain, the Department of Health claims that head lice infestations are on the decrease.[18] A campaigning organisation, Community Hygiene Concern (CHC), disputes this claim as it maintains that reporting of incidents is poor and badly documented. On the basis of its own surveys and analysis of data, CHC states that 'there is a very real head louse problem in our [British] schools affecting pupils of all ages and in every kind of school.'[19] CHC goes on to claim that, due to failures of pesticide control, 'the British school population is on the edge of a head lice explosion.'[20] CHC is critical of the over-dependence on chemical control methods both in terms of their efficacy and safety to the patient. The organisation argues that the key to head lice control is community organisation based on regular inspection and monitoring of school populations and removal of eggs by grooming of hair by qualified personnel.

Bedbugs and Pesticides

The bedbug, *Cimex lectularius*, is another pest species which afflicts millions of people all over the world, and is a health hazard in both temperate and tropical countries. Many people react to bites, which can cause irritation and swelling. Heavy infestations can cause anaemia, particularly in children, due to the loss of iron contained in the blood. There is no evidence that bedbugs transmit diseases to, or between, humans.[21]

The most recent figures available for England and Wales, for example, showed that in 1985–86, local authority environmental health departments carried out 7,771 treatments for bedbugs.[22] Light infestations are usually treated by spraying infected items and rooms with insecticides such as bendiocarb or a synthetic pyrethroid. These insecticides leave long-lasting deposits that kill bugs when they walk over treated areas for up to three months after application.

Fleas and Pesticides

A number of flea species are important pests but, of these, the cat flea, *Ctenocephalides felis*, is by far the most common flea found in domestic premises in Britain. The dog flea, *Ctenocephalides canis*, is a carrier of a tapeworm, whose main host is dogs (though it is sometimes found on cats), and is sometimes transmitted to humans. The human flea, *Pulex irritans*, is sometimes found on dogs and cats, but will breed readily on pigs and hedgehogs; badgers and foxes may also be attacked.[23]

Humans bitten by fleas vary considerably in their reaction and some may become sensitised only after persistent biting. If this continues for long enough desensitisation may occur, after which the bites are hardly noticed.

Chemical control is mainly based on a range of synthetic pyrethroid insecticides such as permethrin and cypermethrin. Bendiocarb can be used, but some cats are susceptible to this chemical. An organophosphorus pesticide, pirimiphos-methyl, is recommended, but this is a hazardous chemical, and has a potential to stain furnishings. One alternative to pesticides is a juvenile hormone, methoprene, which is both persistent and safer. Methoprene acts by preventing flea pupae from turning into adults, so breaking the life cycle. More research needs to be done on non-chemical alternatives.

Rodent Control

The three main rodent pest species in Britain are the Norway rat, the black rat, and the house mouse. Rodents can damage food, crops and buildings, spread disease, and be a general nuisance. Their proximity to the human environment, making use of food and shelter provided by humans, causes problems in terms of chemical control as it limits the range of poisons which can be used, especially as many sub-species of rats and mice are now resistant to some of the common rodenticides.

Rodenticides can be divided into acute rodenticides and anticoagulant materials, and this latter category can itself be sub-divided into first- and second-generation chemicals.

Acute Rodenticides

Acute, single-dose rodenticides were the main control materials until the early 1950s. With rats, which are suspicious of new objects such as baits, a system of 'pre-baiting' is generally used. With this system, plain, unpoisoned bait is laid for a few days before the poisoned bait is used.

- alphachloralose is a narcotic drug which acts by retarding the metabolic processes. It is not recommended either for use against rats or use out of doors.
- calciferol is based on Vitamin D2 and controls both mice and rats. Vitamin D2 causes calcium levels in the blood to rise. If the calcium level in the blood rises excessively then the kidneys are damaged and death occurs from renal failure.[24]
- fluoroacetamide is the most toxic of the currently approved rodenticides and is effective against all three main rodent species. In Britain it is only approved for use in sewers, ships and locked dock premises.[25]
- norbormide is toxic to both the Norway rat and black rat. There have been problems with it due to poor palatability and speed of action.[26]
- zinc phosphide is used for all three species and is quite fast-acting. It should be used in conjunction with pre-baiting for most effective control.[27]

Anticoagulant Rodenticides

Anticoagulant rodenticides, such as warfarin and coumatetralyl were introduced in the 1950s and represented an important step forward in chemical rodent control. They are eaten by mice and rats at low concentrations in bait and, as symptoms of illness are slow to appear, the rodents do not associate them with their food. The bait may need to be eaten for several days before a lethal dose has been consumed, and death may occur much later still. This low-level feeding approach overcomes many of the problems of bait rejection where acute rodenticides have been used. The feeding technique used is referred to as 'saturation baiting' which means laying enough baits to cover the infestation, based on an initial survey. The aim is to ensure that the mice and rats have daily access to the bait points even though they may not feed extensively at any one point or on any one day.

The anticoagulants work principally by reducing the production of a blood-clotting substance known as *prothrombin,* which is vital if blood is to clot quickly when blood vessels and capillaries are broken. Death results either from minor damage to blood capillaries in the rough and tumble of normal, routine activity, from some more severe, localized injury, or from a combination of both. The anticoagulants are toxic to humans, domesticated animals such as cats, dogs and cattle, as well as other mammals, so particular care is needed in their use.[28]

Two of the newest anticoagulant rodenticides, brodifacoum and flocoumafen, are more toxic than other anticoagulants. A rodent can easily absorb enough poison from a single feed, even though death is delayed. Their high toxicity, including risks to non-target species, means their approval is restricted to indoor use only.[29]

Rodents have become resistant to many of the anticoagulant rodenticides, especially first-generation materials. With the house mouse, resistance is widespread to all first-generation anticoagulants. Resistance is also developing to many of the second-generation anticoagulants, though the overall picture is less clear. Resistance to bromadiolone has been reported in urban and rural areas. Resistance in rat populations is growing, but is more variable than with mice. There are populations of 'super-rats' which are resistant to all first-generation anticoagulants in various parts of Britain – especially central Wales and some areas of the Midlands and South East England.[30] There is also well-documented resistance to difenacoum in the rat populations of Berkshire, Hampshire and some areas of Wiltshire.[31]

Grey Squirrels and Tree Damage

Poisoned grain treated with the anticoagulant rodenticide warfarin is also used to kill grey squirrels which damage commercial broadleaved, hardwood tree plantations. Warfarin use for squirrel control is legal in most of England and restricted parts of Wales provided it is used as a bait in special hoppers up which the squirrel must crawl. It is illegal in Scotland, most areas of Wales and a few areas of England where the native red squirrel is still holding out.[32]

Conclusion

Although domestic pesticide users account for only a fairly small proportion of the total pesticide use in Britain, they are likely to be the least trained users, and to use pesticides in places where children, domestic pets and food are likely to be found. Domestic pesticide use should not, therefore, be ignored when the question of pesticide controls is considered.

9

A Global Business

Precise figures are not available, but some estimates suggest that as many as 80,000 people die each year from pesticide poisoning.[1] The overwhelming majority of these unnecessary deaths occur in developing countries. Their occurrence is a savage indictment of the global trade in pesticides. It is a trade which involves the export of hazardous pesticides from the rich, developed world to the poorer, developing nations, a trade in chemicals, many of which are considered too dangerous to use in the West, but somehow are considered suitable for poor farmers. It is a trade where profits are more important than ethics, a trade which exploits illiterate farmers in Brazil and Malaysia;[2] which blatantly advertises some of its more dangerous products on wall murals in Colombia and Ecuador in violation of international codes of conduct; which misrepresents what is required for the safe use of what it sells, and which still seems largely unconcerned about the casualties of its 'business as usual' ethos.

It would be wrong to view everyone in the developing world as a victim in this process, just as it would be inappropriate to castigate all pesticide manufacturers as villains. Sadly there seem to be few manufacturers with clean hands and the scales are tilted in such a way that some 2 million people are poisoned each year. This is a record that some workers' organisations and consumer groups, in developing countries are determined to change. These organisations are beginning to identify abuses themselves and to publicise their findings. Given the lassitude of governments, and the lack of information, workers and consumers have decided that they need to produce their own literature for the public. Examples of what is available are given later in this chapter.

The Growth of the Pesticide Industry

Why this literature is necessary becomes clear when it is set against the growth of the pesticide industry. There is much talk about the advances of genetic engineering, and the development of pest-resistant crops which may not require pesticides for protection, but these are not available yet. In consequence the pesticide business is still expanding. Accurate figures assessing its exact size are difficult to come by. One way of examining the trade is to consider imports. A paper produced by the Worldwatch Institute in 1987 did just that and the results are shown in Table 9.1.[3] Between 1972 and 1984 global imports of pesticides increased more than twofold. Imports into the Soviet Union increased threefold, with the US not far behind. Asia increased its imports 2.6 times with Africa and Latin America showing smaller rises.

Table 9.1: The Value of Regional Pesticide Imports

Region	1972[1] (US$ million, 1985 value)	1984[2]	Change (%)
Soviet Union	132	552	+318
US	142	535	+277
Asia	314	1,132	+261
Europe	824	2,014	+144
Africa	269	522	+94
Oceania	30	47	+57
Latin America	340	503	+48
World	**2,051**	**5,305**	**+159**

[1] Average over 1971–73   [2] Average over 1983–85
Source: UN Food and Agriculture Organisation, *Trade Yearbooks 1977 and 1985 and Worldwatch Paper 79.*

Although imports are the best means we have of estimating the size of the trade they do not take account of chemicals produced in a country, and sold locally. We know that domestic production of pesticides is increasing.[4] India, for example, has a policy of self-sufficiency, and in 1988–89 some 80–85 per cent of the 87,000 tonnes of pesticides used were produced in the country.[5,6] Indonesia, too, produces chemicals locally, and in 1988 the Indonesia Department of Industry estimated production at 51,090 tonnes.[7] The pesticide market in Malaysia, estimated at US$140 million in 1985, is split with about 50 per cent of products made in the country and the rest imported.[8] South Korea produces some 15,000 tonnes of active pesticide ingredients each year, representing about 75 per cent of its own needs. Mexico is another large domestic producer making some two-thirds of its requirements in a market estimated at about US$240 million.[9] Ghana, Costa Rica and the Philippines are other countries where pesticides are made locally.[10–12] Costa Rica's own production is reflected in the fall of imports from 11,000 tonnes in 1981 to 7,000 tonnes in 1988. These figures illustrate the difficulty of using imports alone to estimate the size of the pesticide business.

The growth of the global pesticide trade is no more evident than in India where an ever-increasing quantity of pesticides is being used in agriculture. In 1960 2,000 tonnes of pesticide were used on 6 million hectares of cropland. By 1985 some 72,000 tonnes of pesticides were used on 80 million hectares. The area sprayed has increased 13 times, the quantity of pesticides used increased 36 times, yet food production has only doubled from 80 to 153 million tonnes.[13, 14] Africa, too, has seen a huge increase in the quantity of pesticides used, with a fivefold increase over the last decade.[15, 16]

This bewildering expansion in the trade says a great deal for the persuasive power of the manufacturers. Salespeople from the companies will have earned considerable bonuses for concluding the deals that led to this rapid growth. But it was not all hype. For many, pesticides offered a ray of hope in the war against insect and other pests.[17, 18] This war is still being waged. Nature, however, is fighting back in ever more enterprising ways as one pest after another acquires resistance to the chemicals used against it.[19, 20] For the moment, however, it is appropriate to consider the human casualties in this war. In another setting they might be referred to as 'the poor, bloody infantry'; the foot-soldiers who march up and down the rows of cotton, or citrus fruits, armed with a knapsack sprayer instead of a gun. The elite corps in this army are those in the motorised division who do their spraying from vehicles, using booms attached to tractors.

Pesticide Poisonings Around the World

The depressing story is the same, be it China, Egypt, Mauritius or the Sudan that is under review: people are being poisoned in vast numbers by pesticides. Horrifying as the figures are, they only show part of the picture. A look at the information on most countries listed in Table 9.2 shows that there are few government statistics to rely on. In most cases there is no mechanism for collecting the information. Many developing countries are so starved of funds that they cannot afford the luxury of having a Bureau of Pesticide Surveillance, or some equivalent agency. For others, the priorities are different.

Where government departments attempt to compile records the information they collate, all too often, severely underestimates the extent of the problem. The ideological complexion of the government makes little difference when it comes to making these comparisons. Nicaragua and Paraguay have totally different governments. In Nicaragua, until February 1990, a revolutionary nationalist/socialist government had been in power for ten years, whereas in Paraguay the military run the country. Yet in both countries official figures do not tell the whole truth. The reasons for the discrepancies between the poisonings recorded, and those which happen, are many and varied. Doctors failing to recognise poisoning symptoms; a failure to report cases; deliberate under-reporting, and the absence of someone charged

Table 9.2 Pesticide Poisonings Around the World

Country	Date	Official Statistics	Private Surveys	Poisonings	Deaths (Occupational)	Deaths (Suicides & Accidental)	Chemicals mainly Responsible	Comments
Republic of China,[21]	—	Not available	—	Estimated to be about 100,000 per year	—	—	Organo-phosphates	Estimates based on assumption that 0.1% of rural population poisoned each year
	1981	—	Xue Shou-Zhen Shanghai Medical University	14,632 (occupational) 987 (non-occupational)	90	385		Statistics based on surveys in 1 district
	1982	—	Xue Shou-Zhen	1,749 (occupational) 1,374 (non-occupational)	5	181		Increase in poisonings associated with move away from organochlorines like BHC, DDT and trichlorfon to toxic organophosphates such as omethoate and methanidophos
	1983	—	Xue Shou-Zhen	671(occupational) 842 (non-occupational)	3	82		

Country	Date	Official Statistics	Private Surveys	Poisonings	Deaths (Occupa-tional)	Deaths (Suicides & Accidental)	Chemicals mainly Responsible	Comments
Colom-bia[22]	1978–86	None	Occupa-tional Health Service Official in Antioquia	3,998 20% affected under 14 years old	Average of 63 per year	—	Organo-phosphates, carbamates	No information on long term problems. No esti-mate of cancers. Poison-ings considered to be a gross underestimate of the true figure
Costa Rica[23]	1980–86	None	Pesticide Programme of National University	2,709 requiring hospitalisation	39	—	Paraquat, organo-phosphates, carbamates	Paraquat responsible for 30% of poisonings. National Institute of Insurance Records for 1987 show that only 1 out of every 40 poisoning cases was referred to hospital
Ecuador[24]	1980–89	None	Manabi Province; Ministry of Health Official	2,017 (29% due to pesticides)	—	—	Carbamates, paraquat, organo-phosphates	No official records of poisoning for several of the years covered even though hospital records exist

Country	Date	Official Statistics	Private Surveys	Poisonings	Deaths (Occupational)	Deaths (Suicides & Accidental)	Chemicals mainly Responsible	Comments
Egypt[25]	1966–80	Amin El-Gamin; Ministry of Health 1982		Average of 1,225 per year	Average of 33.4 per year (not clear whether occupational or suicide)	—	Wide range of pesticides in use	No records of long-term health problems attributed to pesticides
India[26]	1978–87	No nationwide figures; Government Forensic Science Laboratory, Bangalore	—	12,302	Not known	Not known	Organo-phosphates	Under 1968 Insecticides Act, State Governments can make pesticide poisonings a notifiable illness – none have done so
	1988		Kabra[27] and Narayanam	—	—	319 fatal poisonings; mainly suicides	Aluminium phosphide fumigant	Fumigant should only be sold to warehouses, but advertised for use in home

Country	Date	Official Statistics	Private Surveys	Poisonings	Deaths (Occupational)	Deaths (Suicides & Accidental)	Chemicals mainly Responsible	Comments
Indonesia[28]	1985–88	None	Agricultural official covering South and Central Kalimantam	81 (75% occupational or accidental)	34, not clear whether occupational, accidental, or suicide	—	—	Frequent press reports of poisoning incidents. 31 deaths in Central Java in 1982 from DDT-contaminated food
Malaysia[29-31]	1979–86	Ministry of Science, Technology and Environment	—	3,051	Not known	Not known	Paraquat, organophosphates	Likely to be underestimated. Two studies suggest 1,200 deaths from paraquat alone each year
			World Health Organisation Report 1987[32]	400–800 agricultural workers out of 4,351 surveyed claimed to have been poisoned at some time during their working life	—	—	—	

Country	Date	Official Statistics	Private Surveys	Poisonings	Deaths (Occupa-tional)	Deaths (Suicides & Accidental)	Chemicals mainly Responsible	Comments
	1985		Ramasamy and Aros[33]	651 out of 1,214 workers and farmers interviewed had symptoms of poisoning	—	—	Paraquat, organo-phosphates	Many workers on large estates sought medical advice from dispensaries on farms. Others drank coconut water or took herbal remedies such as tamarind, for relief
Mauri-tius[34]	1977 1979 1980 1981			194 298 322	63 22 27 44	Not clear whether these are occupational or suicides	Organo-phosphates	
	1983		Association des Con-sommateurs de L'Ile Maurice	28 workers out of 150 interviewed had health prob-lems because of exposure to pesticides				Many farmers go bare-foot and do not even wear shirts when using knapsack sprayers; Many mix pesticides with their bare hands

Country	Date	Official Statistics	Private Surveys	Poisonings	Deaths (Occupational)	Deaths (Suicides & Accidental)	Chemicals mainly Responsible	Comments
	1985 1986 1987		Association des Consommateurs de L'Ile Maurice			47 53 59	Organophosphates	
Mexico[35]	1963–87	Institute of Social Security only no Ministry of Health figures		4,186 (actual) Estimates of between 10,000–20,000 per year	44 (not clear whether occupational, accidental or suicides)			
			Asociación Mexicana des Estudios para la Defensa Consumidor	Estimates 50,000 poisonings per year	Estimates 500 deaths per year			

Country	Date	Official Statistics	Private Surveys	Poisonings	Deaths (Occupa-tional)	Deaths (Suicides & Accidental)	Chemicals mainly Responsible	Comments
Nicara-gua[36]	1976-80		Corrales, D.	321-1,187 per year in cotton growing region of Leon and Chinandega				Pesticide poisoning a notifiable disease since 1979. Poisonings in cotton region established from public and private hospitals and clinics
	1980-83	Ministry of Health		121 cases for 3 years				Hundreds of cases in Leon and Chinandega reported to hospitals over this period, yet Ministry statistics only record 2 cases from the area
Nicaragua	1984		Cole et al[37] in Leon and Chinandega region	396	4	2	Organo-phosphates, particularly parathion-methyl	Official Government statistics only recorded 7 poisonings; principally affected were field workers; those mixing and loading pesticides; and workers applying the chemicals. Most

Country	Date	Official Statistics	Private Surveys	Poisonings	Deaths (Occupa-tional)	Deaths (Suicides & Accidental)	Chemicals mainly Responsible	Comments
								poisoning was through skin contact; protective masks were worn by 38% of those affected, and long sleeved shirts, or overalls, by about 25%
Papua New Guinea[38]	1976–86	None	Mowbray[39]	63		41 deaths due to paraquat; mainly suicides	Paraquat	Other poisonings suspected. No reliable records to provide confir-mation. Education Campaign by government and ICI has led to a reduction in poisonings
Other coun-tries[40] of South Pacific		None		Poisonings reported for countries in region but no comprehensive information				

Country	Date	Official Statistics	Private Surveys	Poisonings	Deaths (Occupa-tional)	Deaths (Suicides & Accidental)	Chemicals mainly Responsible	Comments
Fiji[41]	1980			44 (paraquat)		21		Records of a single hospital
	1980–83			59 (39 suicides, 18 accidental)				
Para-guay[42]	1980	Patchy Health Authori-ties figures		68				Non-governmental Organisation Alter Vida says figures underestimate problem. 45% of deaths not registered. Doctors do not recognise pesticide poisoning. Local press recorded 15 deaths 1987–89 of which 4 were suicides, and the others occupational, or from eating contaminated food
	1981			199				
	1982			77				
	1983			80				
	1984			105				
	1985			73				
	1986			38				
	1987			45				
	1988			57	3 (not clear whether occupational or suicides)			

Country	Date	Official Statistics	Private Surveys	Poisonings	Deaths (Occupational)	Deaths (Suicides & Accidental)	Chemicals mainly Responsible	Comments
Philippines[43]	1980–87	Fertiliser and Pesticide Authority; Ministry of Health		2,694 suicides, 714 accidental, 558 occupational			Organophosphates, organochlorines, carbamates	
	1961–82		Loevinshon[44]	27% Increase in non-traumatic deaths among rural men when insecticide use is high compared with period when use is low				Suggests many poisonings from organochlorines may be confused with a stroke. Likely to be gross under-reporting of poisonings
Senegal		None	Thiam and Dieng[45]	Many accidental poisonings occur No records on overall situation		Some recorded	Over 200 pesticides registered for use including organochlorines and organophosphates	33 pesticides used in Senegal have been banned or restricted in other countries; 11 of these are used with no restrictions[46]

Country	Date	Official Statistics	Private Surveys	Poisonings	Deaths (Occupational)	Deaths (Suicides & Accidental)	Chemicals mainly Responsible	Comments
Sri Lanka[47]	1975	Office of the Judicial Medical Officer		14,653	938	Most deaths are probably suicides; in 1979 1,034 causes of poisoning were examined: 66% were suicide attempts; 23% were occupational; and 8.6% accidental	Organophosphates, paraquat	Pesticide poisoning in 1988 was the third highest single cause of deaths. Between 1983–1985 pesticides were used in nearly 50% of suicides. 75% of all pesticide-related deaths are in the 15–39-year-old age group. Imports of pesticides increased 7-fold between 1976 and 1983.
	1976			13,778	964			
	1977			13,648	1,042			
	1978			14,699	982			
	1979			11,372	1,045			
	1980			11,811	1,112			
	1981			11,103	1,205			
	1982			15,480	1,376			
	1983			16,649	1,521			
Sudan[48]	1988	None		70 people poisoned by bread made with grain treated with seed dressing			Many pesticides available, including organochlorines and organophosphates	Any information about pesticide poisonings has to be collected from the local press

Country	Date	Official Statistics	Private Surveys	Poisonings	Deaths (Occupational)	Deaths (Suicides & Accidental)	Chemicals mainly Responsible	Comments
				167 people poisoned after pesticide (probably malathion) use in cheese process				
Developed World								
California	1986	California Department of Food and Agriculture[49]		1,056 occupational; 146 non-occupational 464 poisonings where information insufficient to confirm pesticide link – but pesticides most likely explanation	8 deaths between 1976–86 related to pesticide use in the field		13,000 pesticide products made up of 800 active ingredients are registered for use	Some 200,000 tonnes of pesticides are used in California each year

Country	Date	Official Statistics	Private Surveys	Poisonings	Deaths (Occupational)	Deaths (Suicides & Accidental)	Chemicals mainly Responsible	Comments
United Kingdom	1981–86		Proudfoot and Dougall[50]	9,000 poisonings recorded; only 54 linked to pesticides		8 suicides, 2 accidental deaths	Fungicides, rodenticides, insecticides and herbicides	Most suicides (6) used paraquat. The other two used phenylmercuric acetate and copper sulphate. Inhalation was the most common route of exposure in the work-related incidents

with compiling accurate information are but a few of the reasons. Hence the statistics can only be regarded as giving an impression, much like a canvas with a few brushstrokes, where the detail is still to follow. However, the figures leave no doubt that there is a serious problem, serious because each number in the table represents a sick person, or a grieving family. Figures are shown in Table 9.2 for California and the UK to emphasise that the problem is indeed global, but also to highlight the far lower death toll in these two countries.

Some of the casualties in Table 9.2 were suicides who found an answer to their unenviable mental isolation by drinking highly toxic organophosphate insecticides or paraquat, a herbicide uniquely capable of despatching those who swallow it. For most of the remainder, contact with pesticides at work poisoned them. There is a prevailing myth that a mask is all it takes to protect against chemicals. If the chemical is not inhaled or swallowed, it follows that all will be well. If only this were true; it is not. The vast majority of poisonings from pesticides are caused by the chemicals passing through the skin.[51, 52] On the job, certain activities are high-risk, and this is as true for Nicaragua as it is for California. In a survey of poisonings in the cotton-growing region of Leon-Chinandega of Nicaragua, three job categories stood out as being dangerous: field workers, pesticide applicators and mixer/loaders. Those mixing and loading were exposed to high concentrations of the undiluted chemical. Of those poisoned, some 30 per cent had worn some type of mask. As for the rest, some 25 per cent relied on long-sleeved shirts, hats, boots or overalls to protect them. Very few had all these items. In consequence, 46 per cent became poisoned simply through skin contact with the chemical, whereas 19 per cent were affected both through skin contact and breathing contaminated air.[53]

The poisoned workers in California were in four categories: those doing the spraying on the ground; diluting concentrated pesticide; in contact with the concentrates; or farmworkers who had to enter fields soon after they had been sprayed.[54]

The need for protective clothing and unambiguous instruction in the 'dos and don'ts' about pesticides seems all too clear after this catalogue of problems. Little progress is being made on protective clothing, and there are some moves, albeit too few, on instruction. In a recent survey in Sri Lanka most smallholders and rice farmers did not use protective clothing.[55] A few of the larger estates had issued protective clothing to their employees, but very few had complemented this with instructions in the proper use of pesticides. Only 11 per cent of estate workers claimed to have been trained before handling pesticides. The overwhelming majority (87 per cent) had to learn from either field supervisors or their more experienced colleagues. Of the smallholders, only 8 per cent received formal training in pesticide usage from government agencies.

Clothing and training are only two of the factors in the equation. If both were provided, most of the giddiness and skin rashes associated with organophosphate contact, and the nail damage and nose bleeds linked with paraquat use, would be greatly diminished.[56] The Sri Lankan survey serves to underline why both protection and instruction are necessary. Of those questioned about their knowledge of the hazards of pesticides, gained by reading the labels on the containers, only 24 per cent said they understood the colour codes and toxicity classes; a disturbing 21 per cent did not even know what they were about to use. On estates many workers claimed that they had not been told about hazards. In some instances the estate management claimed that instruction had been given, but the workers denied this. On one estate the management itself did not understand the significance of what was on the label. Over 80 of those interviewed (about 8 per cent of the sample) said they could not read, and a further 21 maintained that they did not *want* to read the label.

Faced with this official and unofficial indifference to the need to impart information to those who need it most, little can be done to improve the situation. Change will be difficult. Those with vested interests in keeping things as they are, are both rich and powerful. Even so, there is no need to accept all the prescriptions. Indeed, there is evidence that some of the chemical companies themselves recognise that they have to help reduce the number of poisonings in developing countries. Britain's ICI is attempting such a programme in Papua New Guinea in collaboration with the government.[57] This is a small step, long overdue.

The Pesticide Code

Ten years ago the Carter Administration in the US tried another approach, restricting what could be sold abroad. Carter's first action was the signing of Executive Order 12114 requiring any government agency to file an environmental impact statement before starting a new project in foreign countries.[58] Five days before he left office Carter introduced a far more radical control: Executive Order 12264, which Carter hoped would control many abuses, stipulated that the US, before exporting pesticides, had to pass on information about any domestic banning orders; governments in developing countries had to be told about these bans, but more importantly, had to approve any sale before it took place, and before the pesticide was exported.[59] The latter order did not last long. Shortly after taking office the next US President, Ronald Reagan, rescinded Executive Order 12264. But even this relaxation was not enough. More was to follow. In 1985 the US was alone in opposing a United Nations (UN) resolution which called for the continuation of the service which listed restricted and banned pesticides and the areas where these controls were in force.[60] At the moment there are no restrictions on the export of pesticides from the US to developing countries.[61]

The US is not alone in having no restrictions on the export of chemicals deemed too dangerous to use at home. Britain, Germany, France and Switzerland, to name but a few, take a similarly relaxed view. However, some governments have had enough. Chivvied by those surveying the suffering pesticides were causing in developing countries,[62,63] a number of governments decided to seek a code of conduct in an attempt to introduce some controls. Following debates and resolutions at the UN, it was agreed that the UN Food and Agricultural Organisation (FAO) was the most appropriate agency to administer a code. Such a Code was adopted in November 1985 but it had to be voluntary as some might construe it as an interference with free trade.

As written, the Pesticide Code has much to commend it.[64] Some of its more important provisions are shown on page 193. Many require no elaboration. Who could argue against better provision of information, or the supply of pesticides of adequate quality which have been properly tested? Similarly, there ought to be none who would object to guidance and instructions on the treatment of poisonings, proper labelling, truthful advertising, childproof containers, warnings written in the appropriate language, and the use of pictures and pictograms for the illiterate. But there were arguments against some of these clauses, and subdued mumbling behind the scenes. Given that the Code was voluntary and could not be enforced, who would notice if it was ignored? More to the point, no one could prosecute violators or levy fines. The Code was not a law. It might have moral force, but that was all. In view of this it should come as little surprise therefore that many violations have occurred. A few of the violations that have been observed, and the appropriate article of the Code flouted, are shown on page 196.

Most of the credit for monitoring the Pesticide Code and for documenting violations belongs to the Pesticide Action Network (PAN) – an informal coalition of some 300 non-governmental organisations (NGOs) in over 50 countries – which is concerned about the use of pesticides. Three particularly active NGOs are the Pesticide Education and Action Project in San Francisco, the International Organization of Consumer Unions (IOCU) in Penang, Malaysia, and the Pesticides Trust in London. Drawing on the help of their network of consumer, and workers' groups around the globe these three organisations have kept an eye on the trade in pesticides. Two trail-blazing books paved the way for others to follow. *Circle of Poison* by David Weir and Mark Shapiro appeared in 1981.[65] With its haunting pictures and simple, to-the-point text, this book served notice that there was a problem which would not go away. In 1982 David Bull of Oxfam published his thoroughly researched *A Growing Problem*.[66] What Weir and Shapiro achieved with imagery, Bull consolidated with his impeccable and extensive reference material. There could be no doubt after the appearance of these two publications that the pesticide trade, satisfactory though it was to the manufacturers, was a problem for consumers. Disorder reigned. A new order was needed.

International Code of Conduct on the Distribution and use of Pesticides FAO 1986

Some Key Clauses from the Code

What it says
(Numbers refer to the articles of the Code)
The code establishes ... *voluntary* standards of conduct (1.1) which

- encourage responsible ... trade practices (1.5.1)
- assist countries ... to regulate the quality and suitability of pesticides (1.5.2)
- encourage ... safe and efficient use of pesticides (1.5.3)

Who it Affects

The code is designed ... to assist government authorities, pesticide manufacturers, those engaged in trade and any concerned *citizens* to judge whether their actions or the actions of others constitute acceptable practices (1.6)

Governments ... Should

(1) provide technical assistance to other countries to help assess the relevant data on pesticides (3.3.1)
(2) implement a pesticide registration and control scheme ... (5.1.1)
(3) provide guidance and instructions for the treatment of suspected pesticide poisoning for physicians, hospital staff and basic health workers (5.1.3)
(4) establish national or regional poisoning information and control centres ... accessible at all times ... (5.1.4)
(5) ensure that pesticides ... are physically separated from food, medicines, clothings and any other products which might be used for internal consumption or topical application ... (5.1.6)
(6) take action to introduce the necessary legislation ... and make provisions for its effective enforcement, including the provision of appropriate educational, advisory, extension and health care services ... (6.1.1)
(7) collect and record data on the ... import, formulation and use of pesticides (6.1.4)
(8) prohibit ... an extremely toxic product ... if the control measures or good marketing practices are inadequate to ensure that the product can be used safely (7.5)
(9) notify other countries, and importing countries in particular, whenever they ban or severely restrict a pesticide (9.1, 9.3)
(10) prohibit the repackaging, decanting or dispensing of any pesticide in food or beverage containers ... (8.2, 10.4)

The Pesticide Industry ... Should

(1) cooperate fully in the observance of the Code and promote its principles and ethics – irrespective of a government's ability to observe the Code (12.3)

(2) halt sales, and recall products, when safe use does not seem possible by following directions or restrictions ... (5.2.3)

(3) make less toxic formulations available (5.2.2.1)

(4) sell products in ready-to-use packages (5.2.2.2)

(5) use containers that are not attractive for subsequent re-use ... and use containers that are safe ... (for example childproof) (5.2.2.3, 5.2.2.4)

(6) test all pesticide products to assess their safety for humans and the environment (8.1.1)

(7) see that persons involved in the sale of a pesticide are adequately trained ... so that they can provide the buyer with advice on safe and efficient use (8.1.9)

(8) follow international standards for manufacturing and formulating, packaging and storage, labelling and advertising, particularly where the national law is inadequate (3.2, 8.14, 10.1, 3.1, 12.3)

(9) use labels that ... (Article 10)
   (a). are in the appropriate language
   (b). include appropriate symbols or pictograms wherever possible
   (c). show the WHO hazard classification of the contents
   (d). are clear and concise (5.2.2.5)
   (e). include:

   • warnings and precautions
   • warnings against re-use of containers
   • instructions for disposal
   • batch numbers (without codes)
   • the date of formulation
   • storage life

(10) ensure that packaging and repackaging is only carried out on licensed premises ... (10.3.2)

(11) ensure in advertising that ... (Article 11)
   (a)   all statements are capable of technical substantiation
   (b)   restricted use products are not publicly advertised ... unless prominently indicated as restricted use
   (c)   safety claims (e.g. 'safe', 'non-toxic', etc) are not made, with or without phrases like 'when used as directed.'
   (d)   statements comparing safety of different products are not made.
   (e)   no guarantees (for example 'more profits with ... ') are made unless there is definite supporting evidence
   (f)   no visual representation of potentially dangerous practices are used (for example application without sufficient protective clothing ...)

(g)   all advertisements draw attention to warning phrases and symbols on labels

(h)   all staff involved in sales promotion are properly trained ... to present complete and accurate information on the products sold.

Restrictions

Avoid pesticides whose handling and application require the use of uncomfortable and expensive protective clothing and equipment ... (3.5)

Integrated Pest Management

Governments and the pesticide industry should develop and encourage integrated pest management ... (3.8)

The Food and Agriculture Organisation (FAO) refused us permission to quote extracts from the Pesticides Code, so we have paraphrased the relevant sections. The FAO said we could reproduce the *entire* code if we wished, but few publishers have the space to do that. The upshot of this short-sighted approach is that fewer people will hear of the code. This, surely, cannot be the intention.

The Pesticide Code was designed to provide the backbone of what was seen as a new era in the trade in chemicals. As shown on p. 196, that era is still some way off. Three books by the Pesticide Action Network,[67,] [68] and the Pesticides Trust,[69] provide a detailed critique of the Code violations as well as confirmatory evidence in the form of photographs, container labels and supporting testimony (the code violations cited on p. 196 are taken from these three books). The proof that the Code is being ignored should be enough to embarrass the companies into changing their ways. There is no satisfactory excuse for this behaviour. Profits may be one of the reasons advanced, but this is not good enough. Money does not buy lives, and it is totally inadequate as compensation for somebody's health. If the violations were confined to an irresponsible few it would be bad enough. Regrettably the catalogue of violators is a checklist of 'Who's Who' in the chemical industry. Misleading advertising; the use of highly toxic pesticides where safer alternatives would do; poor packaging; unsubstantiated claims about the safety of products, and no visual reminders in the form of pictures or symbols to warn users to take care, are but a few of the techniques that the industry has used and is still using.

Good though it is, the Code was clearly not enough on its own to stop the abuses. A say in what pesticides could be imported would be a significant step in transferring power to the users of chemicals in developing countries. With this in mind, and taking account of UN support for the issue, and after years of lobbying from campaigning organisations such as the Pesticides Action Network, in November

Violations of the FAO Pesticide Code

The following are but a few samples of recent violations of the Pesticide Code together with the particular articles which have been ignored

Article	Country	Details
3.4.1 3.4.2 3.4.3 5.2.2.2 8.1.10 10.1 10.2 10.3 10.4	Mexico (1989) Indonesia (1989) Paraguay (1989)	Repackaging of pesticides in inappropriate containers by traders with only the name of the pesticide displayed – and no warning labels.
5.2.2.3	Senegal	Old pesticide containers recycled to carry food, milk or cooking oil; 19 people die in one village after a cook uses cooking oil from bottle previously containing the toxic insecticide ethyl parathion.
5.2.24	Thailand Philippines Indonesia	Pesticide packaging not childproof: Liquid pesticides on sale in easy-to-open screw cap containers.
10.2.4	Sudan	Literature not in local language: The label for the insecticide Aldrex T (aldrin) sold by Shell was only written in English.

Misleading Advertising

Some of the more flagrant violations are to be found in advertisements for the pesticides. Here the company is being misleading about the product it sells.

Article	Country	Details
11.1.2	Indonesia (1987)	Misleading Safety Claims: Chevron leaflet claims that Monitor 200 (methamidophos) a highly hazardous insecticide is 'not poisonous and does not leave dangerous residues'.
11.1.8		Statements made about safety without qualifications such as 'when used as directed'.
	Brazil (1987)	Roussel Uclaf describes Decis (deltamethrin) as the 'safest insecticide in the world'. Monsanto claims the herbicide Roundup (glyphosate) is 'safe'.

Article	Country	Details
	Indonesia (1989)	ICI Indonesia says the insecticide Ambush 5ULV (permethrin) is 'safer for humans and the environment'.
		Mitsubishi, and the Nippon Kayaku subsidiary, state that the insecticide Mipcin 50WP (isoprocarb) is 'safer for humans and the environment'.
	Colombia (1989)	Dow Chemical insists that its insecticide (chlorpyrifos) can be sprayed from aircraft or backpacks without risk. The product is safe for field staff and those who spray it. (In June 1989 Lorsban 4E leaked from Dow's site into the nearby Cartagena Bay killing countless fish and causing widespread environmental damage.)
	Mexico (1989)	Rhone-Poulenc advertises Mocap 10G — containing the extremely hazardous insecticide, ethoprop, as 'safe and easy to apply'.
	Philippines (1989)	An advertisement written in a local Philippines dialect, Tagalog, claims that Shell's Furadan 3G (carbofuran), a highly hazardous insecticide, is safe to use.
11.1.9		Inappropriate comparisons made about the safety of different products.
	Indonesia (1989)	Sumitomo compares Sumicidin 5EC (fenvalerate) with rival products, claiming it to be of medium toxicity compared with other pesticides.
11.1.11		No guarantees or implied guarantees.
	Indonesia (1989)	An affiliate of Mitsubishi and Nippon Kayaku claims that the insecticide Petroban 200EC (chlorpyrifos) will yield 'healthy crops and a successful harvest'. Hoechst says that the insecticide Thiodan (endosulfan) will 'guarantee your harvest'.
	Philippines (1989)	Sumitomo implies that the insecticide fenitrothion has been the main insecticide used on rice in Japan since 1963, and as a result rice growers had greater yields. Using a local dialect, Monsanto entreats farmers to use the herbicide Lambast (butachlor) 'for added harvest and big profit'.
	Colombia (1989)	Rhone Poulenc tells farmers to 'obtain more and better sorghum [by] applying Methavin', (the trade name for the insecticide methomyl).

Article	Country	Details
11.1.12		No visual representation of dangerous practices such as spraying without protective clothing ...
	Indonesia (1987)	Dupont calendar for 1987 advertising eight insecticides, herbicides and fungicides, shows an attractive female model dressed in a sarong spraying a tobacco crop.
11.1.13		Promotional material draws attention to warning phrases and symbols ...
	Malaysia (1989)	Local company Tiram Kimia Sdn. Bhd. advertising seven Shell pesticides makes no reference to warnings, but offers purchasers of the chemicals an 'exciting opportunity to collect valuable gift vouchers' which will enable them to go on their 'Parkson shopping spree'.
	Philippines (1989)	ICI advertisement depicts a farmer without protective clothing spraying the insecticide Cymbush 5EC (cypermethrin). No warning phrases or symbols are shown on the advertisement.
	Colombia (1989)	Poster advertising Hoechst's insecticide Hostathion (triazephos) – no warnings shown.
	Ecuador (1989)	Mural wall posters for insecticides: (i) Furadan (carbofuran) imported from Cyanamid (US), Bayer (West Germany) and Greece. (ii) Tamaron (methamidophos) imported from Chevron (US) and Bayer (West Germany). (iii) Dimecron (phosphamidon) imported from Ciba-Geigy (Switzerland). None of the murals have any warnings about dangers.
11.1.17		Encourage purchasers and users to read the insecticide labels ...
	Philippines (1989)	Sumitomo advertising the insecticide Sumithion (fenitrothion) says nothing about reading labels on containers.
	Costa Rica (1989)	No warning phrases or symbols, or instructions to read labels on six brochures for the following Bayer pesticides: Disyston 10GR (disulfoton); Nemacur 10GR (phenamiphos); Curater 5GR (carbofuran); Metasystox R500SL (demeton-methyl); Oftanol (isofenphos) and Tamaron (methamidophos). These Bayer products are extremely toxic.

1989 the FAO adopted the principle of Prior Informed Consent (PIC). In future, when companies receive an order for pesticides for export, they must check whether the importing country allows the import of this pesticide. Exporting companies are bound by the Code and have to comply with the importing country's decision. A synopsis of what is involved with PIC is shown on pp. 200–1.[70]

PIC is a considerable step forward, but, as part of the Pesticide Code, it is voluntary. There are no penalties for countries, or exporters, who break the rules. What is now expected of them is a moral stance. PIC is a sort of customary international law where those who disregard the rules are frowned upon, rather than punished. The situation is more akin to pupils (in this case governments and the chemical industry) being kept in line by the stern gaze, rather than the strap, of an old hand at teaching.

Monitoring is essential to assess whether the PIC procedure will contribute to any reduction in the number of poisonings from pesticides. Were agreements on their own enough to stop abuses then adoption of the FAO Pesticide Code in 1985 should have started the salvage operation. This has not happened. In fairness, it must be said that a PIC procedure was necessary to give the FAO Code some teeth. But as neither the Code, nor PIC, are mandatory, we will need to wait to see if the teeth are effective. They need not be just window-dressing. PIC now offers a procedure which, if properly used, could lead to a dramatic reduction in the sale of dangerous chemicals, and particularly those which are associated with so many deaths.

The scheme will be administered jointly by FAO and UNEP through the International Register of Potentially Toxic Chemicals (IRPTC). Given that governments must inform the IRPTC about any bans or restrictions on particular pesticides and that the IRPTC, in turn, will notify all the other participating governments about these actions, we now have an established conduit for the transfer of information. In future, government approval for pesticide imports can be based on details about the health and environmental effects of pesticide use. Ignorance is no longer an excuse. Lack of resources might be. Many governments still lack the trained personnel to implement a PIC scheme. Help will be needed. Both the Pesticide Code and the PIC provisions acknowledge that the more affluent countries – with established toxicological monitoring procedures – will be expected to help their newer, less experienced, trading partners learn the rules.

There are deficiencies in the PIC scheme – apart from its voluntary nature, pesticides which are banned or severely restricted because of problems in handling and application will not be included. Nor is it possible to include pesticides because of their toxicity alone – they must be banned or restricted for health or environmental reasons. This means that pesticides in the WHO categories 1a (extremely hazardous) and 1b (highly hazardous) will not be included (see Table 9.3), unless they are shown to have an impact which meets the agreed definitions.

Prior Informed Consent (PIC)

The principle of 'Prior Informed Consent' which requires exporters of banned, or restricted, hazardous pesticides to obtain the approval of importing countries before shipment, is now considered to be good practice. At its 25th Conference in November 1989 the United Nations Food and Agriculture Organisation (FAO) adopted the principle of PIC. The Conference recognised that exporters of dangerous pesticides should, before exporting, inform purchasing countries of the health and environmental risks of their chemicals. PIC will become part of the 1986 FAO Code of Conduct on the Distribution and Use of Pesticides.

What PIC Means in Practice

Legal Status None. Like the Pesticide Code, PIC is voluntary.

Supervision Joint United Nations Environment Programme (UNEP) and FAO Executing Agency, administered by UNEP's International Register of Potentially Toxic Chemicals (IRPTC) in Geneva.

Participants Member nations of FAO and UNEP invited to take part as exporters and importers.

Designated National Authorities A Designated National Authority (DNA) will be nominated by participating countries. The DNA will notify IRPTC of government control measures; import/export decisions; and will receive IRPTC reports and decisions.

Selection of Pesticides Member governments to inform IRPTC by 31 December 1989 of previous control measures on pesticides. IRPTC will evaluate to ensure that they conform to agreed definitions of banned or severely restricted. The PIC list will be drawn up from:

(1) The initial list which will be made up of pesticides already banned or severely restricted for health or environmental reasons in five or more countries.
(2) Once PIC is operational, pesticides will be covered when one Government bans or severely restricts their use, for health or environmental reasons.

Prior Informed Consent (PIC) *(continued)*

	(3) The FAO/UNEP panel of experts will review 'extremely hazardous pesticides' in the World Health Organisation (WHO) Class 1a category, and others which are found to cause health or environmental hazards under the conditions found in most developing countries, to determine whether more should be covered by PIC. Sixty-one pesticides are currently in Class 1a.
Notification	FAO will notify member governments about controls imposed on pesticides and will send national authorities a Decision Guidance Document to help them reach decisions.
Response	After each notification, importing countries have 90 days to inform FAO whether or not they will continue to allow the pesticide to be exported to them. A response may be a stop-gap measure or a final decision. No response means that exports are forbidden without the express permission of the importing country.
Action	National authorities will be told by IRPTC how each country responds.
Exporting Countries	National authorities of exporting countries will tell exporters about decisions. Governments are expected to ensure that pesticides are not exported if an importing country does not want them.

Source: *Pesticide News*, November 1989.

It is not credible to exclude chemicals from PIC for toxicity reasons. In recognition of this, the expert panel advising on the implementation of PIC has recently implemented a new 'category 1a+'. Pesticides which are banned or restricted because of their toxicity, but which are found to cause health and environmental hazards because of conditions operating in developing countries, will now be scrutinised for possible inclusion in PIC.

As more governments become aware of the risks associated with particular pesticides there will be a move towards the use of safer alternatives. The Pesticide Code actively encourages this shift, but so far there have been few moves in this direction. Had there been more, and if the chemical industry had voluntarily withdrawn its more toxic products, a PIC scheme might not have been necessary. It has become so because the pesticide trade will not act on its own; some coercion is necessary.

The move to alternatives will not go unrewarded. The market for food that is produced without pesticides is growing apace and many consumers are willing to pay more for this produce. Even before this market began to grow, Western governments restricted imports of food considered to be too contaminated. In 1967 the US turned back 300,000 pounds of Nicaraguan boneless beef because of its high DDT content.[71] Fourteen years on, the US was still turning back South American beef. In 1981 US Department of Agriculture officials refused entry for 2,437,268 pounds of beef – about 3 per cent of South American imports – because it was contaminated.[72, 73] From September to December 1980 meat intended for export from Honduras to the US was refused entry because on five occasions it was found to be contaminated with high levels of DDT, dieldrin or heptachlor.[74]

Table 9.3: Recommended Classification of Pesticides by Hazard

| | LD50 for the rat (mg/kg body weight)[a] | | | |
| | Oral | | Dermal | |
Class	Solids[b]	Liquids[b]	Solids[b]	Liquids[b]
Ia Extremely hazardous[c]	5 or less	20 or less	10 or less	40 or less
Ib Highly hazardous[d]	5–50	20–200	10–100	40–400
II Moderately hazardous[e]	50–500	200–2,000	100–1,000	400–4,000
III Slightly hazardous[f]	Over 500	Over 2,000	Over 1,000	Over 4,000
IV Unlikely to present acute hazard[g]				

Notes
[a] The classification refers to the actue toxicity of a pesticide. In some cases account is taken of longer-term chronic toxicity and some pesticides known to cause cancer in animals, or man, are included in Class 1a.
[b] The terms 'solids' and 'liquids' refer to the physical state of the product or formulation being classified.
[c] 61 pesticides listed
[d] 92 pesticides listed
[e] 162 pesticides listed
[f] 146 pesticides listed
[g] some 280 pesticides

Source: *The WHO Recommended Classification of Pesticides by Hazard and Guidelines to Classification 1988–89*, World Health Organization, WHO/VBC/88.953.

What is true for beef also applies to other crops. High pesticide use inevitably leads to contaminated food. In the late 1970s the US Food and Drug Administration estimated that nearly half of all imports of green coffee beans contained detectable amounts of pesticides which had been banned in the US. Many of the beans also had excessive concentrations of pesticides which were regulated in the US.[75] Cocoa beans present a similar picture and Chapter 10 is a case study of the problems associated with just one crop in the developing world – cocoa.

The problem would be complicated enough if the pesticides were restricted to foods that could be dumped. Human breast milk does not come into this category. Milk from the breast is sterile, rich in nutrients and antibodies for the growing infant, and hence it is now acknowledged to be the appropriate food for young babies. Sterile the milk may be, but it is not uncontaminated. In many developed countries breast milk samples have revealed the presence of minute quantities of highly toxic chlorinated dibenzodioxins and dibenzofurans.[76] These chemicals are likely to have been produced in municipal and industrial incinerators as well as in many other situations where organic compounds have been burnt. The WHO acknowledges that this evidence is disturbing but it considers that the advantages of breast feeding still outweigh the disadvantages of a slightly contaminated feed.

Contaminated breast milk is not the preserve of the developed world. Indeed the problem would appear to be much worse in countries where organochlorine pesticides have been used on a large scale. These pesticides are soluble in fat. After gaining entry to the body they remain in fatty tissue for a long time – in the case of DDT, for years. When a mother breast feeds some of her body stores of fat are transferred to the milk to provide it with enough calories. Any pesticides stored in the fat also move into the milk. DDT is one insecticide found in the breast milk of women around the world.[77] At one time human breast milk contamination with DDT and its metabolites in Nicaragua averaged 2.29 ppm for women living in agricultural areas, and 2.12 ppm in women living in the city of Lyon.[78] The WHO recommends that the amount of DDT consumed per day should be no more than 0.005 milligrams for each kilogram of body weight.[79] On this basis Nicaraguan infants in the first few weeks of life would have been consuming some 70 times the quantity of DDT that the WHO considers safe (the calculation assumes a 4 kg infant drinking about 0.6 l of milk per day). Even higher intakes than this have been recorded in Guatemala.[80] In India a survey in 1985 and 1986 of 60 breast milk samples collected from women whose children were less than a week old showed that, on average, children were still consuming 12 times more DDT than WHO considers reasonable.[81] Pesticides clearly do not remain at the farm gate; many are likely to become visitors at the dinner table and long-term residents in our body fat.

Concern about the health implications of using toxic chemicals is one reason why there are moves to regulate the trade in these products. Another reason is the growing resistance of many insects and other pests to a range of pesticides. By 1984 at least 447 insect and mite species had become resistant to pesticides.[82] The resistance is genetic and the new, tougher species have developed immunity to a wide range of pesticides, necessitating alternative methods of dealing with them. In 1954–55 the resistance of the boll weevil to organochlorine pesticides alerted agronomists to the problem. Prior to this, insect resistance had not been taken seriously. Today many species are resistant both to organochlorines and to the new organophosphate insecticides developed as less persistent alternatives. Where organophosphates have been used to control mosquitoes, the insects have not been long in acquiring resistance to these as well.[83] Pesticide resistance is described by some as a treadmill where newer pesticides will always be needed to replace those that the insects eventually make redundant.

Another part of this equation concerns the blanket use of pesticides to keep pests at bay. This 'insurance policy' approach to pest management often means that natural predators are destroyed too. This leads to a further problem known as resurgence, where pest populations rebound to previous or even higher levels, after insecticide use. Linked to this is the phenomenon of secondary pest resurgence where insects which were initially present in very low numbers suddenly increase and become pests following the use of insecticides to kill a different pest.[84] One of the best-known examples of this occurred in 1925–26 when calcium arsenate used to kill boll weevils on cotton, led to an outbreak of cotton aphids.[85] Another example was the outbreak of various citrus pests in the US following the use of DDT and other insecticides which killed the insects which preyed on them. Dr Paul DeBach, an entomologist at the University of California, recorded a 1,250-fold increase in the density of scale insects on trees sprayed with DDT compared with those which were unsprayed. All trees had roughly the same density of scales before the spraying.[86, 87]

One reason for the phenomenon of resurgence is that organic insecticides are often more toxic to the predators than they are to pests which are their prey.[88] Many plant-feeding insects, and particularly those likely to attack crops, have a greater ability to detoxify poisons than the predacious, or parasitic insects likely to be their enemies. The plant-feeding insects have had to adapt to deal with chemicals which plants use to defend themselves, whereas the parasitic and predacious insects which feed on other insects, do not have to deal with the same poisons in their food and therefore lack certain detoxification mechanisms.[89, 90]

There is another phenomenon, not yet fully understood, which also occurs. Referred to as 'stimulation' it is caused by certain insecticides which alter the survival of some insect pests.[91] The brown planthopper is probably the single most important insect pest which attacks rice. Following experiments in 1978 with the organophosphate parathion-methyl scientists at the International Rice Research Institute in the Philippines, discovered that planthopper numbers had increased 30 to 40 times. This resurgence in numbers occurred in tandem with a direct effect on the planthopper's egg-laying habits. Parathion-methyl, in common with four other insecticides – fenitrothion, decamethrin, diazinon and fenthion – caused female planthoppers to lay more eggs.[92] The insecticides also resulted in a decrease in the number of predators, which in turn led to more planthoppers surviving and more eggs being produced. In one insecticide-treated plot egg production increased 2.5 times, egg-to-adult survival by 5.6 times and adult female survival increased 6.3 times.[93, 94]

Integrated Pest Management

These and other observations have led many farmers and agronomists to reconsider the blanket use of pesticides. Not only are the chemicals expensive, but their use can also be counterproductive and lead to even more pests. If pesticides kill the predators then their prey are likely to multiply. For many this approach was ridiculous. Reconsidering this strategy, the concept of using predators together with minimal application of pesticides emerged. Known as Integrated Pest Management (IPM) this has become the single most important new approach to pest control where the controlling habits of predators are harnessed – rather than eliminated – to complement the use of pesticides. Examples of its success are legion and well documented.[95-97] (See also Chapter 12.)

IPM has established its credentials. Indeed some institutions now consider it to be the management tool of choice for dealing with insect pests. The World Bank which funds vast numbers of development projects, has issued its own pesticide guidelines which state that IPM should be official Bank policy. However, a recent review of of some 24 Bank projects revealed that only one – in Sudan – even referred to IPM. The Bank, it seems, is largely ignoring its own guidelines, and seeking to use chemicals alone to control pests.[98] This emphasis on chemicals will lead to far more being used. For example, in the Bank's Second Agricultural Development Project in Egypt, designed to benefit some 650,000 farmers, insecticide use will increase ninefold, fungicide use 9.5-fold and herbicide use 8.6-fold over the six-year programme ending in 1991.[99] In Mexico money has been allocated for pesticides in the Second Tropical Agricultural Development Project which aims to increase the income and agricultural productivity of farmers in the humid, tropical, lowland coastal plain of the country. By the fifth year of the project – earmarked to run from 1986–1994 – spending on pesticides will have increased between 2 and 16.4 times.[100] In the Yemen, pesticides will be used on the Wadi Al-Jawf Agricultural Development Project (1985–1992) where previously there was very little use because of traditional farming practices.[101] In Chad's two-year Emergency Cotton Program which began in 1986 following the slump in the price of cotton, the World Bank programme envisaged chemical-intensive cotton production. Fertilisers and pesticides will be the basis of production and will be used regularly.[102] Sadly, this might be the ruination of the cotton programme in the long run because blanket spraying of cotton with chemicals has led to an increase in pesticide-resistant pests in Peru and Nicaragua. If ever there was a need for IPM it is in the growing of cotton.

A similar picture of over-reliance on pesticides emerges from both a World Bank maize and cowpea project in Nigeria,[103] and a coffee project in Zambia.[104] In Ghana, too, the World Bank restructuring of the cocoa industry envisages the need for increasing use of pesticides (see Chapter 10). Only in the Bank's cotton project in the Sudan does it appear that IPM is being pursued in accordance with the Bank's own guidelines.[105] In the Sudan attention is also being given to adequate storage facilities for pesticides, and a helpful policy on the destruction of old stocks of chemicals. It would appear, however, that the Bank has not given enough consideration to the need to train farmers in the Sudan in the safe use of pesticides, nor thought about protective clothing for those doing the spraying.[106]

Worker Education

Official indifference to the many problems farmers and plantation workers face, has left these groups with little alternative but to organise themselves into unions or co-operatives to enable them to campaign more effectively for change. In Ecuador the Association of Free Farmers (*Federación Nacional de Campesinos Libres del Ecuador*) is educating its members about the dangers of pesticides.[107] In Brazil rural workers are joining unions because they see them as the only way to achieve better wages and safer working conditions.[108] Workers in Ghana join for the same reasons. Often the unions are the only organised bulwark that can prevent the depradations that a World Bank loan might require.[109]

Contrary to popular misconceptions, many organised groups in developing countries do not lack information about the dangers of pesticides. What they often do lack, however, are the resources to pass on what they know to those who are in need of it – the pesticide users. Occasionally government departments help with this and compile valuable handbooks on the 'dos and don'ts' of using pesticides. One of these guides is the excellent *Manual de Sugurança No Uso de Agrotoxicos*, produced by the Ministry of Works in Brazil in 1987.[110] Simple illustrations and a clear text present a convincing case for good hygiene when using chemicals. The booklet shows that pesticides can be inhaled, swallowed if the user doesn't wash before eating and, much more importantly, absorbed through the skin if it, or clothing, becomes contaminated. Protective clothing and masks are vital, the booklet argues, as are clean, secure storage facilities under lock and key to keep toxic chemicals away from the unwary (such as children). Stress is placed in the manual on maintaining spraying equipment in good order, and on keeping it clean to protect subsequent users. Perhaps most important of all, pesticide users are encouraged to read instructions on containers before using a product. Users are reminded that the label will tell them how to dilute in the recommended way and that it will warn against using a product beyond its shelf-life. Out-of-date pesticides need to be safely buried in the ground, says the manual. And for those unfortunate enough to be exposed to pesticides the Ministry of Work's booklet describes the likely symptoms, the need for first aid, and the necessity of getting victims to hospital for proper medical care.

For anyone using pesticides there has rarely been a clearer manual of instruction and the Brazilian Ministry of Works deserves commendation for its effort in producing this guide. Sadly, the praise should not be unqualified; the Ministry comes in for some censure too. For some inexplicable reason only 3,000 copies of the manual were produced and even these were not widely distributed.[111] Regional government officials each received their allocation, but officials made little effort to distribute the booklet any further. This is a palpable waste of resources and time. It borders on the censoring of information. Perhaps the government can claim publicly that it has written a guide, but others need have no fear that there will be a push to educate all users. Booklets will be kept in Ministry filing cabinets where they will cause no difficulties for those who may need to pay money to implement its recommendations.

Where governments show no inclination to warn of the dangers from pesticides, others have stepped in to do the job. The Association of Free Farmers in Ecuador has an enviable track record in this regard. Two of its publications stand out. The first is its manual for users *Riesgos del Trabajo Agrícola*.[112] Relying more on words than pictures this booklet sets out to inform readers about risks. It is intended more for the instruction of those running training courses or talking to small farmers or workers on plantations. Besides covering what the Brazilian Ministry of Works manual addresses, the Ecuadorian publication also focuses attention on a number of chemicals that it considers are causing harm. In particular it draws attention to twelve pesticides identified in 1985 by the Pesticide Action Network as being particularly hazardous. Many countries now either have official bans or restrictions on the 'dirty dozen' as they have come to be called.[113]

For those who may not be able to read well, the Association of Free Farmers also produces a calendar – in the form of a poster – illustrating good and poor pesticide practice.[114] An attractive feature of the calendar is the humour used to convey the message. Taking the FAO Code of Practice as its starting point, the calendar emphasises both the need to restrict imports of toxic pesticides, and the importance of using chemicals safely. A fish crying 'help' and leaping from a stream in which pesticides have been dumped, or equipment washed, underlines the need to protect the wider environment. With its elegant design and clear messages the calendar is one of the best campaigning posters on pesticides that we have seen.

In 1986 an illustrated booklet produced on behalf of the union representing Plantation and Agricultural Workers in the Dominican Republic left little unexplained.[115] Written as a teaching aid, in uncompromising style, and with wonderfully wicked illustrations, the booklet could not fail to persuade its audience about the immorality of sections of the pesticide trade. Interposed between its graphics are highly damning photographs showing workers spraying without protective clothing, and an advert of an attractive female model dressed only in casual clothes with a knapsack sprayer on her back. No subliminal messages are needed here. The advert is blunt: protective clothing is not necessary. Why else would a young model be so attired? Castigating the pesticide trade for its deficiencies is only one part of of the Dominican Republic manual. To balance the negative, the authors also concentrate on positive features of pest control and introduce readers to the concept of IPM and the use of natural predators to keep insects at bay.

Conclusion

Manuals like the three we have discussed will become more evident in many other countries as people recognise that governments on their own will not be able to bring about the changes that are necessary to protect the majority of users. Few governments have the means, or the inclination, to introduce all the controls that are required. For some, public relations appear more important than substance, and if the world can be convinced that certain import bans are in force then the government need do no more. Fortunately many users now see through this window-dressing and they are demanding real safeguards. With the pesticide trade moving into new fields, and investigating genetic engineering techniques to incorporate disease-resistant features into plants, different controls will be required.[116] For the moment the pesticide industry is not considering plants which can be grown without pesticides; this would be a laudable aim. There appears to be more emphasis on producing crops that will tolerate certain pesticides so that these can be used without the plant being damaged. Besides offering the pesticides industry a breathing space, this approach will increase competition between companies as each advertises the merits of the plants it has engineered. In future farmers buying seeds will be able to purchase an integrated pesticide pack at the same time. The seed/pesticide combinations will not reduce the quantity of pesticide used, but may restrict the number of chemicals applied to a crop. Some chemical companies will benefit from this approach, but it is highly doubtful that the consumers, let alone agricultural workers, will be any better off.

The only way forward that offers everyone more safety is the use of IPM. Working with nature rather than against it is, in the area of pest control, a tried and tested approach that works. The challenge will be to ensure that IPM is used on an ever-increasing scale. If the Pesticide Code is used as it was intended, then IPM should flourish, and workers everywhere need not run the risks of dying from the toxic chemicals that so many still use. Governments, industry, farmers, workers, and consumers, all have a role to play in bringing these changes about.

10

Pesticide-coated Chocolate

Introduction

Much of the chocolate that we eat is contaminated with pesticides. The amount of pesticide in a bar of plain or milk chocolate is tiny, and so far there is no evidence to suggest that the levels are dangerous. The question to ask, however, is: 'Do the pesticides need to be there at all?' From the evidence we have it is clear that much of the pesticide contamination is unnecessary. We do not have to have a sandwich of chocolate and pesticides. Forty years ago the cocoa that was processed into chocolate had no synthetic pesticides in it. It is contaminated today because of the intensive use of pesticides on cocoa plantations: chemicals used to control pests on the plantations end up in the cocoa.

Cocoa is a perfect example of what some refer to as the 'circle of poison'; it illustrates how pesticides made for export in Western Europe and the US poison people in developing countries and yet end up in the very food Europeans and North Americans eat. Many of the chemicals used on cocoa plantations have been banned in Europe and the US because they are too dangerous. However, many of the workers using these pesticides on plantations do so in the most unhygienic conditions and without proper training. Conditions of work could not be more dangerous. Advertising, together with more direct pressure from manufacturers to increase the use of pesticides, is making the problem worse. National laws designed to control pesticide use on cocoa plantations are either too weak, or simply ignored because there are not enough inspectors to enforce them.

The economic pressure on Third World governments is considerable, and many find it difficult to resist the demands put upon them by international agencies and transnational companies to allow greater use of chemical pesticides. Many of the chemical companies are based in Europe and make substantial profits from their sales to developing countries. In the unseemly rush to make money it is no wonder that pesticides are promoted extensively without adequate warnings about their dangers. It is not surprising, either, that they are often used unwisely by workers ill-trained in the handling of dangerous substances.

There is some irony in the fact that the consumer in Europe and the US pays a price for this greed through the consumption of chemical residues in chocolate. An added irony is the fact that many of the pesticides are not needed. The crop is grown in many areas without using any chemical pesticides. Perhaps the final irony is the pressure from consumers in Europe, and Switzerland in particular, to have, if not pesticide-free chocolate, then a product with few residues. Switzerland, it needs to be remembered, is home to two of the biggest sellers of pesticides on the international market, Sandoz and Ciba-Geigy. Some Europeans, it seems, want the money, but not the consequences.

The Cocoa Project

Identification of the problems pesticides pose for cocoa growers has been made possible by the work of the Amsterdam-based Transnationals Information Exchange (TIE). The Cocoa Project, one of three that TIE coordinates, brings together people involved in the cocoa chain. TIE convened a meeting in Brazil in May 1989 for representatives from six cocoa-growing countries. In addition to growers from the host country there were farmers from Ecuador, Ghana, Nicaragua, Nigeria and Sierra Leone. Delegates from Malaysia, another major and expanding cocoa producer were invited but were stopped from attending by the Malaysian authorities.

Representing the European end of the cocoa chain were dockworkers from Amsterdam – where most of the cocoa arrives – workers in the British, French and Dutch food processing industry, and trades union health and safety experts from Holland and Britain. The outcome of this meeting – like others TIE has organised – was a realisation of the seriousness of the problem facing the cocoa industry and those who work in it.

Five companies – Mars and Hershey in the US, Cadburys in Britain and Suchard and Nestlé in Switzerland – dominate the world market for cocoa. The five have absorbed most of their competitors, the most recent acquisition being the takeover of the former British company Rowntree Mackintosh, by Nestlé. Competition in the industry is fierce, forcing companies to cut costs. Cocoa prices have been ruthlessly forced down and are at their lowest point for ten years, causing severe economic hardship for countries like Ghana, with 70 per cent of its foreign earnings dependent on the sale of cocoa.[1] Sandwiched between the growers and the chocolate industry is the processing sector, which produces the cocoa from the beans. Here, too, there has been concentration of effort and the market is dominated by a few companies.

Forcing down the price of cocoa benefits the transnational companies in the short term, but the industry is well aware that if prices are too low farmers may simply stop growing cocoa. Many of the large companies, like Mars, Hershey and Nestlé, have established biotechnology programmes to try to produce high-yielding, uniform cocoa plants. It is intended that these new 'superplants' will be more resistant to disease, have an increased number of beans per pod and have a naturally high sugar content to obviate the need to add sugar to chocolate. It is argued that this uniformity will cut costs. However, these new plants will almost certainly need considerable quantities of fertiliser.

Beans grown by tissue culture techniques and planted on selected plantations in Belize and Malaysia have increased the current yield of 400 pounds of cocoa beans per acre to a high of 1,000 pounds per acre. Researchers hope that this yield may eventually be pushed to 2,000 pounds per acre. According to Patrick Aculey, an employee of Ghana's Cocoa Board, Cocobod, better-yielding cocoa plants will give increased yields of cocoa over the first few years without the need for additional fertilisers.[2]

However, after this first phase fertiliser will be essential if yields are to be maintained, for there will not be enough nutrients in the soil to feed the plants. Herbicide, insecticide and fungicide use is also likely to increase for various reasons. First, because it is extremely unlikely that the industry will be able to breed a plant in the near future that is totally resistant to disease. Second, there is a move to concentrate cocoa production on large plantations, and the owners are more likely to reduce labour costs by using herbicides, instead of manpower, to keep weeds and brushwood under control.

Ghana

Herbicides are not always necessary for weed control on cocoa plantations. Provided the cocoa trees are planted at set distances from one another, so that their own leaf canopy prevents light reaching the ground, weeds will be deterred. In Ghana most small farmers use manual labour to clear bush and young trees. But, as Patrick Aculey points out, the chemical companies have been promoting their products to farmers and the Ghanaian government, saying: 'chemicals and machines can cultivate and produce cocoa more cost effectively.' Aculey maintains that there is pressure on the Ghanaian government to shift cocoa production from the country's traditional labour-intensive practice to one that is more capital-intensive.[3] This policy is being pursued in a country where unemployment is high.

The pressure on Ghana to change its policies is being exerted by the International Monetary Fund and The World Bank. In 1982 the World Bank gave Ghana a substantial loan. (When Ghana's current military regime accepted the World Bank loan the exchange rate for the currency (the Cedi) was Cedi 1.75 = US$1. The exchange rate in September 1988 was Cedi 229 = US$1. A loaf of bread currently costs about Cedi 100.) In return for the money, Ghana reduced the number of employees in state organisations. Twenty thousand employees have already been made redundant by Cocobod; more will follow.[4] An estimated 45,000 employees, many of whom advise small farmers on cocoa-growing practices, will eventually lose their jobs in the restructuring of Ghana's cocoa industry. The redundant workers, many of whom were employed in the industry's extension services, did most of the pesticide spraying. Properly equipped with personal protective clothing and respirators, they applied the limited chemicals that Cocobod approves. The Cocobod tests all products for their suitability and is the only organisation in Ghana currently allowed to advertise pesticides. So far it has resisted pressure from the chemical companies to allow more pesticides to be used on plantations. In future, however, small farmers in Ghana will have to do their own spraying.[5] But like their counterparts in other developing countries, most farmers in Ghana are illiterate, and therefore unable to read instructions on pesticide containers. Accidents, and poisonings are inevitable because few are able to interpret the limited safety instructions that are printed on pesticide container labels. As for protective clothing and face masks, only a few of the richer owners will be able to afford these.

Restructuring of Ghana's economy has meant that the government has had to sell 55 of Cocobod's 95 plantations.[6] The plantations are far too expensive for most Ghanaians to buy and there is a real fear that they will be purchased by foreign nationals or transnational companies.[7] The shift to private ownership of the cocoa industry in Ghana has run in parallel with a move to encourage the use of chemicals. Neither change is in Ghana's interest for, says Aculey, 'machines and their spare parts must be imported, and become more expensive each year. All chemicals used on cocoa farms must be imported, and these are also very expensive. Above all, the almighty foreign exchange is needed.'[8]

Although Ghana's economic problems are considerable they are in no way unique. Other cocoa-growing countries also rely on the sale of cocoa for foreign earnings. Their economies, however, are more robust than Ghana's. Yet, in spite of these difficulties, Cocobod has managed to limit the number of pesticides used on cocoa. It is impossible to predict how long this policy will be maintained if the pressure on Cocobod to sanction other pesticides is kept up. Fortunately, there are some who argue that rather than increasing pesticide use, farmers should experiment with using fewer chemicals, and Patrick Aculey is one. Lindane, he says, was first introduced to kill insects when cocoa production was increasing. Spraying has now become the norm, and is equated with keeping insects at bay – a sort of insurance scheme.[9]

Lindane and Unden (a carbamate) are the two insecticides currently used on cocoa plantations; growers use these chemicals on alternate years. In the north, growers use lindane for two years and then Unden for two.[10] A similar biennial rotation occurs in the south, but in a different sequence so that when lindane is used in the north, Unden is used in the south. In any one year, therefore, each chemical is only used on half the plantations. In this way Ghana hopes to prevent insect resistance to the chemicals for a while longer.

Aculey believes that it would probably be difficult to stop insecticide use altogether, but that it may be possible to reduce it. Instead of the blanket spraying each year it would be feasible to experiment by spraying for 2–3 years, and then stopping for a few years. Spraying could be started again if pests return. But if the insects do not re-appear, it would justify increasing the period between sprayings. Use of insecticides would certainly fall if this policy was adopted.

Brazil

Fungal diseases are another serious problem on cocoa plantations. One fungus called 'black pod' is particularly contagious and can do untold damage. The name black pod is apt because the infected pod eventually blackens. The fungus grows on discarded cocoa pods and if any of these have been left under healthy trees the fungus will be carried to them and infect the ripening healthy pods above. The infected pod may not always appear damaged; inside, however, the white gelatinous coating around each cocoa bean turns to a brown, dry, fibrous material. There is no mistaking the condition. The drying out of the bean in this way means that it can no longer be processed. Normally, when cocoa is ripe for harvesting the pods are collected from the trees, cut open, and the beans extracted and allowed to ferment for a few days before they are dried (either in the sun or over a stove). After drying, the beans are transported to the factory for processing into the cocoa consumers recognise. By 'drying out' the beans, black pod disease prevents fermentation – a necessary step in the transformation of cocoa into an edible food.

Chemical treatment of fungal disease has a long history. Some of the more common treatments for fungal diseases are based on the use of copper. Bordeaux mixture is a copper sulphate solution which has been used to control fungal diseases on grapes for nearly a century.[11] On cocoa plantations in Brazil copper fungicides are much in evidence. No single formulation is used. On any one plantation there may be copper oxide, copper hydroxide or copper oxychloride in the pesticide store. All three formulations are retained to combat black pod.

Perhaps because copper is an essential element in the body there is a widely shared myth that copper fungicides are not toxic – they are. Eye, lung and heart problems are documented in the grape industry and directly attributable to spraying with Bordeaux mixture (see Chapter 5). It was with some distress therefore that in May 1989 one of us (Alastair Hay) watched five unprotected men spraying copper fungicides onto cocoa pods under the dense, leafy canopy of a plantation some 30 miles from the centre of the cocoa-growing region of the state of Bahia. The men had no protection other than rubber boots to prevent snake bites. Three of the men carrying motorised backpack sprayers doused the fruit with spray from a four-foot lance while their two colleagues refuelled a fourth sprayer. On the ground where the men were working the hot, moist air, trapped by the leaves above, took on a brown hue as the spray settled to the plantation floor. The smell was choking, and the concentration sufficient to sting nose and mouth. For the men spraying, it was 'business as usual' and with no masks, they had no option but to breathe in the fumes. Their skins, partially covered by torn working clothes, displayed greeny-pink stains from fungicide that had leaked from the sprayer and run down the lance onto their hands and arms.

The counselling these men received about protection demonstrates how badly advised they were. Some were told that a drink of milk before spraying would protect them; in another instance a doctor recommended a tot of rum before work.[12] Both recommendations are wrong; they are also dangerous. Swallowing milk or rum does not stop people inhaling chemicals, or absorbing them through the skin – the route responsible for most pesticide poisoning. The suggestion that milk or rum will prevent chemicals being absorbed in the gut is life-threatening nonsense and is not an alternative to providing protective clothing. Wilson de Silva Andrade, employed on the Fazenda Santo Antonio displays a degree of resignation about his situation: 'We know that our health is being damaged by these chemicals but we have to do the job as it pays just a little more: that way we can eat better' he says.[13] Wilson receives £1.60 a day and a weekly bonus of £2.70 when spraying.

Conditions are better at the State Research Institute for Cocoa (CEPLAC) where plantations are maintained by Institute employees. The only man spraying on the day Alastair Hay visited CEPLAC was not wearing protective clothing. Masks and overalls were provided he said, but: 'they are not always available and there were none today.' As for wearing a mask, he said he usually did this, but not always, because it was difficult to breathe. It offered little protection anyway because it did not fit tightly. Never ill himself, he claimed to know of others on different plantations who had had hospital treatment. Singling out a copper fungicide as the worst product he used, he said he disliked it because it 'makes me cough up blood'. On every occasion he used the fungicide he would inhale it. Others Alastair Hay met also claimed to have coughed up blood after using pesticides, but it was not always clear what chemical was being referred to. It was clear from the majority of workers he spoke to that they recognised that pesticides could be a threat to their health, but not everyone was so concerned. On the Fazenda Yalon, a plantation near the coastal town of Camamu where some 150,000 kg of cocoa is produced each year, one man did most of the spraying. Although he was able to read the safety instructions on containers he said he never bothered to do this, the spraying had done him no harm and he had never suffered from ill-health which could be attributed to the chemicals. In the rowdy meeting where these comments were made it was apparent that others were not of the same opinion. They spoke of headaches and throat and skin irritation arising from use of the insecticide BHC (hexachlorocyclohexane) officially banned in Brazil in 1981, but still available under-the-counter from many shops selling agrochemicals.[14]

Evidence that pesticides are causing chronic health problems in individuals is not hard to find. Nestor Bispo da Silva, aged 55, has heart, chest and eyesight problems which he says are common among older workers who have been spraying BHC without protection for nearly 20 years. Nestor started working with pesticides when he was 16, but he had been unemployed for nearly a year. Four months after becoming ill with his heart and chest complaints his employer dismissed him without compensation. With no social security payment from the government Nestor now depends on handouts from individuals and charity. Pneumonia is another common complaint related to inhalation of pesticide dust. Aureliano Pereira dos Santos said that his doctor had told him that he nearly died during a three-month bout of pneumonia aggravated by high blood pressure.[15]

Individual cases like these are supported by some evidence collected at a district level in a 1987 survey of 283 farmworkers in three Bahia townships – one in a cocoa region – by Fundacentro, a health ministry department. According to the survey 77 per cent of those interviewed reported symptoms following exposure to pesticides and a startling 48 per cent had absorbed 'intense levels' of chemicals.[16] Statistics such as these are few and far between. CEPLAC used to collate details about poisoning incidents on cocoa plantations but has not done so for the last few years, according to Dr Paulo Fernando Nunes da Cruz.[17] The statistics end in 1984 because there is no one responsible for collecting the information. In the past Dr da Cruz did this, but for the past few years he has worked as an administrator in CEPLAC on crop protection, and no one has been detailed to replace him.

Few doubt that there is a serious health problem from pesticides in Brazil. The country has eight poison centres that offer treatment and advice to doctors on what to do when poisoning has occurred. The centre in Salvador (*Centro Antiveneno da Bahia*) serves the state of Bahia and, according to its records, between January 1986 and June 1988 some 330 cases of pesticide poisoning were recorded, of whom 14 died.

The Salvador statistics are unlikely to be the whole story. Unlike the chemical sprayer on the Fazenda Yalon most cocoa workers in Brazil are illiterate. Most of those Alastair Hay spoke to said that they could not read and therefore could not follow instructions on the labels of pesticide containers. Most had not been shown how to use pesticides safely, although CEPLAC, according to Dr da Cruz, sent out specially trained operators to teach plantation workers how to do this. Dr Da Cruz did confirm, however, that most of the workers CEPLAC instructed were illiterate.[18] He said that CEPLAC encouraged the use of protective clothing, but in the hot humid conditions of a cocoa plantation many workers did not wear any protection. When asked why they did not follow these recommendations, workers on the Fazenda Yalon claimed that it was too uncomfortable to wear the clothing: 'Why don't they make safer pesticides and then we won't have to wear protective clothing?' one man interjected.

CEPLAC, according to Dr da Cruz, is trying its best to improve conditions for workers using pesticides. The organisation has made two fundamental changes, he said. In the past, farmers were encouraged to spray, but CEPLAC now only recommends using pesticides if 'pests constitute an epidemic. If this is not the case then we do not recommend that chemicals are used.' The second change is to promote the use of safer insecticide alternatives such as 'malathion, permethrin and the carbamate, sevin,' instead of the more toxic BHC. It was regrettable that many farmers were used to BHC, he said, and that they still used a formulation of the pesticide made illegally in Brazil. As far as CEPLAC is concerned BHC is dangerous. In the March 1989 issue of the institute newsletter a two-page feature, headlined 'BHC used to harm your health but now it harms your pocket', pointed out that current formulations of the pesticide only contained 0.4 per cent BHC instead of the recommended 1.5 per cent. According to the article buyers were putting their health at risk and being swindled.

For the literate few the article would be helpful. For most plantation workers it would have little impact. Most doctors in the region have little knowledge of the health problems pesticides cause. Dr José Carlos Ribeiro is Director of the Roberto Santos Hospital in Ilhéus and trained in public health. He admits to knowing very little about pesticide poisoning. In his opinion many cases of poisoning will not be recognised simply because doctors are not familiar with the symptoms. This inability to recognise poisoning symptoms means that doctors do not implicate chemicals as the cause of an illness and they make no records in the register.[19] Cases of poisoning are therefore underestimated. Many are also undetected. On some of the plantations Alastair Hay visited, workers said that no doctor visited them. The fact that a doctor might make such a visit was the cause of much amusement to squatters occupying land at Marinbondo, a remote camp some 40 miles inland from the Bahian town of Camamu. 'Why should a doctor be interested in us?', they asked. One family had lost a 3-year-old girl a few days before Alastair Hay arrived. According to her mother, the girl had not appeared very ill at first, and a doctor may well have saved her. But, lacking transport, her distraught parents could only nurse her and watch her die.

Combating plant diseases is not as traumatic. Pesticides are often used for control, but they are not always required.. A novel approach to dealing with black pod disease illustrates this. The fungus that carries the disease grows on discarded cocoa pods. If these are left under the cocoa tree – as often happens – spores from the fungus will be carried onto the healthy cocoa pod, thereby spreading the infection. To break this cycle often all that is necessary is to ensure that no empty cocoa pods are left on the ground near healthy trees. This is simply good management. In parts of Ghana and on a few plantations in Brazil they clear away the cocoa pods and shred them for use later as fertiliser.[20, 21]

On the Fazenda Nogueira, some 30 miles from Ilhéus, Alastair Hay observed empty cocoa pods being mechanically shredded and left in a pile to ferment for two months. Fermentation increases the temperature considerably, and the heat generated destroys any fungus. After fermentation the shredded pods are mixed with raw manure from mules or cattle and used as fodder for worms – especially imported from California – for a few months. The worms digest everything leaving a fine mulch which is used as compost for cocoa trees. Unfortunately only the Fazenda Nogueira uses this process at the moment. Other small-holders around Ilhéus are considering setting up a number of demon-stration plots to show more cocoa growers how to generate this compost. The workers on Nogueira have made compost for several years and in that time they have not had to purchase more worms (as they breed at a phenomenal rate) nor resort to the use of fungicides. Their good housekeeping practices have saved them money.

In Ghana, farmers take additional steps to control the spread of black pod. By selective pruning of trees farmers allow more sunlight and warmth to reach the ground and help keep the area drier. They believe that these simple hygienic measures developed over the years help to keep black pod in check.[22, 23] Some viral diseases affect the embryonic cocoa pod, and for these there are no chemical solutions. Removing the infected shoot is the only answer.[24]

Brazil, like Ghana, relies on cocoa for foreign earnings. Fifth in the league table of exports, cocoa is the most important commodity in the State of Bahia. Some 92,000 square kilometres – an area the size of the Netherlands – is given over to the cultivation of cocoa, representing some 95 per cent of Brazil's production: 2.3 million people in the state rely on the crop for their livelihood. Most of Brazil's cocoa is produced on large, privately-owned plantations; smallholders only produce 19 per cent of the crop. Technical information about how to grow cocoa, what diseases affect the crop and how to control them is provided by CEPLAC. Funded by a 10 per cent tax on exports, CEPLAC is reputed to be far more receptive to the needs of the large plantation owners than it is to those of smallholders. Given that most of its income comes from the sale of cocoa from the plantations, this claim is probably not unreasonable, although Arisual Vesper, Regional Co-ordinator for CEPLAC, maintains that the organisation is there to advise everyone.[25] CEPLAC tests pesticides for use on cocoa plantations.

A large number of chemicals are recommended by CEPLAC for this purpose some of which are listed in Table 10.1. Some of the chemicals used for the same purpose in Ghana and Malaysia are also listed. The table makes depressing reading for it includes very toxic chemicals that have been banned in European countries, Japan and North America. The CEPLAC list may eventually be more selective, as the organisation is aware of the growing concern in the population about the health risks from pesticides. The continued use of all of these chemicals on cocoa plantations must be questioned. Those that cause immediate concern are aldrin, chlordane, dieldrin, heptachlor and lindane. These five chemicals cause cancer in animals and may do the same in humans.[26] Although aldrin is not recommended by CEPLAC for direct use on cocoa it is still used on many plantations to control termites.

Table 10.1: Pesticides Used on Cocoa Plantations

Brazil	Malaysia	Ghana

HERBICIDES

Brazil	Malaysia	Ghana
dalapon (BASF; Dow Chemicals)* diuron (Hoechst) 2,4,5-T+2,4-D (BASF; Dow) paraquat (ICI) picloram + 2,4-D (Dow) simazine (Ciba-Geigy)	paraquat (marketed under 56 different brand names: ICI, Malaysia) 2,4,5-T (ICI, Malaysia)	None approved

FUNGICIDES

Brazil	Malaysia	Ghana
benomyl (Du Pont) mancozeb (Rohm and Hass) copper oxychloride + maneb + zineb (Sandoz)		Fungicides under investigation by Cocoa Research Institute Kocide ⎱ All Redomile ⎰ copper Cocoabre ⎰ based (Sandoz; and local company, Reiss and Co.) Copy flow ⎱ Antioplus ⎰ (Ciba- Copper Nodox⎰ Geigy) Champion ⎰

INSECTICIDES

Brazil	Malaysia	Ghana
aldrin (Royal Dutch Petroleum Co [Shell]; Rhone Poulenc) lindane carbaryl (Sevin) dicrotophos demeton methyl malathion (Cyanamid) parathion-methyl (Bayer) endosulfan (Hoechst)	chlordane (Wesco Chemicals; Velsicol Corporation) heptachlor (F.E. Zuellig; Asean Chemical Fertiliser Sdn. Bhd.) dieldrin (ICI Agriculture, Malaysia; Australian Chemicals Co.; Shell, Malaysia) lindane (ICI Agriculture, Malaysia; Australian Chemicals Co.)	lindane (ICI) Unden 200 (a methyl carbamate formulated in Ghana under licence from Bayer)

* where known the names of chemical manufacturer(s), or supplier(s), are listed and bracketed.

As for paraquat and parathion-methyl, these are two extremely toxic pesticides. Many deaths in developing countries are attributed to exposure to parathion-methyl.[27] Details about paraquat are more scarce. With children often working alongside adults on plantations, it is wrong that products as toxic as this are promoted. Yet they are, and extensive promotion of a vast range of pesticides is evident in Brazil. Of the three countries we have listed in Table 10.1, Ghana provides the most information about the need for safety when using chemicals. Malaysia and Brazil do far less to protect workers. In addition to the men spraying without protective clothing, Alastair Hay also saw evidence of a Brazilian plantation worker using his bare hands and a cup to scoop chemicals out of a sack. A small dust mask was all that was provided, and the man was required to mix the chemicals in a drum with a stick. Mixing necessitated him having to force the stick round and round the drum. When he had finished, his arms were wet with chemicals. The mixture was poured from the drum into a backpack sprayer. Strapping the sprayer onto his back, the man walked to the cocoa tree to begin spraying. No protective clothing was provided for this job other than the dust mask which was totally inadequate to protect against chemical aerosols. This plantation is not unusual.

Back pack sprayers leak. Workers using them in Malaysia and Brazil complain that the chemicals often drip onto their skin and Alastair Hay observed this at first hand in Brazil. The problem is compounded because the pesticide has to be sprayed upwards to reach the cocoa pod and drifts down onto the worker. This is a far cry from the situation depicted in many chemical company advertisements showing workers kitted out in full protective gear.[28] Gramoxone, the trade name for the herbicide paraquat, is sold by ICI Brazil, and is marketed as a grass and weed killer for use on cocoa plantations. In the promotional brochure the man carrying out the spraying is wearing a protective hood, mask, gloves, trousers and boots. The glossy promotional booklet for Cobre Sandoz – the copper fungicide sold by Sandoz to treat black pod disease – says nothing about protective clothing.[29]

Apart from depicting visually, and referring in the text to the proportion of powder to mix with water, the Sandoz booklet stresses that it is important to ensure that the pesticide spray covers the whole cocoa fruit. It is important, the brochure says, not to use too little as this could be a false economy. In pictures showing how to spray, the booklet portrays a man in everyday clothes wearing a small dust mask, gloves and boots. Accompanying this is another picture of a man spraying without any protective clothing. In the second picture the visual and textual messages are that the spray needs to be directed at the fruit. The picture does not refer to the need to wear protective clothing, and the impression given is that it is not essential. Many people using the fungicide will not even see this literature, as the chemical is sold in brown paper sacks without any illustrations, and few instructions.

Pesticides and Health

Pesticides are on sale in local shops in both Brazil and Malaysia. Malaysian cocoa growers maintain that the chemicals are not stored securely. They frequently complain that empty, or half-used, containers are not properly discarded but re-used by villagers after a cursory wash. Brazil has legislation to deal with pesticides but lacks the resources to enforce it. Three ministries, Agriculture, Health, and the Interior, control pesticide use in the country. They deal with licensing, use, and protection of workers and the environment. Attempts to introduce tough regulations on the use of pesticides by the Brazilian state of Rio Grande do Sul were fiercely resisted at first by transnational companies.[30] Required to disclose details about the toxicity of their products, the companies refused to hand this information over, fearing that other states would follow suit. Only after the threat of a £350 a day fine did the companies come into line.

Rio Grande Do Sul's law would have prohibited the use of all organochlorine pesticides. Sadly, it would appear that these restrictions do not apply throughout Brazil. Although BHC was not visible on the shelves of stores retailing pesticides many others were on sale. Aldrin, a toxic, persistent, carcinogenic insecticide was on sale without restriction in six shops Alastair Hay visited. Indeed, in one grocers, the owner highlighted the range of goods on sale by holding aloft a carton of aldrin in one hand and a packet of biscuits in the other. This sad display of wares illustrates the difficulty facing any authority trying to impose some control over the use of chemicals. By common consent, State authorities in Brazil lack the resources and staff to implement national pesticide regulations: CEPLAC employees, plantation workers and small farmers alike appear resigned to the situation for the moment and see little hope of any immediate changes.[31]

Everyone, however, is heartened by the increasing attention being paid by the public, and the media, to the dangers from pesticides. Knowledge about the danger is being publicised in Bahia. Local newspapers in the Salvador region recently featured articles with the headlines: 'Agrochemicals in water kill children', 'Agrochemicals poisoning rural workers'.[32] Politicians were reportedly called upon to act on the issue.[33] Trades unionists have been concerned for some time. Aware of the dangers of over-dependence on pesticides, and the health risks involved, the Association of Professional Agricultural Technicians (STAC) – the union representing CEPLAC workers – has encouraged farmers to try to do without chemicals, and made grants available for this purpose. STAC's campaign has angered CEPLAC. In 1988, to weaken the union's resolve, CEPLAC transferred eleven STAC trades unionists to remote areas of the country.[34] This crude attempt at victimisation failed and was resisted in the courts. All eleven workers have now been reinstated. Fortunately, appeals for clemency for those campaigning against the excessive use of pesticides is not yet required in Europe or the US.

Fumigants in Transport

To stop weevils eating cocoa beans during shipment, ship holds are fumigated. In 1986 dockworkers in Amsterdam handling a cargo of cocoa fell ill. The men noticed a garlic-like odour when the hold was first opened but no one had any idea what it was, nor could anyone explain their illness. The answer appeared three months later, when workers unloading an Indonesian vessel were violently sick. Tablets of aluminium phosphide were found in the hold. Moisture reacts with this chemical to give off the toxic gas phosphine, which is widely used as a fumigant. Manufacturers claimed that the aluminium phosphide would no longer be active by the time the ships docked and that there would be no phosphine emission. The Dutch Institute for Working Conditions discovered, however, that the chemical was still highly reactive and that when exposed to water, it rapidly gave off phosphine.

On investigation it became apparent that, contrary to expectation, aluminium phosphide was not exhausted when exposed to moisture in the hold. Although most of the chemical had reacted with water some had been covered by a protective crust forming on the surface of the remainder. When the sacks of cocoa were lifted out of the hold by the dockers, the aluminium phosphide canisters were dislodged, breaking the crust in the process and allowing the rest of the chemical to generate phosphine.[35] Aluminium phosphide containers are still found in ships carrying cocoa. Three years ago any suspicious-looking grey powder was sufficient reason for Dutch dockers to down tools. To deal with the problem the dock authorities agreed in 1988 to finance a study to identify the origin of ships carrying the powder and to design a suitable safety procedure for dealing with phosphine.

Most cocoa-growing countries use fumigants. An agreement between the buyer and seller determines what is used. It is clear that changes must be made to reduce risks. Magnesium phosphide is a safer source of phosphine. The chemical breaks down more rapidly and there is unlikely to be any left when a ship docks. The rapid release of phosphine could, however, create problems for those loading the vessels and might simply shift problems to other docks. A more realistic alternative would be fumigation, and thorough ventilation, before loading the cocoa. Any insects in the hold would be killed in the process. Cocoa beans could be similarly fumigated, and aerated, before loading. In Ghana, fumigation of both the beans and the ship's hold, is carried out before shipment, as opposed to during the voyage and this is an inherently safer procedure. If this were done everywhere it would remove any remaining uncertainty about which holds are dangerous. At the moment it is not possible to predict whether there will be phosphine in a hold.

Pesticide Residues in Cocoa

Ridding the docks of a fumigation hazard is vital. A less clear-cut issue is whether the pesticide residues in cocoa pose a health hazard. The cocoa processing industry regularly checks cocoa for pesticides, bacteria and moulds. Cocoa de Zaan, the largest processing company in Amsterdam, processes some 30,000 tonnes of cocoa per year. Tests are carried out on every 1.5 tonnes. This high frequency of sampling, according to the laboratory manager, Dr H. Brinkman,[36] ensures that the high biological quality of the cocoa is maintained.

As for pesticide residues, Cocoa de Zaan screens for organochlorine pesticides. Fifty to sixty per cent of any residue ends up in the husk surrounding the cocoa, the rest remains in the cocoa nib – the raw cocoa inside the shell. Brinkman says that organochlorine pesticides are the principal residues. In consequence, Cocoa de Zaan screens mainly for DDT, lindane and contaminants in lindane. He claims that only some 1–2 per cent of cocoa has residues above the legal limit. Most of the contamination is caused by lindane. Other chemicals used in cocoa growing, such as the organophosphate insecticide parathion-methyl, and carbamates, Brinkman claims, are destroyed on the vegetation, during storage, and in processing. (Parathion residues are known to persist on grapes in transit so breakdown in storage does not always occur.) Fumigation could also contaminate cocoa, but as Cocoa de Zaan does not permit fumigation in its own warehouses with methyl bromide, the cocoa it processes is free of this pesticide.[37] 'Poor spraying practices' in the cocoa-growing countries is the reason Brinkman puts forward for pesticides being found in cocoa. Cocoa de Zaan sells to the Swiss chocolate manufacturers, who have the most exacting standards for residues. Any single batch of cocoa with pesticide residues above the prescribed limit is diluted with another containing fewer residues to ensure compliance with the standard required by the Swiss buyers. Cocoa de Zaan uses considerable resources and equipment to check the quality of the cocoa it sells.

Samples of the cocoa processed in the UK are screened for pesticides by a laboratory financed by the trade association representing chocolate manufacturers. Cocoa beans from a wide variety of countries, principally in Africa, the Caribbean and Malaysia are screened under the auspices of the Biscuit, Cake, Chocolate and Confectionary Alliance. According to a spokesman for the Alliance it has been 'monitoring the levels of pesticide residues in cocoa beans' for 14 years. Screening tests cover 32 pesticides and

> our results over the years have shown that the organophosphates and carbamates are almost completely metabolised by the cocoa tree and that only very small traces of their breakdown products are found. As might be expected, the organochlorines are more persistent, but are found only in very low concentrations.[38]

In addition to the screening carried out by the Alliance, the Nestlé Company screens for organochlorine pesticides in cocoa at least twice a year.[39] According to the company, previous experience has demonstrated that this frequency of testing is sufficient as pesticide residues are always lower than the maximum residue limits recommended by Codex Alimentaris (the body jointly sponsored by the two UN organisations, Food and Agriculture and World Health to make pronouncements on food). Nestlé's own screening has revealed hexachlorobenzene, dieldrin, DDE, and aldrin at concentrations ranging from 0–2 parts per billion (ppb). Lindane is present in concentrations ranging from 9–41 ppb. Similar quantities were found by the Alliance laboratory. Of the 179 samples analysed in recent years (a frequency of testing many times lower than practised by Cocoa de Zaan) only 13 samples (7.3 per cent) had lindane residues above 100 ppb, with a maximum of 600 ppb being recorded. The maximum residue recommended by Codex is 1,000 ppb. In view of these results and their monitoring exercise, Nestlé remains confident that 'levels of organochlorine pesticides are within the recommended limits' and, as such, safe.

From the above it must be obvious that consumer standards in the developed world are far stricter than pesticide user standards in the developing world. The promotion of pesticides in developing countries might be reduced by the growing consumer pressure in the developed world for pesticide-free food. In some countries cocoa is grown without using pesticides simply because farmers cannot afford to buy the chemicals. Small farmers in Brazil, Ecuador, Ghana and Sierra Leone find that most pesticides are far too expensive to use. One farmer from Sierra Leone complained about the cost of chemicals, maintaining that his productivity would improve if he could afford to use them.[40] Other farmers from Ghana and Nigeria were not convinced by this argument.[41, 42] There is no evidence to suggest that the residue levels of pesticides in cocoa pose any health risk to consumers. Nevertheless, there is a recognition on the part of the growers that if they were able to produce a crop with little, or no pesticide residues, that consumers would prefer this.

Conclusions

Ghana, with its crippled economy, has its Cocobod valiantly trying to limit the encroachments of the pesticide industry which is keen to promote increased use of chemicals. How long Cocobod can continue to hold out remains to be seen. The omens are not good given that Ghana's financial lifeline to the International Monetary Fund calls for some sacrifices, the net result of which is that Cocobod will lose much of its control over the cocoa industry.

Plantation workers everywhere, but particularly in Brazil and Malaysia, would be the direct beneficiaries of a policy to use fewer pesticides. Their working conditions are poor and cry out for attention. Pressure needs to be applied to the pesticide manufacturers, who appear to be more eager to increase sales of their products than concern themselves about the health problems users experience. The continued use of pesticides that are banned in the developed world is inexcusable. Perhaps the fact that some of these banned organochlorine pesticides end up in our chocolate may help make consumers more aware of the plight of plantation workers.

How long plantation workers will be required to use dangerous chemicals is far from clear. If it is left up to the chocolate industry and pesticide manufacturers to institute changes it is likely to be some time before conditions improve. The investment of the chocolate manufacturers in genetic engineering to develop a cocoa 'supertree' that is higher yielding, disease-resistant, and which will not require pesticides is not likely to bear fruit for years – if at all. Hence only direct pressure on the industry from consumers and workers' organisations will bring about the changes in working conditions that are needed.

11

Pesticide Law

Introduction

Because pesticides are toxic chemicals, and therefore hazardous, public policy dictates that their approval and registration, manufacture, formulation, marketing, sale, supply, advertising, transport, storage, labelling, packaging, use and disposal should be regulated. This regulation varies from country to country. Some aspects of regulation are made the subject of laws, with sanctions if they are broken. Other aspects are the subject of codes of practice or voluntary agreements. International regulation tends to be by means of conventions or agreements. The monitoring and enforcement of regulations is a problem everywhere.

Agribusiness concerns and the chemical industry lobby all governments and international agencies. In general, the system of regulations and enforcement that exist in developed countries do not properly represent the interests of trades unions, community, consumer, or environmental groups; the problem is far worse in developing countries. There are no international enforcement agencies, although there have been proposals to set one up for the control of the nuclear industry and hazardous substances and wastes, including pesticides. The workings of the international agencies that do exist will be examined later. The European Community imposes sanctions for breaches of community law, but the international agencies can only operate on the basis of consensus and political pressure.

Often laws, where they exist, are shrouded in secrecy. Basic toxicity data on pesticides and evaluation criteria are protected by 'commercial confidentiality'. Manufacturers or governments often have no legal obligation to make data public for independent reviews. This is true of the UK where the Advisory Committee on Pesticides (ACP) is a closed system, made up of civil servants and government-appointed 'independent' experts. There is no direct representation of industry, trades unions, or environmental, consumer and public interest groups. In other countries, such as the US, Canada, Sweden and Denmark, freedom of information laws enable citizens to find out what has been decided and on what basis.

Systems of regulation that apply fall into three broad areas:

- International Codes of Practice and Guidance
- EC: European Community law
- national legislation

Each country regulates pesticides differently, under different approval and enforcement agencies, according to its own system of law and government. UK law is dealt with more fully on pages 232 to 254.

International

The development of international regulations and control for all forms of hazardous chemicals has been significantly influenced by organisations such as the UN Food and Agriculture Organisation (FAO), the UN Environment Programme (UNEP), the International Labour Organisation (ILO), the World Health Organisation (WHO), the Organisation for Economic Cooperation and Development (OECD) and the European Community (EC). The programmes, codes of practice, guidelines and so on they have drawn up or implemented, while not legally binding, provide a framework and minimum standards for the development of international controls.

United Nations Organisations and Programmes

World Health Organisation

The WHO, in cooperation with UNEP and ILO, has been responsible for developing the International Programme on Chemical Safety (IPCS). Within the IPCS, the Pesticide Development and Safe Use Unit has developed a WHO Pesticide Evaluation Scheme and published *The WHO Recommended Classification of Pesticides by Hazards and Guidelines to Classification 1988–89*.[1] The IPCS publishes the results of its evaluations as Environmental Health Criteria Documents.

United Nations Food and Agriculture Organisation

The FAO published the *International Code of Conduct on the Distribution and Use of Pesticides* in 1986.[2] The FAO Code is voluntary, but establishes for the first time a minimum international standard for monitoring the standards of governments and the pesticide industry; it lays down the responsibilities of governments, manufacturers and distributors.

The issues covered by the Code include the regulation, availability, distribution and health aspects of pesticide use, as well as advertising, labelling, packaging, storage and disposal. In 1989, after years of lobbying from Third World and Non-Governmental Organisations (NGOs), the FAO included the principle of Prior Informed Consent (PIC) (see Chapter 9). Under PIC provisions, which began operating in 1991, a list has been drawn up of pesticides which have been banned or severely restricted for health or environmental reasons in five or more countries (eventually a ban in one country for these reasons will act as a trigger). Governments participating in the scheme (at present over 76) state, through an appointed national authority, whether they will allow the import of pesticides on the PIC list. Exporters must consult the list, and agree not to export if a country has indicated it does not allow the import of a particular pesticide. PIC is a joint scheme with UNEP, and is administered through UNEP's International Register of Potentially Toxic Chemicals (IRPTC) in Geneva.

United Nations Environment Programme (UNEP)

UNEP's IRPTC is responsible for the exchange of information on hazardous chemicals, and is developing a data base to make the information more accessible. The IRPTC routinely distributes general information on chemicals via newsletters and circulates notices of national regulatory actions affecting particular chemicals. Individual governments and organisations can request regulatory or scientific data on over 700 specific chemicals.

In June 1987, UNEP adopted a code on information exchange, *The London Guidelines for the Exchange of Information on Chemicals in International Trade*, known as the June 1987 'London Guidelines', because the agreement was made in London.[3] The London Guidelines included Prior Informed Consent provisions, and UNEP and the FAO agreed to unify their PIC conditions and operate these jointly through the IRPTC. The EC has made a Directive to cover a list of banned or restricted substance which require the exporter to inform importers of the hazards of these chemicals.[4] However, these provisions are weaker than the FAO/UNEP PIC scheme, in that consent need not be given prior to export.[5] In Britain, the Department of Environment (DoE) is the designated national authority for the PIC and the EC schemes, which will be run through the Health and Safety Executive. It appears unlikely that more resources will be devoted to policing or monitoring the schemes.

Codex Alimentarius Commission

The Codex Alimentarius Commission is a joint effort by the WHO and FAO. The Commission attempts to get governments to agree on international minimum standards for pesticide and chemical residues in food. It publishes the *Codex Alimentarius List* (Codex) which forms the basis of international Maximum Residue Limits (MRLS) set by the FAO.[6]

International Labour Organisation

The ILO lays down standards and conditions on work and employment, including health and safety, in the form of non-binding Conventions. A Convention on Safety in the Use of Chemicals at Work was adopted in June 1990.[7] This includes safety at pesticide manufacturing sites.

The UN Consolidated List

The UN Consolidated List of Products whose consumption and/or use have been Banned, Withdrawn, Severely Restricted or Not Approved by Governments is a very useful directory published by the United Nations General Assembly intermittently, and includes trade and regulatory data from all countries on over 500 hazardous products, one-third of which are pesticides. Most countries, including the UK, operate some list of banned or severely restricted chemical substances. The substances are categorised, for convenience, as pesticides, industrial chemicals or dual usage products.

The Organisation for Economic Cooperation and Development

The OECD functions as the main forum for international cooperation between the US, Europe and Japan on chemical control legislation. The major programmes of the OECD in the field of chemical controls are set out in the *Guiding Principles on Information Exchange Related to Export of Banned or Severely Restricted Chemicals*.[8] These Guiding Principles provide for one-time notification whenever a government takes action to control a pesticide, and the first time a banned or severely restricted chemical is about to be exported. Information on why the pesticide is restricted is only supplied if requested by the importing country. The OECD now recommends compliance with the FAO/UNEP PIC procedure.

The OECD *International Glossary of Key Terms in Chemicals Control Legislation* has helped define legal terms in order to avoid international inconsistencies and barriers to trade.[9]

In 1989 the OECD announced a scheme to investigate the occupational and environmental toxicity of almost 1500 chemical compounds for which little or no toxicological data exists, the Heavy Volumes Project. The 1,500 compounds account for 95 per cent by volume of all the chemicals used around the world. The project focuses on chemicals that are produced in bulk, including pesticides, as these are the ones to which humans and the environment are most likely to be exposed. The compounds under investigation were in common use before governments first introduced laws forcing companies to screen new compounds. The chemicals therefore escaped toxicological scrutiny and as an OECD spokesperson stated: 'the chemicals could be extremely safe or highly toxic; we simply do not know.'[10]

In summary, the three international schemes agreed as codes for the production, import, export and use of pesticides are the FAO Code, the UNEP London Guidelines, and the joint Prior Informed Consent provisions. International control, however, remains weak, and lack of export controls on products and hazardous technologies is an unresolved problem.

European Community Law

European law is having an increasing impact on the national legislation and regulatory frameworks of the twelve Member States. The creation of the Single European Market in 1992, based on the Single European Act 1987, is likely to further accelerate this process. The central aim of the EC is the creation of a free market for the movement of goods and services between Member States. Health, safety and environmental laws cannot be used as a barrier to free trade within the EC. As a result there is a potential source of conflict between health, safety and environmental requirements and those of free trade. The harmonisation of these requirements could mean that the lowest standards of the most poorly regulated member state are adopted as the norm. Whether this will be the case in practice remains to be seen.

The main categories of EC legislation are:

Regulations – are directly binding on all Member States and take precedence over any existing or conflicting national laws

Directives – are binding on Member States in terms of the results to be achieved, but it is left to each State to implement the Directive in its own way through national legislation.

Decisions – are binding on Member States, or the companies, institutions, individuals to whom they are addressed.

Recommendations and *Opinions* – are guidelines which do not have to be translated into legislation and are not legally enforceable; in the strict sense they are not really legislation at all.

Pesticides and the EC

The EC's own legislation on pesticides and related chemicals is based on a series of Directives, some of which have then been translated into national legislation. The Directives can be divided into five broad subject areas: registration and approval; use; residues; information, and exports.

Registration

The first attempt at EC harmonisation of pesticide approvals and testing took place in 1976, when a 'tandem' Directive was issued, aiming to establish one list of approved products and another list containing prohibited substances. In the event, only the half of the Directive (the Prohibitions Directive), containing the list of banned substances was agreed.[11] The Prohibitions Directive has now been replaced by the 1989 EC Directive on plant protection products, which only applies to pesticides used in agriculture and horticulture.[12] This draft Directive establishes a base list of authorised substances used as active ingredients in pesticides and requires Member States to agree common standards and protocols for testing pesticides and sharing information. A procedure is set out for applicants to add new ingredients to the list of authorised substances, and for reviewing all substances within ten years of the implementation of the Directive. The essential proposals are:

(1) establishment of an EC authorised list of pesticide active ingredients – used only for plant protection – which will be tested and approved to EC safety and efficacy standards;

(2) vetting of toxicological data and registration of pesticide active ingredients to be carried out by any of the national pesticide registration authorities, which will remain in existence;

(3) safety review of older pesticide active ingredients over a ten-year period; only ingredients meeting modern safety testing standards will be placed on the authorised list;

(4) sale of the 420 active ingredients and products made from them and currently available in EC countries to still be allowed during this ten-year review period;

(5) 'mutual acceptance' of authorised plant protection products in trade between Member States;

(6) Member States will *not* be able to refuse the import or use of authorised pesticide products on health and safety or environmental grounds; the only criteria for refusing to accept an authorised pesticide product is on special grounds of agricultural or climatic conditions;

(7) an EC Standing Committee on Plant Protection will monitor the registration and authorization of pesticide active ingredients and products in EC countries;

(8) there will be a ten-year period of protection for manufacturers' safety data for new active ingredients, and a five-year protection period for review information;

(9) there is controversy between EC countries over the confidentiality of safety data. Some countries want full public access to basic toxicological, and review data, while other governments, and sections of the pesticide industry, are arguing for strict confidentiality and safeguards;

(10) the marketing of genetically-engineered micro-organisms used as pesticides can be authorised, provided that they have first received a clearance for release under Biotechnology Directive 90/220/EEC on the deliberate release into the environment of genetically modified organisms;

(11) wood preservatives and non-agricultural pesticides are excluded, along with veterinary pesticides. However, a proposal for a Directive on the harmonisation of authorization of non-agricultural pesticides is due to be published in 1991.[13]

Use

There have been no specific EC Directives on pesticide use. However, the Control of Substances Hazardous to Health Regulations (COSHH) covers occupational pesticide manufacture, formulation and use in Britain.[14] This legislation originated from an EC Directive.

Residues

The EC has introduced a framework Directive which aims to establish a set of mandatory MRLs for fruit and vegetables, and certain other crops such as oilseeds and hops.[15] This Directive will also make it easier to prescribe and alter MRLs in food. Two earlier Directives which already set mandatory MRLs for animal products and cereals (incorporated into British Standards) will be reviewed as part of this process.

The Directive will establish a priority list for crops and pesticide active ingredients where there is currently no MRL. The UK government has proposed the pesticide active ingredients maleic hydrazide and pirimiphos-methyl as its top priorities for inclusion in this EC list. Existing MRLs for specific pesticides will be systematically reviewed to see whether the levels should be altered.

Freedom of Information

A Directive on freedom of access to information on the environment will enable individuals to obtain information about the environment held by public authorities, subject to safeguards in respect of trade and industrial confidentiality.[16] Governments will also be required to publish at least every three years, from 1 January 1992,

> a report on the state of the environment containing a general analysis of the national situation, the state of water, air, soil, flora, fauna, and natural sites, and a description of the principal measures taken, or planned to preserve, protect, and improve the quality of the environment and to repair any damage caused.[17]

Pesticide Exports

The EC, as opposed to the OECD, has now become the forum where the pesticide-exporting Member States debate how much responsibility and regulation they are willing to assume for the international pesticide trade. EC countries account for 60 per cent of world pesticide exports, and so have a considerable influence. While the OECD has recommended that its members adopt the FAO/UNEP PIC provisions, the EC has developed a separate set of guidelines. Its guidelines to Member States will be weaker than PIC: importing countries must be informed of hazards of chemicals on the EC list, but need not consent to their import before the goods are exported.[18] However, the EC participates in the FAO/UNEP discussions relating to codes on information exchange, distribution, and use, including those relating to PIC.

United Kingdom Law

A bewildering array of Acts, Regulations, Codes of Practice, Orders and Consents govern pesticides in Great Britain. In addition to the HSE and MAFF, the other government departments directly concerned with pesticides include the Scottish and Welsh Offices, the Departments of Environment, Health, Employment, and Trade and Industry. Special arrangements apply to Northern Ireland, as discussed on page 239.

Pesticides have only recently been subject to the type of control and laws which have regulated other hazardous substances. In the past, voluntary schemes such as the Pesticides Safety Precautions Scheme and some regulations of limited scope were used to control pesticide testing and safety.[19] The inadequacy of the voluntary schemes combined with growing public pressure forced the government to legislate. The laws on pesticides are best explained by reference to four areas: pesticide approvals; pesticide use and related areas; environmental and consumer protection; and enforcement.

Pesticide Approvals

It is now illegal to sell, supply, store, advertise or use a pesticide unless it has been granted government approval under a statutory framework regulated by the Food and Environment Protection Act 1985 (FEPA),[20] and the Control of Pesticide Regulations 1986 (COPR).[21] All approvals are granted by the government in response to an application from a manufacturer, formulator, importer or distributor (or in certain circumstances, a user) supported by the necessary data on safety, efficacy and humaneness. The pesticide ingredient and/or formulated, ready-to-use product is approved by the ACP,[22] which reviews the data, supported by the HSE and MAFF Secretariats. There is no independent government toxicity testing scheme.

The statutory framework for the approval of pesticides is laid down in Part III of FEPA and COPR. Part III of FEPA, Section 16 (1) states:

The provisions of this Act shall have effect:

(a) with a view to the continuous development of means
   • to protect the health of human beings, creatures and plants
   • to safeguard the environment
   • to secure safe, efficient, and humane methods of controlling pests
(b) with a view to making information about pesticides available to the public.

The specific requirements governing approval are laid down in COPR, made under the authority of FEPA. COPR explains what approvals are required before a pesticide can be sold, stored, supplied, used or advertised. COPR also covers testing and labelling requirements. The procedures to be followed are set out in legal Consents, which are updated and amended as required. Consents are published in the official gazettes, in the *Pesticides Register*,[23] and reproduced each year in the HSE/MAFF Reference Book 500, *Pesticides*.[24] The following Consents have been issued to date:

- Consent A: Advertisement of Pesticides
- Consent B: Sale, Supply and Storage of Pesticides
- Consent C: (i) Use of Pesticides; (ii) Aerial Application of Pesticides

A FEPA-specific Code of Practice covering suppliers of pesticides to agriculture, horticulture and forestry has also been agreed.[25]

Legal Definitions of Pesticides

The legal definition of a pesticide for the purpose of approval, use, storage, sale, supply and advertisement (but not manufacture, formulation or disposal) is given by COPR. According to COPR, pesticides are any substance, preparation, or organism prepared or used for:

- protecting plants or wood from disease-causing organisms or harmful creatures;
- regulating plant growth;
- controlling harmful organisms in water systems, buildings or other structures, or on manufactured products;
- protecting animals against external parasites (ectoparasites).

COPR does NOT apply to:

- organisms other than bacteria, protozoa, fungi, viruses and mycoplasmas used for destroying or controlling pests;
- substances whose use or sale within the United Kingdom is controlled under any of the following laws – the Medicines Act 1968, the Agriculture Act 1970 Part IV, the Food Act 1990, the Food and Drugs (Scotland) Act 1956, and the Cosmetic Products (Safety) Regulations 1984.

Also exempt are:

- disinfectants, bleaching agents or sterilants or any substance (including water) other than soils, compost or other growing medium;
- products used for the micropropagation of plants or substances used in the production of novel foods;
- plant growth stimulants;
- products used in anaerobic fermentation of silage;
- pesticides in adhesive pastes, decorative paper, textiles, metal

working fluids, paints, insect repellents, water supply systems, swimming pools, those intended solely for export, or used as part of a manufacturing process.

Pesticide active ingredients and products falling outside the COPR definition and therefore the Approval scheme, have to be labelled in accordance with the requirements of The Classification, Packaging and Labelling of Dangerous Substances Regulations 1984 (CPL) which apply to all types of industrial chemicals and preparations, including chemical intermediates and pesticides used in manufactured products.[26] They also have to meet the safety testing requirements of The Notification of New Substances Regulations 1982 (NNSR).[27] Pesticides, as defined by COPR, are exempt from the CPL and NNSR regulations.

Labelling Requirements/Conditions of Approval

Approvals are normally granted for individual products and only for specified uses. Exceptions to this rule are the approval of commodity chemicals, such as formaldehyde, methyl bromide, sulphuric acid and urea, which have a variety of pesticide and non-pesticide uses, and are generally sold or supplied as substances rather than pesticide products. These substances can be legally used as pesticides (but not sold, supplied or advertised as such).

Conditions of approval differ slightly from product to product but are normally set out on the label (apart from off-label approvals). Labels may include special storage or supply conditions, but these are rare. Conditions are legal requirements which usually cover and define:

(1) field of use – whether it is use within broad categories such as agriculture, horticulture, or forestry, wood preservation, public hygiene, home, garden or the like, or more specific areas within these;

(2) crops, plants, or surfaces on which it may be used;

(3) the maximum application rate;

(4) the maximum number of treatments;

(5) the latest time of application;

(6) any limitation on area or quantity allowed to be treated;

(7) any statements about operator protection or requirement for operator training;

(8) any statements about environmental protection.[28]

(N.B. Other specific prohibitions may be set depending upon the product concerned.)

Approvals for off-label uses allow pesticide users to apply to have the conditions of approval for certain pesticides extended beyond those stated on the product labels. Extended approvals may have additional specific conditions. Off-label approval is undertaken at the user's choosing, and the manufacturer is not liable for damages.[29]

Veterinary Products

Pesticides used on animals are classed as animal medicines and, along with other veterinary products, are deemed 'medicinal products' under section 130 of the Medicines Act 1968.[30] The manufacture, sale, supply and importation, but not use, of veterinary products is covered by the Medicines Act (and not FEPA). The Licensing Authority is the Minister of Agriculture and the Health Minister acting together who take advice on efficacy and safety from the government-appointed Veterinary Products Committee (VPC) – an equivalent body to the Advisory Committee on Pesticides.[31]

Once a veterinary product has been granted a licence it must be labelled under the Medicines (Labelling) Regulations 1976.[32] In broad terms these are comparable to the Health and Safety: Classification, Packaging and Labelling Regulations in that they require the identification of the active ingredient, directions for safe use, warnings, precautions and information about any special requirements for storage or handling.

Use of veterinary products is not regulated under the Medicines Act as vets are at liberty to alter dose rates or use on animals according to their clinical judgement. The use of veterinary products on farms, riding stables, dog kennels etc. is subject to the Health and Safety at Work Act (HSWA) 1974,[33] and COSHH. Under COSHH, veterinary products are now subject to specific controls relating to use and protection of the user. The HSE and MAFF are currently reviewing amendments to veterinary product labels to indicate where the COSHH Regulations apply.

Approval System Data Requirements

To obtain approval for a pesticide, the manufacturer has to provide toxicological data and other scientific information to the government. The information usually covers 4–10 years of toxicological testing under controlled laboratory conditions. This information includes animal studies on the short- and long-term toxicity of the pesticide active ingredients and other formulation products, cumulative effects, and any others which might emerge after a period of time. The requirements for data on safety, efficacy (efficiency) and humaneness are set out in a government document, *Data Requirements for Approval Under the COPR*.[34] The data is supplied by the pesticide manufacturers and vetted by the government. Examples of the types of data required are given on page 237.

Advisory Committee on Pesticides

This data is submitted to the government-appointed ACP and its Scientific Sub-Committee, who review it. The ACP and the Scientific Sub-Committee both have medical and scientific experts who assess risks to humans, animals and the environment, provide advice and make recommendations to the government on approvals, reviews and revocations (withdrawals). The ACP has no representation from the trade unions, consumer and environmental groups, or industry, all of which have a legitimate interest in pesticide safety.

The ACP has, however, set up a number of expert advisory panels to examine various issues in more depth to which the Trades Union Congress (TUC) and Confederation of British Industries (CBI) can nominate representatives. There are currently four main expert panels (plus working groups) and these are:

- Pesticide and Application Technology Panel
- Pesticide Container and Labelling Panel
- Medical and Toxicological Panel
- Environment Panel – restricted to government appointees only.[35]

Stages of Approval

Different stages of approval can be granted as follows :

- Experimental Permit – when the pesticide is still under trial and development;
- Provisional Approval – where further data is required;
- Full Approval – full commercial clearance as per the label requirements.

The ACP also carries out safety reviews and evaluations of pesticides and can recommend revocation or amendment of approvals.

Pesticide Use and Related Areas

The two main laws regulating pesticide use in Britain are the Control of Substances Hazardous to Health Regulations, 1988 (COSHH), and the Food and Environmental Protection Act, 1985.

The COSHH regulations are occupational health and safety laws which apply to pesticide manufacture, formulation, storage and use. COSHH covers more than just pesticides; it applies to other hazardous industrial chemicals, dusts and disease-causing micro-organisms. The COSHH regulations are made under the authority of the Health and Safety at Work Act 1974, so all the general legal requirements of this earlier legislation also apply. The Regulations are accompanied by a COSHH Approved Code of Practice which lays out in more detail what is required of employers, the self-employed and employees.[36]

FEPA (The Food and Environment Protection Act 1985) and COPR (The Control of Pesticides Regulations 1986) made under its authority, regulates the use, approval, sale, supply, storage and advertising of pesticides and sets down environmental standards.

In the area of pesticide use the COSHH and FEPA laws overlap and have different legal requirements, based as they are on different philosophies of control of hazardous substances. The COSHH

Information Required on the Toxicity of Pesticides in Animals Under the 1985 Food and Environment Protection Act[37]

The following list of tests is not necessary for all pesticides; some sections may not be relevant. Other pesticides may, however, require more extensive investigation if, for instance, there was evidence of damage to the nervous or immune systems.

(For acute toxicity in mammals, species and laboratory strain must be stated as well as a brief statement of the mode of action, for example uncoupler, inhibition of cholinesterase.)

- oral single-dose toxicity LD 50
- percutaneous toxicity (toxicity following skin absorption)
- inhalation toxicity
- delayed effects in animals which have recovered from a near-lethal dose by other routes, for example directly into abdominal cavity, intravenous, into the eye etc.
- skin and eye irritation
- short-term toxicity
- oral administration (up to 90 days)
- allergic sensitization
- inhalation
- other routes
- supplementary toxicological studies
- toxic effects of metabolites from treated plants and of impurities
- metabolic studies
- accumulation of compound in tissue and cumulative effects
- neurotoxicity (damage to nervous system)
- ability to cause mutation
- reproduction studies including effect on fertility etc. A risk of malformation or toxicity to the developing foetus
- long-term toxicity and/or carcinogenicity
- potentiation (does it increase the risk of exposure to other chemicals)
- effects on livestock, poultry etc.?

Toxicity classification

- adequate toxicity data to enable classification of the product

Human toxicology

- modes of action in humans
- health records of occupational exposed workers; in both industry and agriculture
- accidental, suicidal etc. poisoning cases
- observations on exposure of the general population
- experiments with volunteers

Diagnosis and treatment

- symptoms, specific signs of poisoning
- laboratory findings, clinical tests, forensic methods
- proposed treatment: first aid measures, antidotes, medical treatment prognosis
- information on previously reported cases

Information on environment and wildlife hazards

- toxicity to birds
- acute toxicity
- other information about harm to bird species , for example population studies, impaired fertility, residues in tissues of dead or living birds and their eggs

Toxicity to fish

- acute toxicity
- other information on direct harm to fish, for example fish kills, population studies, residue studies
- effects on water quality, fish food and other aquatic organisms

Harmful effects on other wildlife, mammals, amphibia, reptiles

- toxicity to honeybees.
- toxicity to beneficial insects etc.
- field trials and observations.
- toxicity to earthworms and other soil invertebrates, changes in soil ecology, micro-organisms etc.
- information on existing approvals in other countries
- tolerances.

Industry carries out this testing at present and presents the results to MAFF. All of the data remains secret. Demands on MAFF to publish the information it uses for assessment are considerable and likely to continue.

Very few of the pesticides that have been sold for 15 years or more have been rigorously tested. It is essential that they are all tested properly. With its limited resources it will take MAFF more than 20 years to review the safety of the pesticides in use today.

Regulations have been drawn up by the HSE while FEPA/COPR is the responsibility of MAFF.

Because of the overlap of COSHH with FEPA, there are two main Codes of Practice covering pesticide use; a third Code on fumigants may also apply. These are:

Pesticides in Northern Ireland

Different provisions apply to Northern Ireland. The Health and Safety at Work Order (Northern Ireland) 1978, closely follows the 1974 Health and Safety at Work Act in Great Britain, but there are differences in the administrative structure which can have important practical implications.[38] The Northern Ireland Order set up the Health and Safety Agency, but it is mainly an advisory body unlike its British counterpart, the Health and Safety Commission, which has considerably more power and independence.[39] This difference is largely because the Health and Safety Inspectorate in Northern Ireland (the British equivalent is the Health and Safety Executive) is part of the Department of Economic Development, as is the Health and Safety Agency itself. The Inspectorate tends to see its role as essentially one of promotion rather than enforcement.[40]

- The joint COSHH/FEPA Code of Practice for the Safe Use of Pesticides on Farms and Holdings (but not forestry);[41]
- The COSHH Approved Code of Practice for the Control of Exposure to Pesticides at Work covers non-agricultural use in industry, by local authorities, wood preservative treatments and so on;[42]
- The COSHH Approved Code of Practice: Control of Substances Hazardous to Health in Fumigation Operations.[43]

This duplication of legislation confuses the user and adds to the cost of regulation. The situation is made even more confusing because pesticides used in agriculture and horticulture are not regulated in the same way as those used in forestry and for industrial purposes. The joint COSHH/FEPA Pesticide Code only regulates the agricultural and horticultural use of pesticides,[44] whilst the COSHH Code covers non-agricultural use of pesticides.[45] There is concern that agricultural and horticultural users of pesticides may have lower standards of protection than their industrial counterparts.[46] From the user point of view the COSHH legislation provides higher standards of protection as it emphasises the control of hazards at source.

Under COSHH, employers have to carry out a risk assessment before allowing employees to be exposed to pesticides. Prevention and control measures based on technical and engineering controls have to be used to reduce exposure before the use of personal protective equipment (PPE) can be considered.

By contrast, FEPA relies heavily on PPE as the main method of operator protection: a less desirable and effective control method. As discussed in Chapter 4, PPE only provides partial protection and is uncomfortable to wear for long periods. PPE should only be used to supplement other control measures and for a short time.

What is 'Approved'?

NOTE: The word 'Approved' is important in health and safety law. Codes of Practice, 'approved' by the government-appointed Health and Safety Commission (HSC), are not *laws* but they are *admissible* in law as evidence of recommended practice.

A summary of the main differences in terms of user protection between the COSHH and FEPA legislation is given in Table 11.2.

Other Legal Requirements Concerning Use

User Training

The COSHH Regulations, especially as they are made under the Health and Safety at Work Act, place a general duty on employers to provide adequate training for employees using or exposed to pesticides, and to pay for it.

Training is vital not only for the safety of the operator but also to minimise risk to the public, wildlife and environment from poor spraying techniques. As well as in-house training, there may be a need to use qualified instructors from outside bodies like the Agricultural Training Board for either basic or refresher courses.[47]

In terms of training, FEPA is much weaker as it only requires employers to provide adequate instruction and guidance to their employees using pesticides. This is not the same as a legal requirement for training, and will mean variable standards of competence and proficiency amongst spray operators, as some employers may choose not to train their workforce in safe procedures. Under FEPA the *user* will be blamed if pesticides are not used safely.

Training standards for pesticides used in agriculture and horticulture are laid down in the joint COSHH/FEPA Code of Practice.[48]

HSC has also published training standards for non-agricultural pesticide users which cover wood preservatives, public hygiene pest control, rodenticides, surface biocides, anti-fouling paints (including fish farms), fumigation, and food storage.[49]

Certification

FEPA, via COPR, does require certain groups to obtain Certificates of Competence, however, and this implies the need for training.[50] The Consent on Use specifies who needs a certificate and lists the following, unless they are working under the direct, personal supervision of a certificate holder:

- agricultural, horticultural and forestry contractors and anyone carrying out a commercial service not on their own (regardless of age or experience);

Table 11.2: Prevention and Control Checklist Requirements

REQUIREMENTS

COSHH	FEPA
RISK ASSESSMENT The employer has to carry out a risk assessment, using a competent person, before exposing employees to substances hazardous to health. The risk assessment should cover all the areas listed below and, in particular, specify what prevention and control measures are required.	**RISK ASSESSMENT** No requirement on employers to carry out a risk assessment.
PREVENTION **Elimination** – the employer has first of all to consider if use of/exposure to the substance can be eliminated altogether. 'Does the pesticide have to be used at all?' should be the first question.	**PREVENTION** **Elimination** – there is no specific emphasis in FEPA on eliminating or reducing pesticide use.
Substitution – if elimination is not possible, can a safer, substitute chemical be used; or even a safer formulation of the existing chemical?	**Substitution** – once again there is no specific emphasis on substitution.
CONTROL If it is not possible to prevent exposure by elimination or substitution, then the employer must ensure there is 'adequate control' of exposure. The list of control measures should be implemented in the order that they are written, starting with:	**CONTROL** There is no emphasis on a hierarchy of control measures starting with technical and engineering controls. The main method of operator protection under FEPA remains the use of personal protective equipment – the least effective and desirable control method.
Technical and Engineering Controls – should be the main control method when dealing with chemicals, dusts etc. Such controls include local exhaust ventilation, sealed mixing and filling systems, forced air filtration units in tractor cabs etc.	**Technical and Engineering Controls** – there is no specific requirement for employer to use such controls as the first step. However, from 1992 all pesticide labels approved under COPR will require the use of such controls to be considered.[51]

Operational Controls – the employer must also consider safe working practices and systems to reduce exposure. Measures such as keeping all non-essential personnel away, reducing length of exposure etc.

Operational Controls – there is no specific emphasis on safe systems/methods of work.

Personal Protective Equipment (PPE) – is the least desirable and effective control method. PPE should only be used to supplement the control measures listed above. It should not be the first line of operator protection.

Personal Protective Equipment (PPE) – is the main control method under FEPA which is why it is poorer legislation from the viewpoint of operator protection.

MAINTENANCE OF CONTROL MEASURES
Employers have to maintain the control measures. Planned maintenance is essential for technical and engineering controls. Respiratory protective equipment has to be maintained. Records of maintenance have to be kept by the employer.

MAINTENANCE OF CONTROL MEASURES
There is no specific requirement on employers to maintain controls or keep records.

MONITORING
Monitoring has to be carried out periodically to check that the control measures are still working effectively. Employers have to work out what kind of monitoring of airborne concentrations of substances hazardous to health is required, how often it should be done and by whom.

MONITORING
There is no requirement on employers to monitor for pesticides.

HEALTH SURVEILLANCE
For certain categories of substances, including organophosphorus pesticides, COSHH specifies that health surveillance measures must be carried out.

HEALTH SURVEILLANCE
There is no requirement for employers to carry out health surveillance of their workforce.

RECORD-KEEPING
Under COSHH, employers have to keep records of maintenance, monitoring, health surveillance and training.

RECORD-KEEPING
There are no requirements for employers to keep records or make them available to their employees.

INFORMATION

Employers have to provide employees with the following categories of information:

- label information
- product safety data sheets
- maintenance records
- monitoring records
- health surveillance (personal records only)
- training (individual)
- conclusions of any risk assessments and, in particular, what is expected of the workforce

TRAINING

Employers must provide 'suitable and sufficient' training for their employees so they can effectively apply and use:

- control methods
- personal protective equipment
- emergency measures

Management and supervision must receive specific training where they have been designated as 'competent persons'

INFORMATION

Information provision requirements are not spelled out in any detail.

TRAINING

Employers have to provide 'adequate instruction and guidance' for their employees. The need for training is implied but there is no specific legal requirement in FEPA either for employers to provide training or to pay for it.

- agricultural, horticultural and forestry spray operators who were under 25 on 1 January 1989 (that is, born after 31 December, 1964);
- anyone selling or supplying pesticides approved for use in agriculture, horticulture and forestry;
- anyone storing approved pesticides for the purpose of sale or supply. The storage certificate provided by the British Agrochemicals Standard Inspection Scheme (BASIS) is a recognised standard.[52]

Pesticide Sale and Supply

COPR imposes general legal obligations on anyone selling or supplying pesticides to take all reasonable precautions to protect the health of human beings, creatures and plants, and to safeguard the environment. They must also be qualified and will have to obtain a Certificate of Competence based on standards laid down in the FEPA Code of Practice for Suppliers of Pesticides.[53] A special Certificate of Competence for the sale of forestry pesticides is available for those advising in this area who do not require the range of other subjects covered by the existing certificate.[54]

Pesticide Storage

Storage by users on farms, in horticulture, by local authorities, contractors and the like is dealt with in an HSE Guidance Note.[55] The standards for storage sheds are set out in detail in Chapter 4. For smaller quantities, a chest, bin, vault or cabinet may be used, though these should be kept in secure, well-ventilated and well-lit premises.

A 1989 HSE survey of 870 farm pesticide stores found that more than 60 per cent did not fully meet the specified standards; in 73 per cent of the inspections some form of enforcement was taken by health and safety inspectors.[56]

Advertising

Consent A under FEPA regulates the advertising of pesticides.[57] Only provisional, or fully approved, pesticides may be advertised and only in relation to their approved uses. All printed and broadcast advertisements must mention the active ingredients in the pesticides, include a general warning phrase, and include any specific warning phrases considered necessary.

Aerial Spraying

Aerial spraying is also regulated under FEPA. Operators must hold an aerial application certificate granted under Article 42 (ii) of the Air Navigation Order 1985.[58] Only pesticides specifically approved for aerial application may be applied from the air, whether from a fixed-wing aircraft or helicopters. The Civil Aviation Authority provides a list of pesticides approved for aerial application.[59]

Under FEPA Consent C (ii) the following legal requirements apply to aerial spraying:

- only pesticides specifically approved for aerial application may be applied from the air;
- spraying should not be carried out within 200 feet of an occupied building or garden, nor below 250 feet over any motorway or below 100 feet over any public highway;
- at least one day before spraying, contractors must notify: the occupants of each building and the owner or his agent of any livestock or crops within 75 feet of the boundary of the land where spraying takes place; any hospital, school or other institution (including the grounds) any part of which lies within 500 feet of the intended flight path
- contractors must consult the Nature Conservancy Council if spraying near Nature Reserves or Sites of Special Scientific Interest or the Water Authority if spraying near water;
- the local beekeepers' spray warning scheme should also be consulted at least two days prior to aerial spraying.

Pesticide Manufacture and Formulation

Pesticide production and formulation is regulated by COSHH, and pesticides are treated as general industrial chemicals and regulated to the same standards.

To minimize the risks of major chemical accidents, certain pesticide production and/or formulation plants, along with other chemical factories and warehouses, are regulated by the Control of Industrial Major Accident Hazard Regulations 1984 (CIMAH).[60] The CIMAH Regulations require notification of sites to the Health and Safety Executive where certain dangerous substances are manufactured, formulated, or stored. The persons running, or responsible for, these sites have to prepare on-site and off-site emergency plans to prevent or minimize the consequences of accidents which could harm the workforce, local community, or environment. The CIMAH Regulations implement an EC Council Directive on major accident hazards, also known as the 'Seveso Directive',[61] introduced following the release of sodium trichlorophenate and the most toxic dioxin (2,3,7,8 – tetra-chlorodibenzodioxin) from a chemical plant over a residential area in the Italian town of Seveso in 1976.[62]

Pesticide Transport by Road

Various health and safety regulations, made under the authority of the Health and Safety at Work Act, control the transport and delivery of pesticides and other dangerous substances. The regulations which apply depend on the quantity of pesticide being transported, and in what form:

(1) Regulations on bulk transport in road tankers and tank containers require vehicles to display hazard warning panels and for driver/operators to be properly trained and to carry health and safety emergency information with them in the vehicle.[63] Agricultural crop sprayers, travelling on public roads, are covered by these regulations if the sprayer unit is structurally attached to the main vehicle.

(2) Regulations on transport of pesticides in packages set out the duties of operators and drivers of vehicles carrying packages of more than 200 litres.[64]

Harbours and Pesticides

Another set of dangerous substances regulations provide a uniform and comprehensive framework of controls over the carriage, loading, unloading and storage of dangerous substances, including pesticides, in harbours and harbour areas.[65]

The main features of these regulations are:

- ports must be given prior notice of arrival of dangerous substances from inland destinations; the transport operator is required to inform the Harbour Master at least 24 hours before arrival;
- vessels carrying certain dangerous substances must show a red flag or light;
- freight containers and portable tanks must be suitable for the purpose, and be properly packed and labelled;
- every harbour must prepare an emergency plan.

Environment and Consumer Protection

Environmental controls for the protection of plants and wildlife are laid down under FEPA and COPR. Environmental controls are often a condition of pesticide approval and are usually specified on the label. Detailed requirements for professional pesticide users are contained in the COSHH/FEPA Code of Practice on Pesticides. The idea is that employers, self-employed and other pesticide users should assess the environmental risks from pesticides. The environmental risk assessment should identify the problems and what prevention and/or control measures are needed to avoid, or minimize, damage to wildlife, or pollution of land, or water.

The COSHH/FEPA Code highlights the following reasonable precautions:

- do not exceed the maximum, legal dose when spraying;
- avoid spraying round field margins which are important plant and wildlife habitats, particularly if hedges and/or ditches are present;
- advise local beekeepers 48 hours in advance of any intention to spray;
- warn local neighbours in advance so they can take precautions such as keeping pets and livestock under cover;
- clean and decontaminate carefully spraying equipment using proper soakaways and avoiding watercourses;
- make proper arrangements for the disposal of waste pesticide and empty, washed containers;
- consult the Nature Conservancy Council if it is proposed to apply pesticides in the vicinity of a Site of Special Scientific Interest, National Nature Reserve or other area of nature conservation interest.

Pesticide Residues in Food

The Pesticide (Maximum Residues Levels in Food) Regulations 1988, made under FEPA, introduced legal residue limits for some 60 or so pesticides used on specific crops, as discussed in greater detail in Chapter 6.

Pesticide Residues in Water

Unlike food, pesticide residues in water are regulated under EC rules as incorporated in the Water Act 1989, not under FEPA (see Chapter 6)

Pesticide Disposal

Safe disposal of pesticides and pesticide containers is still a grey area and the source of considerable confusion. We have given guidance on required practice in Chapter 4. Various types of disposal are involved:

Disposal of revoked pesticide: this arises when the government has withdrawn (revoked) the approval for a pesticide on safety grounds, or because the manufacturer has not provided certain information. In either case, the pesticide has to be disposed of safely but it is not made clear who should do this. There is a good case to be made that the responsibility for safe disposal of banned pesticide should lie with the manufacturer or supplier who should be legally required to collect stocks of revoked pesticides for safe disposal. This process is referred to as 'reverse chain' disposal. The responsibilities of manufacturers and suppliers in this respect have still to be clarified. The COSHH/FEPA Code refers to safe disposal but does not specify the duties of manufacturers or suppliers in regard to reverse chain disposal.

Disposal of approved surplus or waste pesticide concentrate, and of empty pesticide containers is another problem where there is conflicting advice and regulation. Concentrated pesticide waste, in whatever form, should be safely disposed of by reverse chain disposal, authorised waste disposal contractors, or by delivery to specially licensed, local authority-controlled waste tips. Guidance on disposal is given in the COSHH/FEPA Code and has been written in conjunction with the Department of the Environment, which has overall responsibility for pollution control and chemical waste disposal via the following laws:

The Control of Pollution Act 1974,[66] and the Special Waste Regulations 1981 made under this Act,[67] define two main categories of waste:

(1) controlled waste from industrial, commercial and household sources, and
(2) special waste, where the controlled waste is especially hazardous to human health; extra requirements apply for this waste; most pesticides are classified as special waste under the Regulations and Act.

Special waste must be disposed of at special waste sites licensed and operated by the local authority. Local authorities have the details of sites licensed to accept special waste. However, waste from premises used for agriculture (within the meaning of the 1947 Agriculture Act) is excluded from the definition of controlled waste, and hence special waste.[68] Agricultural waste is interpreted as waste produced on a farm by a farmer, and disposed of on her/his own land. Lower standards govern the disposal of agricultural waste and farm pesticide waste than waste arising from industrial and commercial sources.

Disposal of waste and dilute pesticide: once again rules for safe disposal have not been finalised, though guidance is given in the COSHH/FEPA Code. At the time of writing, it is unclear if the Environment Protection Act 1990 will modify any of the requirements on pesticide disposal.

Genetic Modification (Biotechnology) and Pesticides

The main regulations controlling genetic modifications are:

(1) the Genetic Manipulation Regulations 1989 which require the HSE to be notified of work involving genetic modification,[69] those undertaking such work must assess the risks using an approved method and set up a local genetic modification committee to advise on these assessments;[70]

(2) Part 6 of the Environment Protection Act 1990 deals with biotechnology and controls the planned release to the environment of genetically modified plants, animals, microorganisms and medicines;[71]

(3) FEPA also applies to genetically modified pesticides, and the ACP will review approvals.

Enforcement

Laws require enforcement if they are to have any meaning and achieve their purpose. Responsibility for enforcement of pesticide laws is divided between the HSE and its various inspectorates, and local authorities. In recent years both have suffered cutbacks in the number of enforcement officers, support staff and resources while, paradoxically, the number and complexity of laws of enforcement have increased. The result has been fewer health and safety inspections. In view of this trend there must be doubts over the effective implementation of the latest pesticide laws.

The Health and Safety Executive (HSE)

The HSE is the government organisation responsible for workplace health and safety in Britain, as well as preventing harm to the public arising from work activities. In conjunction with the HSC, the HSE drew up and implements COSHH; it also issues technical guidance and advice. HSE inspectors visit workplaces to check that laws are being implemented. Two inspectorates in the HSE deal with pesticides: the Agricultural (AI) and Factory Inspectorates.

HSE Agricultural Inspectorate

Most pesticide enforcement fall within the remit of the AI. Some 160 Agricultural Inspectors carry out inspections on farms, in horticulture and in forestry.[72] These inspectors enforce all health and safety laws, not simply pesticides. They are also responsible for investigating public complaints over pesticide spraying incidents, including aerial spraying, where there is alleged to have been damage to human health or the environment.

The trades union representing health and safety inspectors, the Institution of Professionals, Managers and Specialists (IPMS), takes the view that the number of inspectors is completely inadequate to cover nearly 300,000 work premises and a workforce of 664,000, half of them self-employed farmers and their families, for which the AI has responsibility.[73]

Accident figures seem to support the IPMS claim. According to the HSE's own figures, serious accidents, let alone poisoning incidents, increased by 34 per cent in agriculture between 1981–85.[74] The upward trend seems to be continuing as, for example, the serious injury rate for agricultural employees rose from 331.7 per 100,000 employees in 1986/87 to 525.9 per 100,000 employees in 1989/90, an increase of 58 per cent.[75] These figures are likely to be an underestimate, as even the Director of the HSC acknowledges that in agriculture there is major under-reporting of accidents by employers, by perhaps 50 per cent or more.[76]

With a sharp fall in the number of inspectors since the early 1980s, routine visits to farms, horticultural premises and forests take place at longer and longer intervals.[77] The 1989 IPMS Report states: 'given that each inspector has a target of some 220 inspections per year and assuming that each labour-employing workplace is visited on a rotational basis, the inspector would call only once in every 9.8 years.'[78] This rate of inspection is too low to ensure compliance with the law and protection of the workforce and public. In fact, even the above figures do not reveal the full picture. Most routine inspections take place on labour-employing farms and premises, yet some 400,000 self-employed persons work in agriculture in nearly 200,000 work premises. They are often medium to large farms, and major users of pesticides, and their use of these chemicals poses risks not only to themselves but to the public and environment as well. Yet because of the shortage of staff they can expect a routine visit from an Agricultural Inspector only once in every 30 years – once every generation.[79]

HSE Factory Inspectorate

Factory Inspectors enforce health and safety in pesticide manufacturing, formulation, and use; use includes manufactured products such as carpets, textiles and marine paints. Under COSHH/FEPA, inspectors now have additional powers to control use by local authorities, by water, gas, electricity and dock boards, by industrial pest control firms, including timber treatments, and for vermin control in industrial food storage facilities. Factory inspectors also deal with dangers to the public arising from pesticide use in, for example, parks and gardens, and on playing fields, as well as incidents arising from pest control treatments in schools, public buildings and the home. Some 550 inspectors have to cover 16 million employees in 400,000 registered premises.[80]

Local Authorities:

Local authorities also enforce health and safety legislation in offices, shops, warehouses and commercial premises. The bulk of health and safety enforcement work, including COSHH/FEPA work, in these premises is carried out by local authority Environmental Health Officers (EHOs). They are also under-resourced, too few in number, and lack support staff to enforce health and safety as well as food hygiene legislation. The situation has worsened as local authority expenditure has been cut back.[81] As discussed in Chapter 6, there are too few local authority Public Analysts to effectively monitor food for pesticide and other chemical residues.[82]

EHOs enforce both COSHH and FEPA. The Pesticides (Fees and Enforcement) Act 1989, made under the authority of FEPA, allows government Ministers to specify which local authority officials may act as enforcement officers. The 1989 Act also allows local authorities and Port Health Authorities to authorise the individual officers concerned.

Wildlife Enforcement

MAFF – and in particular its Agricultural Development and Advisory Servic (ADAS) in England, WOADS in Wales and DAFS in Scotland – has responsibility for monitoring and enforcing wildlife incidents to mammals, birds and honeybees. MAFF has established a Wildlife Incident Investigation Scheme (WIIS) to ensure the collaboration of individuals and departments under FEPA. Field investigations are conducted by MAFF regional staff and coordinated by the MAFF Regional Wildlife and Storage Biologist in England and Wales and by Area Principal Agricultural Officers in Scotland. Specimens can be examined for disease by veterinarians, and biochemical investigations performed by staff at either the Tolworth Laboratory, MAFF, or the Agricultural Scientific Service Laboratory, DAFS. *Post mortem* examination can be carried out at the local Veterinary Investigation Centre, which forwards samples to the relevant laboratory.

The Politics of Approval, Control and Enforcement

The UK pesticide approval system is slow, inefficient and secretive. It is badly under-resourced, and lacks adequate scientific and technical staff to vet new pesticides, as well as review the older ones. There is also a strong conflict of interest involved, as MAFF tries to regulate both food production and food safety. As a result, the system has attracted widespread criticism from the pesticide industry, the NFU, trade unions and consumer, environmental and countryside groups.

Conflict of Interest

The UK Pesticide Approval system involves six government departments, so six government ministers have responsibility for pesticide registration and regulation, those for Agriculture, Employment, Environment and Health and the Secretaries of State for Scotland and Northern Ireland. They are not all equal partners, and in reality, although the government often denies this, the dominant department is MAFF.[83] MAFF runs the approval system for agricultural and horticultural pesticides; provides the main secretariat to the ACP; and draws up legislation. Historically, MAFF has been closely allied with the pesticide industry and large farmers, and this alliance has not always worked in the public interest. Relying on MAFF to help maximize food production and at the same time regulate pesticide health and safety in favour of the workforce, consumers and the public is like asking the fox to guard the chicken coop.

The HSE is the junior partner to MAFF in the approvals scheme, though it denies this is the case. The HSE is fully responsible for controlling all types of hazardous substances except pesticides, but within the approvals scheme it only deals with non-agricultural pesticides. Also the HSE is not a government department but a quango under the control of the Department of Employment, so it does not have the same direct access to Ministers as MAFF. Having both HSE and MAFF dealing with pesticide control has also led to a duplication of user legislation, namely COSHH and FEPA, which does not help the end-user.[84]

Pesticides are both an occupational and environmental hazard, and this poses problems in developing an effective regulatory structure. However, within the present regulatory framework and pending the creation of a US-style Environment Protection Agency, responsibility for pesticide approval should be fully transferred to the Health and Safety Executive, and away from the control of the Ministry of Agriculture. On the safety side, the link between pesticides and agriculture, which arose because pesticides were first called 'agrochemicals', needs to be broken. Pesticides should be regulated in the same way as other forms of hazardous substances, and by a government body which has no vested interest in food production.

Staffing and Resources

The approval system for pesticides is badly under-resourced, even though the government promised more support when FEPA became law in 1985. As a result the British Agrochemicals Association (BAA) – the pesticide manufacturers trade association – claimed that it was taking them up to three years to get a new pesticide fully approved, compared to less than a year in most other EC countries. The BAA claimed that this delay costs the industry millions of pounds in lost orders – £22 million in 1986 alone.[85] This criticism prompted the government to announce the provision of more staff and resources to speed up approvals and reviews. In May 1989, the government claimed that the waiting time for the processing of pesticide applications had gone down over the previous twelve months from 90 weeks to 47 weeks.[86] In 1990 the government announced further increases in scientific and technical staff for the approval scheme.[87]

The safety review of older pesticides, many of which would not meet modern toxicity testing requirements, is behind schedule. In 1985 the government promised to review all older pesticides within ten years. Four years later, in April 1989, the government finally published a limited review timetable, far short of the systematic review promised.[88] IPMS has estimated that, given current staffing levels, it could take 20–30 years or longer to carry out the full review promised.[89] As a result, potentially dangerous pesticides continue to be used in the UK even though their uses have been banned or phased out in other countries. In November 1990 the government announced that safety reviews of 38 pesticide active ingredients were under way as part of the routine review which extends to the 296 pesticide active ingredients approved before 1981.[90]

Secrecy

The approval system is also highly secretive and there is no public access to information. The government argues that the toxicity data, on which approvals are based, remains the legal property of the manufacturers, so it cannot be publicly released without their permission. Thus the toxicity data and the standards by which they are evaluated are not open to full public and scientific scrutiny. It is hardly surprising, therefore, that there is suspicion and lack of confidence in the system.

This suspicion is in contrast to countries where Freedom of Information Acts operate, such as the US, Canada, Australia, Denmark and Sweden. In these countries a member of the public has a right to see basic toxicity information. The secrecy in the UK stifles discussion and hampers decision-making on pesticide safety. It is indeed ironic that the government and pesticide industry often accuse critics of being 'hysterical' about pesticides, yet continue to deny them the very information that would allow a proper, calmer debate.

Enforcement and Lack of Inspectors

Owing to government under-resourcing the number of Agricultural and Factory Inspectors fell by some 20–25 per cent since 1980, though the decline in numbers levelled off towards the end of the decade.[91] As a result serious accidents rates have soared in all industries including agriculture – a 30 per cent increase in general manufacturing industry and a 45 per cent increase in construction.[92] The government claims that the HSE received the funds it asked for. The 1989 IPMS report disputes this claim:

An examination of the Treasury's Supply Estimates 1990–91 reveals that the Government is being disingenuous in its claims to have increased HSE's resources. In 1988-89 the total grant to the HSE was £122.2 million and by 1989–90 this figure had risen to £125.7 million. For 1990–91, however, the figure shown is £116.6 million – a cut of more than £9 million![93]

These figures are not reassuring. The HSE has received no extra inspectors to enforce either FEPA or COSHH (which covers all forms of hazardous substances including pesticides). The Chief Agricultural Inspector acknowledged that:

> FEPA work increases the range of knowledge and skills which inspectors use in the course of their work and it will add to the heavy demand that the investigation of pesticide complaints and suspected poisonings already places on the Inspectorate's limited resources.[94]

In 1988, the first full year of FEPA enforcement only 60 enforcement notices were served and only a handful of prosecutions planned. Furthermore, only 1,909.9 inspector days were expended on FEPA work: equivalent to fewer than ten full-time inspectors working full-time.[95]

Until the number of inspectors is significantly increased no adequate preventative inspection programme for pesticides and other areas of health and safety is possible. IPMS states:

> we also support the view that the HSE is a preventative body first and an enforcement and prosecuting body second. It is quite clear that the priority must be to prevent the accident or incident happening rather than to prosecute after the event. Only an increasing number of preventative inspections from an increasing number of experienced inspectors can achieve this aim.[96]

IPMS goes on to say: 'to meet responsibilities under the Health & Safety at Work Act, FEPA and COSHH, the number of agricultural inspectors should be increased by a minimum of 100 to 260.' Even the pesticide manufacturing industry supports this demand.[97]

This concern is now being echoed by the Health and Safety Commission and its Executive. Speaking at the press conference marking the launch of the HSC 1989/90 Annual Report, the Chairman of the HSC stated:

> this Christmas [1990] finds me cataloguing the sad facts that several hundred families were bereaved during the year by industrial accidents; over 30,000 people suffered major work-related injuries and around 160,000 workers were hurt severely enough to stay away from work for at least three days.[98]

In introducing the 1987/88 Annual Report, the same HSC Chairman stated:

> in the face of evidence of poor safety management in sectors such as construction, agriculture and quarrying and several major disasters [Zeebrugge ferry, Kings Cross fire, Piper Alpha oil platform], I cannot conclude that 1987/88 was a satisfactory year for industrial health and safety.[99]

Campaign Groups and Enforcement

As a result of these shortcomings in the regulatory system, an unprecedented alliance of organisations has come together to put pressure on the government to increase resources devoted to the approval, monitoring and enforcement of pesticides. In August 1989 the British Agrochemicals Association (BAA), Friends of the Earth, the Green Alliance, the National Federation of Women's Institutes, the Pesticides Trust and the Transport and General Workers' Union published a joint letter to Ministers responsible for pesticides.[100]

The organisations called for:

- completion of safety reviews of older pesticides by 1992;
- speeding up of the introduction of new chemicals by reducing the waiting time for approval of new pesticides to one year;
- recruitment of an extra 100 health and safety inspectors to help ensure that pesticide regulations are properly enforced;
- more frequent testing for pesticide residues, over a broader range of foodstuffs;
- a National Pesticide Incident Monitoring Scheme to replace the present four different systems for reporting incidents.

The BAA has publicly stated that it is willing to pay higher registration fees to speed up approvals and re-approvals and to provide more resources for effective enforcement of pesticide laws.[101] The government reply stated that many of the issues raised were now being effectively and speedily dealt with.[102]

What to do About a Poisoning or Pollution Incident

There is no pesticide investigation or incident centre in the UK and no central referral or monitoring service, and this can make the task of finding out which organisation or authority to inform a difficult one. Here are some suggestions on possible courses of action. In all cases keep a record of *dates*, *place* and *time* and put the matter *in writing*, keeping copies of any letters or correspondence.[103]

Types of Incidents

Poisoning or Contamination of People at Work or Home
Action

- If at home then contact either the Health and Safety Executive (HSE), your local authority environmental health department (EHD), or community health council (see telephone directory). The police can also be contacted and they should be able to put you in touch with the relevant authorities.
- Contact your doctor or the hospital. Ask for an immediate blood test. There are also specialised Poisons Centres which can be contacted.
- If you suspect a court action may result – for example against a company who may have been liable for negligence – try to get the doctor involved to make a simple written statement that poisoning seems to have occurred.
- If at work, make sure you report the accident or incident to your employer or management, and ensure that it is recorded in the accident book. If the accident or incident results in either hospitalization or your being off work for more than three days, then your employer is legally obliged to report what has happened to the relevant authorities – either the HSE or EHD.[104] Also contact your trades union – telephone your local trade union officer or office and ask about procedures.

Aerial Spraying and Ground Crop Spraying
(Including Fruit and in Greenhouses)
Action

- Obtain a witness and photograph if possible, and get the number of the aircraft or the make of the ground sprayer or tractor (or its registration number).
- Check the windspeed and record the location of the incident.
- GPs may contact the Poison Units for advice on pesticide poisoning symptoms.
- Reporting: both aerial and ground spraying complaints should be reported to the HSE Agricultural Inspectorate stating that there has been a risk to human health from spraying activity and asking them to investigate the incident. Confirm it in writing and keep a copy. Also report it to your local authority EHD. The HSE have a legal obligation to make publicly available copies of improvement and prohibition notices issued by inspectors.[105]

Local Authority Spraying

Action

- Contact your council engineers, parks and highways or similar department as appropriate and ask what pesticide(s) were involved.
- Ask if the council has a Pesticide Management Policy.
- Lobby your Councillor and MP.
- Confirm it in writing.
- Reporting: Incidents arising from the use of pesticides by local authorities in public areas such as parks and gardens, roadside verges, pavements and pest control treatments in schools, nurseries, council homes and other public buildings should be reported to the HSE Factory Inspectorate. As with the agricultural inspectorate, ask them to investigate the incident stating that you believe there has been a risk to human health. Confirm it in writing.

Walking in the Countryside

Action

- If a path is ploughed or planted with crops make sure you report it to the highway authority (county, metropolitan district or London borough council).
- If you find a pesticide on a crop over a path, especially if you see it being sprayed, report it to the local HSE agricultural inspector. If you actually suffer directly yourself, try and get your doctor to confirm the effect of the pesticides, as well as reporting to the agricultural inspector.
- Write to your MP telling her or him of your concern about the government's failure to ensure the protection of walkers in the countryside, both by failing to control the use of pesticides near paths, and by failing to stop crops being planted across them.
- Let the Ramblers' Association know if you find a pesticide on a crop over a path, or if you find a path ploughed, or planted with crops.[106]

Mammal, Bird and Bee Corpses

Action

- Veterinary surgeons should send corpses of wild or domestic animals to the local MAFF Veterinary Investigation Centre, which are part of the Wildlife Incident Investigation Scheme. Ask for the results again in writing. The RSPCA or RSPB may also be able to help.
- With bee poisonings also contact your local beekeepers' spray warning scheme and notify them of the incident.

- If aerial spraying is involved, check with your local EHO if two days' notice of aerial spraying in your area was given.

Other Contacts

Other contacts include the press; local newspapers; TV; radio; Women's Institutes; the local offices of the National Farmers' Union (NFU); Friends of the Earth; the Soil Association and the Freedom of Information Campaign (see also 'Contacts').

Conclusion

Laws governing pesticide approval, use, environmental and consumer protection, and enforcement vary greatly from country to country. We have only looked at the UK laws. In the EC, however, there are now moves to harmonize laws on approvals, use and residues. At an international level, regulations are voluntary. Enforcement of regulations on pesticide exports is weak, and Third World countries bear the brunt of this as they do not have the necessary infrastructure to regulate hazardous and toxic pesticides for health and safety. Nevertheless the law is one important way of improving health, safety and environmental standards for pesticides.

12

Alternatives to Pesticides

Introduction

The UK Royal Commission on Environmental Pollution (7th Report, September 1979) stated:

> Some farmers feel themselves to be on a treadmill with regards to pesticide usage, compelled by circumstances to depend on chemicals to an extent that they as countrymen, intuitively find disturbing. We believe that positive action aimed at reducing, or at least questioning, pesticide usage would be welcomed by many.[1]

Evidence is also mounting to show that pesticides, far from solving pest problems, may aggravate existing ones and even create new pests. Regular pesticide use can lead to dependence, where more and more pesticide or newer and newer active ingredients have to be used to maintain pest control levels. This phenomenon is often popularly referred to as the 'pesticide treadmill effect'.

Problems which often arise from regular and intensive pesticide usage are:

Pesticide resistance: Pesticide resistance is a major and rapidly growing problem worldwide and this, rather than worries over health, is the main impetus behind the demand for alternative pest control methods.[2] Resistant strains of weeds, insects and fungi are now common and the list is getting steadily longer (see Chapter 2). The more intensive the pesticide use the quicker the resistance develops. Pest resistance to one chemical often means resistance to other pesticides in the same chemical group.[3, 4]

Pesticide manufacturers, having spent tens of millions of pounds developing a new pesticide active ingredient, are concerned that its economic life, and hence their profits, will be cut short by pests becoming rapidly resistant to it. There is major economic pressure from manufacturers and users for alternative pest control methods, if only to lengthen the effective life of current pesticides.[5]

Creation of secondary pests: As 'pests' along with beneficial and non-harmful species are often indiscriminately controlled by broad-spectrum pesticides the 'natural balance' is upset. Previously minor pests, freed from competition by these other species, can fill the ecological niche, or gap, created and become pests in their own right.

Control of beneficial species: Insecticides and fungicides, in particular, are often non-selective and control friend as well as foe. They reduce or eliminate the control of pest species by beneficial predators, parasites and diseases.[6]

While natural controls are by no means perfect, further weakening them by use of non-selective pesticides can lead to a rapid increase of the partially controlled pest species, a phenomenon known as 'pest resurgence'. Aphid populations (blackfly, greenfly etc) can multiply much more rapidly than beneficial controlling species such as ladybird larvae, and all the more so if these natural enemies have been decimated by the aphicide.[7]

Control is also based on the notion that there is a safe level below which a chemical causes insignificant biological damage. There is growing evidence to suggest that such a threshold does not exist and that exposure of humans and wildlife to very low levels of pollutants over a long period can be as biologically damaging as exposure to very high levels over a short period. However, when dealing with chemicals that may cause cancer, reproductive or neuro-behavioural damage it is extremely difficult to establish 'safe' levels.

Alternatives

The fact that pesticides may create, or aggravate pest problems, and lead to user dependence on them, coupled with the difficulty of establishing safe levels and methods of control, has lead to the development of alternative forms of pesticides and pesticide use, and alternative systems of agriculture. The problems of control leads quite naturally to the question of alternatives. The main alternatives can be broadly categorised as: Integrated Pest Management (IPM); biotechnology; organic farming.

Integrated Pest Management

There is no one system of IPM. It is a combination of techniques used in a harmonized way. IPM is not aimed at elimination or annihilation of species but at the reduction or maintenance of pest populations at a level below those causing economic injury and loss of money. IPM is applicable to all areas of pesticide use – agriculture, public health, amenity, industrial pest control and even domestic use. IPM is sometimes referred to by other names, such as Integrated Pest Control (IPC).

The attraction of chemical control is that it provides a rapid, simple and, for a very limited period, effective solution to pest problems. However, problems soon build up in the form of pest resistance, secondary pests becoming major pests, and lower levels of natural biological control. This results in an increase in the number and frequency of sprays, and pesticides become a major factor in production costs. The extreme result is a fall in yields and chemical costs spiralling upwards.

In contrast, an effective programme of IPM is necessarily complex. It demands detailed studies of each field to determine which organisms are the key pests and which are secondary. Before any control measures are applied, growers and farmers have to sample fields or orchards, counting the number of pests, and predators, to reach an estimate of the 'economic threshold' of the major pests. This threshold is the density of a pest population below which it costs more to control the pest than accept the losses caused by it. Only if major pest numbers exceed the 'economic threshold' is treatment going to be economically worthwhile.[8]

Unlike chemical control, no two IPM treatments are likely to be identical. There can be a variety of ways to control major pests, depending on the nature of the pest and the ecological, environmental and climatic conditions in which it has become a problem. The main components of an IPM approach can include biological control, host plant resistance, cultural control, autocidal (sterile) insect control, along with the use of chemicals such as pheromones or insect growth regulators which modify the behaviour of insect pests. IPM can never replace insecticides completely but it makes it possible to use slower-acting, less persistent and less potent chemicals.

Despite its complexity, IPM can and does work for small growers and farmers. Some success stories are given on pages 261 to 269. However, there are certain prerequisites if it is to work:

(1) government agricultural departments have to be committed to its principles;

(2) farmers must somehow be convinced that IPM can solve their pest problems; this is easier said than done; the hardest thing to do when your crops are attacked by pests is nothing!;

(3) the pesticide industry has to be willing to tackle the problems of manufacturing biocontrol products; the industry is interested in biological control only if the target is a pest of worldwide importance;[9]

(4) IPM also benefits from high technology; Peruvian scientists, for example, are trying to incorporate into potatoes the gene of the natural pesticide *Bacillus thuringiensis*, which produces a protein that is toxic to many insects.[10]

Economic and Environmental Benefits of IPM

Pests are only controlled when they start to cause economic damage, so overall control costs are reduced.

Costly and harmful practices such as cosmetic spraying and preventative (calendar) spraying are minimized, or eliminated.

Crop damage from chemical pesticides is reduced, or eliminated. In glasshouse crops such as tomatoes, yields increase in IPM programmes as compared to conventional chemical treatments.

Pesticide use is reduced through lower dose rates and frequency of spraying or even complete elimination of chemicals.

Less toxic materials or more selective chemicals and formulations are substituted for more toxic or total pesticides.

Natural controls by predators, parasites, pathogens and crop competition are maximized.

Examples of IPM Programmes

IPM is applicable to all areas of pesticide use – agriculture, local authorities, industrial weed control and even domestic use. There are over 200 examples of successful biological control/IPM programmes worldwide, ranging from control of insect-type pests by other insects or mites to the use of viruses or fungal diseases to control weeds and insects. This number grows each year.

Some examples of successful national or regional IPM schemes including the use of biological control are given below.

Integrated Mosquito Control in India

India has the biggest anti-malaria programme in the world, spending half its health budget on malaria prevention. Intensive insecticide use led to resistance in malaria-carrying mosquitoes, as well as unacceptable pesticide residue levels in human and bottled milk. As a result the Indian Council of Medical Research abandoned the use of insecticides for the control of mosquito-borne diseases in favour of community participation schemes based on environmental management and larvae-eating fish. The aim was to stabilize mosquito populations at low levels where disease transmission stopped.

In Pondicherry a health education campaign enlisted public co-operation in clearing up thousands of cesspits, ponds, wells, open drains and swamp where mosquitoes bred. In two years much surface water was covered or eliminated, wells were stocked with fish, and the swamp converted to a park. In 1986 the number of mosquito bites had dropped by 90 per cent and there were only three cases of filariasis (a disease caused by filarial worms also transmitted by mosquitoes).[11] The scheme is being extended to other industrial towns, and the cost of the whole project is only a fraction of the economic loss to industry due to malaria.

IPM Techniques

Economic threshold levels are worked out to determine the level of economic damage at which it becomes cost effective to control the pest species. This technique is used for predictable pest damage where control measures can be effective even after the crop has been attacked.

Pest forecasting systems: where pest damage is on a more random seasonal basis, and prevention is the only effective control method, forecasting systems are used to predict the likelihood and timing of a pest attack. Pesticides need only be applied if pest damage is predicted and at the right time, rather than blanket preventative spraying.

Substitution: an important part of any IPM programme, is careful choice of less toxic and/or persistent pesticides to maximise natural controls. The aim is to use selective pesticides which control the pest species but leave beneficial and other species unharmed.

Methods and timing of application: the correct method of application linked to choice of formulation can help maximise efficient use of the pesticide and minimise unwanted side effects. For example, spot treatments as opposed to overall spraying can reduce the amount of pesticide used and save money.

   Correct timing of applications, say early in the growing season when the pest is active but before natural enemies have built up in large numbers, is also important, especially if biological control techniques form part of the IPM programme.

Biological control is defined in a number of ways, but is used here to mean the deliberate introduction of predators, parasites or pathogens to control pests. These killers are often specially bred and, in the UK for example, are widely used in glasshouse crop IPM programmes on tomatoes and cucumbers (see pages 267–8).

Plant breeding for resistance is a developing area of pest control and one increasingly linked to biotechnology. The UK National Institute for Agricultural Botany now employs minimum standards of disease resistance for new crop varieties. Even high-yielding crop varieties will not be recommended unless disease resistance standards are met.

Crop rotation: monoculture leads to a build-up of pest species, whether weeds, insects or diseases, especially if the same crop is grown continuously in the same field. Regular crop rotation between fields, linked to more varied crops being grown can reduce dependence on chemical pest control. A better balance between arable and pasture land would reduce pesticide consumption, as arable farming uses more pesticides.

Timing of planting: early or late planting of crops often helps reduce pest damage by either avoiding the pest attack, or ensuring the crop is sufficiently developed to be less harmed by attack.

Cultivation: the way the soil is cultivated has a direct effect on pests, and the amount of chemical application, and the three basic techniques involved are:

- direct drilling – this is *not* a good IPM technique as the soil surface is left completely undisturbed, special drills being used to sow the crop; the weeds and crop stubbles are then controlled purely by pesticides;
- minimum cultivation – only the top 75–100 cm of the soil is disturbed and cultivated, forming a shallow seed bed or 'tilth' into which to sow crops; build-up of weeds from season to season is a problem here, and the system relies on pesticides, but less heavily than direct drilling techniques;
- ploughing – inverting the soil profile buries stubble, disease infected trash, weeds, seeds and destroys insect larvae and pupae, as well as generally aerating the soil and maintaining structure; it is an important part of many IPM programmes, and ploughing is definitely back in favour as an agricultural technique.

Crop hygiene: destroying unwanted crops left over after harvesting, which act as a 'green bridge' for pests until the next growing season, is especially important.

Sterile insect control involves the release of irradiated or chemically sterilised (in the future, genetically sterilised) insects so that 'false', unproductive mating occurs. This technique has, for example, been used in IPM programmes for tsetse fly control.

Insect pheromones are synthetic chemicals that mimic the natural pheromones released naturally as sex attractants by insects. The synthetic pheromones attract pest species to traps, or particular areas, where they can be selectively sprayed.

Insect antifeedants are likely to be part of future IPM programmes. A focus of current scientific research is the commercial development of synthetic 'anti-feedant' chemicals, which inhibit feeding so insects may die of starvation.[12]

Biotechnology, including genetic modification techniques such as gene splicing, gene cloning and gene synthesis which breed resistant genes into plants, could form part of integrated management programmes in the future (See pages 270–5).

Biological Control of Harmful Fungi in Wood and Trees

Two naturally-occurring fungi which have been commercially developed as biological control (biocontrol) preparations for the treatment of wood and trees against harmful fungi are *Peniophora gigantea* spore tablets, and *Trichoderma* pellets, or wettable powder. Both preparations are cost-effective and safe for the consumer, the workforce and the environment. The major use of *Trichoderma* species is for the control of silverleaf disease in fruit trees. Potential uses against wood-rotting fungi in telegraph poles and marine timbers are also being considered.[13]

Skeleton Weed Control

In Australia skeleton weed was successfully controlled by the introduction of the rust fungus *Puccinia chondrillina* from the Mediterranean region where the weed originated,[14] saving an estimated $26 million (Australian) per annum in terms of chemical control.[15]

Mycoherbicides and Weed Control

In the US the use of mycoherbicides – herbicides based on natural fungal diseases – for weed control has proved biologically feasible. Commercial applications have taken place in at least three cases, for control of water hyacinth, strangler vine control in citrus fruit groves, and northern joint vetch control in rice and soya bean.[16]

Biological Control of Water Weed in New Guinea

In the Sepik flood plain in Papua, New Guinea the water weed *Salvinia molesta* has been successfully controlled following the introduction of an insect weevil *Cyrtobagus salviniae*. In 1980 the water weed occupied some 500 square kilometres of waterways, and villagers could not fish or travel to collect food. The weed was literally destroying their way of life. The first release of the insect weevil took place in 1982, and by 1985 the area occupied by the water weed had decreased tenfold to 50 square kilometres of waterway.[17]

Biological Control and Conflict of Interest

There can be a conflict of interest in some instances over biological control programmes. A bitter eight-year battle in Australia between government, scientists, beekeepers and farmers over the biological control of a purple-flowered weed, *Echium plantagineum*, ended in 1988 with the release of a species of moth in New South Wales and Victoria. The beekeepers had legally blocked the release of the moth, *Dialectica scalariella*, claiming that the weed provided much of the pollen on which their bees depended. The farmers in turn claimed the weed poisoned livestock and drastically reduced the productivity of farmland.[18]

IPM and Grain Store Pest Control

The UK produces some 20 million tonnes of grain per year, about two-thirds of which is kept in long-term storage on the farm and in large commercial stores. The stored grain is often routinely sprayed for pests as a preventative measure, as up to now there have been no simple methods of monitoring pest levels. This routine use of pesticides has lead to residue problems in grain, and grain products such as flour and bread.[19] Researchers in the US have now developed a new way to monitor grain for storage pests, using a simple 'pitfall trap' that is proving to be up to ten times more efficient than current methods. The trap is placed in the grain and the number of insects falling into the trap is recorded, which allows the timing and severity of pest attack to be calculated. These traps give early warning of pests, avoiding the need for blanket treatment of the stored grain with pesticides.[20]

Nicaragua and IPM

In 1983 the Nicaraguan government established a Commission of Occupational Health to investigate pesticide-related health problems. The Commission wanted to reduce the level of pesticide use through the development of alternative pest control methods, and planned an education programme to develop awareness of the pesticide problem and the steps which could be taken by workers to reduce poisonings. Nicaragua had become trapped on a 'pesticide treadmill', using more and more pesticides while yields fell and costs spiralled upwards. Pesticide contamination and poisoning was widespread. The Nicaraguan Health Fund reported that DDT in women's breast milk in the Leon region was 40 per cent above the safe level recommended by the WHO.[21] Nearly 400 people died from pesticide poisoning during the 1969/70 harvest season.[22]

The Nicaraguan agricultural ministry is helping to encourage IPM programmes. A bilateral research project between the University of Leon, Nicaragua and the Institute of Virology, Oxford, UK is investigating the use of viruses to control food crop pests such as the soya bean looper and fall army worm. The project is training Nicaraguan scientists and agricultural technicians in IPM techniques. The main problem at the minute is the lack of trained people.[23]

Bacillus Thuringiensis and the Blandford Fly

The River Stour in Blandford, Dorset is the main breeding ground of the black fly, *Simulium posticatum*, which bites hundreds of people each summer in southern England. Some of the bites are as severe as a burn. As part of a scientific experiment to control this insect, the pesticide *Bacillus thuringiensis*, variety *Israeliensia* (BTI), which is based on a naturally occurring micro-organism, was released on a small scale into the river at a concentration of one part per million. The spores of *B. thuringiensis* kill the larvae of the fly after they are swallowed. The unusually alkaline conditions in the gut of the larvae prompt the ingested spores to release a toxic protein which destroys the gut wall of the host. Drift samples taken after treatment revealed large numbers of dead black fly larvae while other invertebrates were unharmed.[24]

Enzyme Tests and Pesticide Resistance

A new technique, which can easily be carried out by the farmer or grower in the field, could save time and money wasted on spraying pesticides onto resistant insects. The technique uses specific enzyme reactions to detect the biochemical basis of resistance in a single squashed insect.[25] The tests are particularly easy to carry out in the field. The field worker crushes an insect and dilutes the resulting material, some of which is placed in the wells of a microtitre plate a small plastic plate with wells containing various chemical reagents. If the reagent changes colour it reveals that the insect is resistant to a particular pesticide. A worker can do some 30 different tests on a single insect to check its resistance to different pesticides.

For example, health workers can now check a mosquito's resistance to the organophosphate insecticides by adding a reagent to the crushed insect to test for the presence of acetylcholinesterase. If the insect is sensitive to organophosphates the test will be negative because the pesticide will have inhibited its action. If the insect is resistant to organophosphates because their acetylcholinesterase enzyme does not respond to the pesticide, the test will detect the enzyme and the reagent will turn a bright yellow.

The micotitre plate technique can give the farmer or grower an answer in a couple of hours. Conventional bioassay tests would take up to a couple of days. The new technique is more sensitive than bioassays, which often fail to pick up resistance until 60 per cent or more of the insects are affected. The new technique can detect resistance in fewer than 20 per cent of the population, while there is still time to do something about it. The new tests are also much cheaper, costing about US$38 compared to US$232 for the WHO's bioassay kits. They are likely to help farmers and growers avoid needless and costly spraying of pesticides, so they can form part of an IPM approach.

Sex Pheromones and Cotton Bollworm

Farmers in Pakistan have been involved in trials of a plastic twist tie containing pheromones to protect cotton against cotton bollworm pests. The twist ties are similar to those used for fastening rubbish bags and have two strands, one containing pheromone, the other allowing it to be fastened to a cotton stem.[26]

Male moths, of bollworm and other species, find their mates by scent. Just before mating the female moths release a sex pheromone to tell the males where they are. The pheromone in the twist ties mimics the natural insect pheromone and evaporates into the air in such quantities that there is too much scent in the air, making it impossible for males to locate females. The twists ties, carrying a mixture of the pheromone of all three major bollworm pests, cut the number of insecticide sprays to maximum of two, sufficient to protect the cotton from bollworm infestation for the entire growing season. Pheromones are safe to use and harmless to living organisms.[27]

Vacuum-cleaning and Strawberry Pests

Strawberry growers in California, now use tractor-mounted vacuum-cleaners, called 'Bugvacs', on their fields to get rid of harmful insects. The bugvac is drawn over the rows of strawberries, and insects in the upper leaves accelerate to a speed of about 30 miles per hour by the time they hit the fan inside the machine which 'blow-kills' them. Their dead bodies then blow back out over the field.

Results show that vacuuming the strawberry fields keeps the lygus bug, a pest which lives in the upper leaves of the strawberry plants, down to manageable levels without spraying the plants. Predatory insects, without insecticides which would kill them off, can keep the western flower thrips and remaining lygus bugs in check, while surviving predatory mites control the spider mite pests.[28]

Natural Plant-derived Pesticides

At least 2,000 plants are known to have pesticidal properties, yet only pyrethrum, tobacco (nicotine) and derris (rotenone) are exploited commercially on a large scale. The development of IPM programmes is, however, causing new interest in natural plant pesticides.[29]

The US government has approved the use of a pesticide extracted from the Neem tree, a native of India, but now widely distributed throughout South-East Asia, East Africa, sub-Saharan Africa and parts of Central America. Neem has insecticidal, repellent, anti-feeding, growth-inhibiting, fungicidal and nematicidal (anti-eelworm) properties. It can be used against a wide range of important pests including aphids, mites, cutworms, some grass-, leaf- and planthoppers, locusts, some cabbage worms, the Mediterranean fruit fly, whitefly, several stalk borers and beetles. The entire plant can be used, but the seeds contain the highest concentration of active ingredients.[30]

Pyrethrum is a well-known insect poison derived from the daisy species *Chrysanthemum cinerariaefolium*. The flowers are pulverised to obtain the active ingredient, pyrethine, which has a quick 'knock-down' action on aphids, beetles, locusts, mites, thrips, moths etc. It also has anti-feedant and repellent properties. Pyrethrum production is limited to certain areas of the world, primarily East Africa, and owing to its high price as an insecticide in comparison with chemical chlorinated hydrocarbon insecticides its cultivation is not widespread.[31]

The synthetic pyrethroid group of chemicals are modelled on the pyrethrum molecule, so they have similar properties.

The UK Glasshouse Industry

Biological control within the context of IPM programmes has found successful commercial application in the more controlled environment of glasshouses.

The Glasshouse Crops Research Institute estimated in 1985 that approximately 75–80 per cent of the heated cucumber area and 15 per cent of the heated tomato area used an artificially-introduced Chilean predator mite, *Phytoseilulus persimilis*, to control an important pest, the glasshouse red spider mite, *Tetranychus urticae*.[32] Since then its use has been maintained close to the 1985 level.[33] *Phytoseilulus* is very efficient at searching for its prey, and each female will devour up to 5 adult or 20 young spider mites per day. Unlike its host (the spider mite), the *Phytoseilulus* predator will not normally survive the winter, and must be reintroduced into glasshouses each year as required. Like most parasites and predators, *Phytoseilulus* is very sensitive to many toxic chemicals in current use, therefore careful choice and timing of pesticide use is essential.[34]

The greenhouse whitefly, *Trialeurodes vaporarium*, is controlled by a small parasitic wasp-like insect, *Encarsia formosa*. *Encarsia* lays its eggs in one of the developing whitefly scale stages. Its larvae feed off the whitefly scale before emerging as adult wasps.[35] The Glasshouse Crops Research Institute estimates that approximately 70 per cent of the heated tomato area and 25–30 per cent of the heated cucumber area is controlled with *Encarsia*.[36]

Other biological agents are successfully used in the UK glasshouse industry. For example, biological control of aphids has been developed on chrysanthemums using a parasitic aphid, *Aphidus matricariae*, which is an efficient parasite of the important aphid pest *Myzus persicae* (one of the major 'greenfly' pest species).[37]

In the UK control of outdoor agricultural and horticultural crops using biological agents is more problematic. *Bacillus thuringiensis* has been successfully used to control caterpillar larvae, being applied in spray form as a natural pesticide. Yet there are promising developments even in this area. Predatory nematodes, antagonistic fungi (for example *Verticillium lecanii*), and mycoherbicides are all being developed for commercial application.[38]

UK Apple, Pear and Hop Orchards

Intensive, broad-spectrum chemical spray programmes in apple, pear and hop orchards have led to serious pest resistance problems. Apple orchards, for example, are treated on average with 17 fungicide and 5–6 insecticide/acaricide applications per growing season.[39] IPM programmes, based on allowing naturally-occurring predators to survive and multiply, have been introduced; these require careful monitoring of pest build-up, correct choice of chemicals, and timing of application, if a decision to spray is made.[40]

For all the major pests of apples – moth larvae, aphids, mites and apple suckers – there are natural predators which can help provide effective control as long as their populations survive the pesticides used. The most important general predators are *Anothocorid* and *Mirid* insect species. Other important predators are *phytoseiid* mite species such as *Typhlodromus pyri*, especially where these mites have themselves become resistant to organophosphorus insecticides. This resistance is exploited in IPM programmes, in which OP insecticides are used to control aphids and moth larvae, leaving *T. pyri* mites to survive and regulate pest species like the fruit tree red spider mite, *Panonychus ulmi.*[41]

In pears, the main pest is the pear sucker, *Psylla pyricola.* IPM techniques in pear cultivation centre on preserving populations of the predator insect *Anthocoris nemoralis* by restricting chemical applications to occasions when treatment thresholds are exceeded, and then choosing materials that cause the least possible damage to *A. nemoralis.*[42]

The two most important pests of hops are the damson hop aphid, *Phorodon humuli,* and the two-spotted mite, *Tetranychus urticae.* In hops IPM is based on combining the use of synthetic pyrethroid insecticides, to which the two pests are less resistant, with the survival of predator species like *A. nemoralis.*[43]

Salmon Farming and 'Cleaner Fish'

The sea louse, a type of crustacean parasite, is a serious problem for farmed salmon. Fish kept at high stocking rates in sea cages are prone to the parasite which can cause extensive damage to the skin and, in extreme cases, penetrate the brain and kill the fish.

Fish farmers currently control the parasite by dosing the cages with the organophosphate insecticide dichlorvos and treatments have to be made every three to four weeks. Dichlorvos is a serious pollutant of other marine organisms and has, for example, been linked to a sudden upsurge in the incidence of eye cataracts in wild fish.[44] It is also expensive, costing up to £100 per year to treat one cage with pesticide. So fish farmers and government scientists have begun looking for environmentally safer, as well as cheaper, alternatives.[45] Following successful trials in Norway, selected UK fish farmers are experimenting with the use of a 'cleaner fish' species, the golden sinny wrasse, *Ctenolabrus rupestris,* as a means of removing parasitic sea lice from farmed salmon. The wrasse is a small, orange-coloured fish some 12 cm. long and it picks off external parasites from other fish.

Improved Grain Fumigation Techniques

Australian scientists have developed a simple air pump, called the Siroflo, which could help fumigate grain silos more effectively using much less pesticide. The technique involves pumping low concentrations of the fumigant phosphine, mixed with either nitrogen or carbon dioxide, through the grain at a constant rate under positive pressure, allowing the phosphine to penetrate the grain mass more effectively, killing more insects with less pesticide. As the air flow is continuous, silos do not have to be fully sealed, so old silos, which would normally be abandoned, can now be used again.[46]

Conventional fumigation techniques mainly use sachets of aluminium phosphide which react with moisture in the atmosphere to give off phosphine (see Chapter 10). This method often gives mixed results as it normally eliminates the pests in two stages of their development, at the larval and adult stages, but not as eggs and pupae. The surviving eggs and pupae are difficult to detect, and high fumigant concentrations may be needed to control them. Silos also need to be sealed effectively as the concentration of phosphine gas falls with distance away from the sachet. Another problem is 'phosphine narcosis'. Insects exposed to very high concentrations, about 60 times as high as the normal lethal dose, enter a torpid state where they are resistant to the phosphine. It is not known why they do this.

The Siroflo technique uses much less phosphine to protect stored crops, so it is much more effective, and potentially safer to operators. In addition, farmers and food processors do not need to use other insecticides, such as the organophosphate fenitrothion, which leaves residues in the grain.[47]

Cockroaches and Temperature Control

A pest control company has found a way to exploit insects' vulnerability to heat in order to kill common house pests. The Californian company, Isothermics, encases a house in a canvas tent, installs propane burners and fans inside, and raises the air temperature to around 66°C. After four hours at the temperature, every part of the building will have reached more than 50°C, which is enough to kill cockroaches, ants, flies, ticks, moths and even termites.[48]

The pest control industry has known for some time that heat can kill insects, but until recently it was content to use chemical pesticides. According to Isothermics, there are several articles in scientific journals from as long ago as 1916 showing that insects can die at temperatures quite bearable to humans: 'Cockroaches can withstand a higher dose of radiation than humans can, but they cannot thermoregulate, so heat that we can stand – because we sweat to keep our body temperature down – will kill them,' says Richard Brenner, an expert on cockroaches from the US Agricultural Research Service.[49]

Biotechnology and Pesticides

Biotechnology covers all uses of living organisms to make or modify products for human use. It is now used more specifically to refer to genetic modification (engineering) of animals, plants and microorganisms by modern techniques such as gene splicing (recombinant DNA technology).

Biotechnology has traditionally been based on the use of naturally-occurring microbes modified and improved by conventional breeding and reproduction techniques. Micro-organisms are already widely used on an industrial scale to produce, for example, antibiotics such as penicillins, other organic chemicals, and a limited range of microbial pesticides like *B. thuringiensis*. With the advent of genetic modification, a new type of biotechnology is available and being rapidly developed on an industrial and commercial scale in agriculture to produce new types of animals and plants, and even pesticides.

Microbial Pesticides

Microbial pesticides are non-chemical pesticides based on the use of naturally-occurring micro-organisms. 'Microbial' is used in the broad sense to include viruses, which are technically not alive and, therefore, not organisms.

A problem with the development of microbial pesticides is that they are likely to have a small market. Only six or so species are available worldwide and they account for less than 1 per cent of the total pesticide market,[50] and 90 per cent of that figure is *Bacillus thuringiensis*.[51] Nevertheless, further development of microbial pesticides is an important part of any broad IPM strategy.

Some 1,500 naturally-occurring micro-organisms or microbial products have been identified as potential biological control agents against fungal and viral diseases, insects, and even weeds. These agents include bacteria, fungi, nematodes, protozoa and viruses.[52]

Bacterial Insecticides

Over 100 species of bacteria are known to infect insects but so far only four of these have been closely examined as control agents:[53]

- *Bacillus thuringiensis*
- *Bacillus popilliae*
- *Bacillus lentimorbus*
- *Bacillus sphaericus*

Of these four, only *Bacillus thuringiensis* has been developed as a commercial insecticide. Various strains of *B. thuringiensis* have been used in biological pest control since the 1950's. Its insecticidal properties are limited to certain insect groups, primarily caterpillar pests. The subspecies *Israeliensis* (also called *B. thuringiensis* pathotype B) is effective against mosquitoes, especially at the larval stage, so it can be used to control the carriers of important diseases such as malaria and yellow fever.

Bacillus thuringiensis kills the insect host by production of a toxic poison. The highly alkaline conditions in the gut of the larvae prompt the ingested *B. thuringiensis* spores to release a toxic protein which destroys the gut wall of the host.[54] Because it is highly host-specific, *B. thuringiensis* is harmless to honeybees, humans and animals. It is sold commercially, including for garden use, under product names such as Thuricide, Dipel and Bactospeine.

Viral Insecticides

Six groups of viruses are known to infect insects. One group, the baculoviruses, are now used in commercial spray preparations, primarily to control caterpillar pests; especially the baculovirus sub-groups, the nuclear polyhedrosis viruses and granulosis viruses. This is because baculoviruses only attack invertebrate species such as butterfly caterpillars, and are therefore considered safer than other insect viruses related to vertebrate or plant viruses. They are also fairly host-specific and lethal. The virus is contained in the form of stable, protein-encapsulated crystals which have a long commercial shelf-life and are compatible in formulations with other chemical pesticides. Examples of nuclear polyhedrosis viruses include cotton bollworm virus, Douglas tussock fir moth virus, alfalfa looper virus, and European pine sawfly virus. An example of a granulosis virus is codling moth virus, used in apples.

Baculovirus preparations, containing dormant forms of the virus, are sprayed onto the plant and ingested by young caterpillars as they eat the foliage. Once ingested the virus is activated and multiplies in the host (the caterpillar) and, approximately 3–4 days after spraying the caterpillars become sick and die.

Baculoviruses have been used against a wide range of pests such as turnip moth, cabbage looper, lucerne butterfly, army worms, cabbage worms, American bollworm, corn earworm and soya caterpillar.

Fungal Insecticides

More than 500 species of fungi are capable of infecting insects and there are vulnerable insect species in all the major families of *Insecta*.[55] Yet, in spite of their potential as microbial control agents, worldwide use of fungi on a commercial basis is limited. This is because their control is often variable and easily inhibited by climatic or environmental factors. In particular, the high relative humidity required for successful germination of fungal spores limits their use.[56] Some successful examples, however, include:

- *Verticillium lecanii:* aphid/insect control in UK glasshouse crops
- *Beauvaria basiana:* Colorado potato beetle control in the USSR and European corn borer control in China
- *Hirsutella thompsonii:* citrus rust mite control

Plant-beneficial Bacteria

There is growing interest in the idea that bacteria associated with the underground parts of plants (the rhizosphere) may have beneficial effects on plant growth, either by providing nutrients and growth factors, or by producing antibiotics which may help protect the plant from pathogenic fungi or bacteria.

In the future, plant growth-promoting bacteria may be used as bio-fertilisers or microbial disease control agents. Genetic modification is likely to be used to improve strains of plant growth-promoting bacteria.[57]

Genetic modification techniques (biotechnology) may be used in the future to increase the virulence of micro-organisms, by making them more tolerant of physical and chemical conditions, as well as perhaps widening their host range. For example, the toxin gene of the *B. thuringiensis*, (subspecies *Kurstaki*), has been isolated by means of recombinant DNA technology.

Belgian scientists have now transferred the gene into the cells of the tobacco plant, and shown that it successfully protects the plant from insect predators. The cells grew into normal adult plants which were highly resistant to insect larvae. Larvae feeding on the genetically altered plants became paralysed after 48 hours and died within 3 days. Plants pass on genes for insect resistance to future generations.[58]

Genetic modification makes it possible to transfer genetic material from one species of animal, plant, insect, or micro-organism to another and alter the genetic make-up in a way that would not arise from conventional breeding methods. The genetic material transferred is DNA, which contains the genetic information controlling species type and individual characteristics. Hence the term 'recombinant DNA technology'.[59]

Biotechnology raises fundamental ethical and moral questions about the right to create new species or alter individual characteristics. There are also many questions as to the inherent safety of these techniques for human health and the environment. The public is wary of genetic modification and the possible risks associated with it. They fear that enthusiasts and supporters for this new industry, with massive amounts invested and precious little sign of any return, will play down the dangers. Poor risk-assessment and regulation has increased public concern, especially as there have been clear examples of regulations covering genetic releases being ignored or by-passed.[60]

Biotechnology and Agriculture

One of the early promises of biotechnology was that it would provide environmentally preferable alternatives to intensive agricultural practices, and especially alternatives to chemical pesticides. Plants could be altered to be more resistant to insects and diseases, to be able to withstand drought and frost and to fix their own nitrogen from the air. They could also be made to produce compounds toxic to insects and diseases, and disease-suppressing micro-organisms could be developed.[61] The application of biotechnology to crop production and crop protection could, in fact, help reduce the use of chemical fertilisers and pesticides, as well as improving IPM programmes and alternative crop production methods.

The direction that commercial biotechnology has taken is somewhat different and could lead to continued chemical dependency in agriculture. Faced with the fact that such advances in genetic modification could, within the next 20 years or so, consign much of the pesticide and fertiliser industry to the commercial dustbin, the large multinational oil, chemical and even food companies have reacted by buying up seed companies worldwide, to gain control of the basic genetic plant material. They can produce plants with characteristics suited to their commercial operations, and protected by plant and seed patent rights. One consequence of this is that herbicide-tolerant crop plants are being developed which are resistant to certain chemical pesticides.

Herbicide-tolerant Plants

Multinational companies and even governments are now engaged in large-scale commercial development of genetically modified, herbicide-tolerant plants which promises to prolong and even increase chemical pesticide use. Herbicides are still the largest category of pesticides used, so there is a huge financial incentive for companies to develop and patent herbicide-tolerant crops which can be marketed worldwide.

Far from phasing out chemical herbicides, these types of biotechnological developments mean continued or even increased dependence on chemicals, with corresponding risks to human health and the environment. There are also fears that the genetic herbicide tolerance trait may be transferred by cross-pollination into weeds closely related to crop plants, thus increasing the number of weeds resistant to herbicides.[62]

Herbicide-tolerant plants can grow in the presence of amounts of herbicide that harm or kill non-tolerant plants. Conventional plant breeding has attempted to breed herbicide tolerance into crop plants but with only limited success. Conventional plant breeders are limited to tolerance traits they can find in plants closely related to the plant they want to modify. With modern genetic modification techniques this constraint has now been removed. Once a trait for herbicide tolerance in plants has been found, that trait can potentially be transferred to a variety of crop plants from widely differing parent plants of different genetic make up. All the major food crops are now the target for herbicide tolerance research – corn, potato, rice, sorghum, soya bean, tomato and wheat. Other projects involve trees, vegetables, alfalfa, cotton, oilseeds, flax, oats, sugar beet, sunflowers and tobacco.

All the world's major pesticide companies have initiated herbicide-tolerant plant research – Bayer, Ciba-Geigy, ICI, Rhone Poulenc, Dow/Elanco, Monsanto, Hoechst and DuPont. Likewise, many of the world's largest seed corporations sponsor similar research – Ciba-Geigy, Dekalb-Pfizer, ICI, La Farge/Rhone Poulenc, Pioneer Hi-Bred, Sandoz-Hilleshog and Upjohn. Field tests have already been carried out on genetically modified plants resistant to herbicides such as bromoxynil, glufinosate, glyphosate and sulfonyl ureas. Such plants could be on the market by the mid-1990s. Older pesticides may also have an extended life. Research is in progress to develop tolerance to 2,4-D and atrazine, two herbicides linked to cancer. Atrazine is also a major contaminant of ground and surface waters in Europe. Monsanto, for example, has found the plant gene which confers resistance to their herbicide, glyphosate. The company is now trying to introduce this gene to plants previously not tolerant of glyphosate, or make existing crops able to tolerate higher, or more frequent applications.[63]

Deliberate Release of Genetically Modified Pesticides

Trial, planned or deliberate releases to the environment of genetically modified pesticides, based on micro-organisms, have taken place in the UK, the US and elsewhere, and the pace of development is accelerating rapidly.

The first release in the UK of a genetically modified pesticide took place in 1986, when the National Environmental Research Council Institute of Virology, Oxford, released a small amount of the *Autographica california* nuclearpolyhedrosis virus for the control of the pine beauty moth, which is a serious pest of pine trees in Scotland.[64] The Institute has since conducted other small-scale experimental releases using the insect virus *Autographica california*, which was released for the control of three species of moth – lime hawk moth, small mottled willow, and the cabbage looper.[65] The viruses were genetically modified to include a 'marker' which allows their presence to be detected, and which also ensures they self-destruct within a short period.

The world's first commercial pesticide based on a live, genetically engineered organism is now on sale in New South Wales, Australia. The product, called 'No Gall' protects stone fruits, nuts and roses from crown gall disease. It contains a benign strain of the bacterium *Agrobacterium tumefaciens* which causes the gall disease. The benign strain of *A. tumefaciens* produces an antibiotic which kills its disease-causing relative. The benign strain has been engineered to remove a plasmid – a loop of genetic material – which confers resistance to the antibiotic.[66]

In the US planned release of genetically engineered organisms has a varied history since the first approvals in 1986. First, the private company receiving approval to test a genetically engineered bacterium in a strawberry patch admitted that it had already tested the organism in the open without telling anyone.[67] Then the US Department of Agriculture mishandled the trial approval of a genetically engineered pig vaccine. The department quietly approved the proposal without telling its own experts, but the US environmental biotechnology lobby found out.[68] Following this, a US plant pathologist from Montana State University tried to protect trees against Dutch elm disease using a genetically engineered bacterium without informing the US Environmental Protection Agency.[69]

Organic Agriculture: the Non-pesticide Option

This book has assumed that pesticides are an essential, if sometimes unwelcome, part of modern agriculture, horticulture and land management. For a growing number of organic farmers and growers this is no longer the case. There is increasing interest in the practice of organic farming and gardening methods in Britain. Organic cultivation avoids the use of both soluble nitrate fertilisers and artificial chemical pesticides.

Organic growing does not just mean 'going back' to traditional methods which were in use before the advent of artificial pesticides. Used professionally it is a sophisticated technique which relies on a detailed understanding of ecology, soil science and crop breeding. At the same time, the basic steps can be used by any gardener. Important elements of organic growing are:

- *crop rotation* to build up soil fertility and break the life cycle of crop-specific pests. One element in the rotation is a leguminous plant (a member of the pea and bean family); these plants have bacteria in their root nodules that can 'fix' nitrogen from the air and store it in plant tissues. Ploughing in the remains of the legume allows nitrate levels to be maintained in the soil.
- *recycling nutrients* through composting and careful use of animal manure. These methods top-up nitrate levels where necessary, and also increase the amounts of trace elements in the soil. Manure is usually composted before application to make sure that its nutrients are not leached away by rain.
- *non-chemical pest and weed control:* this includes biological control; cultural control (through planting strategies etc.) and use of mechanical weeders, flamers etc. These methods are examined in more detail below.
- *mixed farming or gardening* practised in an extensive system. Ideally, organic farms have fairly small fields and a mixture of crops and livestock.
- *care for wildlife and the countryside* while producing food. It has been shown by research in both Denmark and Britain that wild bird species increase on organic farms. The Soil Association has incorporated specific environment and conservation standards into its Soil Association Symbol, to increase the wildlife potential of organic farms.
- *humane treatment of animals,* allowing them access to pasture, and no unnecessary exposure to drugs or hormones. Animal welfare organisations, including the Royal Society for the Prevention of Cruelty to Animals and the Farm and Food Society, back organic farming because of its improved treatment of livestock.

The History of the Organic Movement

The pioneers of organic growing began forming their ideas during the 1930s, in reaction to the intensification of agriculture. In 1943 Lady Eve Balfour published *The Living Soil*, which described her philosophy of organic agriculture, and the work of her experimental farm.[70] As a result of the considerable public response, she started the Soil Association in Suffolk in 1946 to explore the links between soil-plant-animal-human and the practice of organic methods on an agricultural scale.

In those days the fledgling organic movement was deeply influenced by the enormous problems of soil erosion, which had occurred widely as a result of incorrect farming practices. Soil erosion had formed 'dustbowls' over huge areas of Australia and the western US, with a range of social effects probably best described in John Steinbeck's modern classic *The Grapes of Wrath*.[71] There was no particular political philosophy embodied in the movement, and early Soil Association

Council members represented opinions ranging from the extreme right to the far left.

The Soil Association continued its experimental work for some decades, but was eventually forced to sell its farm after 30 years when continually rising prices were not, at that time, matched with public enthusiasm. Nevertheless the Association remained intimately involved in the development of organic agriculture, and the setting and maintaining of appropriate standards.

Today, the Soil Association is probably best known for running the Soil Association Standards for Organic Agriculture, which define good organic practice for farmers and growers and are a consumer guarantee of genuine, high-quality organic food. The Association also campaigns on a wide range of issues relating to food, health and the environment, lobbies parliament and other bodies involved with agriculture and food, and carries out policy research. It is currently running the '20% Organic by the Year 2000' campaign. It is a membership organisation, with a quarterly magazine, *The Living Earth*.[72]

Soon after the founding of the Soil Association, Lawrence Hills set up the Henry Doubleday Research Association (HDRA) at Braintree, in Essex, to carry out practical research on organic gardening. In the years since, HDRA has also moved to near Coventry, where it runs the National Centre for Organic Gardening, which is open to the public. HDRA also produces the *All Muck and Magic* television series on Channel 4. It is a membership organisation, with a quarterly journal and a large mail order service for organic gardening products, books and pamphlets.[73] HDRA also runs a successful reforestation project in Cape Verde and carries out research into organic gardening methods, often through the cooperation of its own membership.

In the early years of the movement, organic growing remained the interest of a minority, who were often regarded as impractical romantics. The development of environmental interest during the 1970s and 1980s led to many more professional organic farms being set up, and organic food started appearing in shops.

Several other organic organisations have appeared in the years since. British Organic Farmers and the Organic Growers' Association are the main organisations representing professional organic producers in Britain. They offer advice, run a journal, publish technical material, run farm walks and represent the needs of organic producers to agricultural policy makers.[74] Elm Farm Research Centre has been established as a charitable trust and research organisation, based on a working organic farm in Berkshire. Elm Farm Research Centre has carried on the original work of the Soil Association in terms of research into organic methods.

More recently, regional organisations have been established. In 1989 a Scottish centre, based near Edinburgh, was set up to research and develop the market for organic meat. It was part-funded by the retail chain Safeways, which is interested in increasing the market potential of organic meat in Britain. A year later the Aberystwyth Centre for Organic Husbandry was established as a liaison operation between several Welsh universities and research organisations, to carry out research into organic farming within Wales.

In 1989 the government also increased its links with organic farming, by setting up the United Kingdom Register of Organic Food Standards to oversee standards for organic food, and ensure that food was not sold fraudulently. This is becoming increasingly necessary as demand grows and premiums are paid for organic produce. The government also made limited amounts of money available for farmers wishing to convert to organic methods, through the set-aside and beef and sheep extensification policies of the EC, under the aegis of the CAP. These policies, introduced to reduce food surpluses in the EC, are controlled by the European Commission, but allow individual Member States leeway in how they are enacted. *Extensification*, in effect, is reversing the trend towards *intensification*, which has been in operation since the Second World War. Indeed, the European Community is itself becoming increasingly involved in organic agriculture, and published its own Draft Standards at the end of 1989.[75] These are supposed to provide Europe-wide continuity in the practices of organic farming and quality of organic food.

By the time Lady Eve Balfour died, aged 91, in January 1990, all major retail chains stocked organic food and the government had acknowledged her vital contribution by awarding her the OBE. It is ironic that the day after she died the first grant to help British farmers convert to organic methods was announced. Organic farming enjoys support from more than half the population, according to a survey carried out by the National Farmers' Union.[76]

Controlling Pests on an Organic Farm

From the perspective of this book, the ways in which organic farmers and growers control pests is of key importance. Organic methods of pest control include several non-chemical methods. The Soil Association Standards also allow occasional use of a small number of plant-based insecticides and fungicides, including natural pyrethrum, and rotenone. The Standards regarding pest control are described on p. 279.[77]

The key to pest control on organic farms is the system as a whole, rather than any individual steps. Chemical farmers say, quite rightly, that if they cut out all chemicals overnight they would be over-run by pests. The conversion period needed to qualify for the Soil Association Symbol, or any other organic food standard, is not just set to 'detoxify' the soil of any agrochemicals, but is also necessary to allow the growing area to come into nutrient balance again, build up predators of pests, and so on. When organic farmers meet to talk about the problems of farming, they do not usually have pests or weeds at the top of their list, although for a chemical farmer locked into the 'pesticide treadmill' this is sometimes difficult to believe.

Organic Pest and Disease Control – Standards

The British Organic Standards Committee (the 'Symbol Committee'), which determines the Soil Association Standards for Organic Production, has set a series of clear recommendations and obligations concerning pest and disease control. The following quotation from the *Standards for Organic Agriculture* makes these clear:

> Pest and disease control in organic agriculture is primarily preventative rather than curative. At present, because of the lack of technical development, currently available remedies for pest and disease control are often inefficient, expensive or both. In addition to good husbandry and hygiene, the key factors of pest and disease control are:
>
> (a) balanced rotational cropping to break the pest and disease cycles;
> (b) balanced supply of plant nutrients;
> (c) the creation of an ecosystem in and around the crop which encourages predators, utilising, where appropriate, hedgerows or mixed plant breaks within fields, companion planting, undersowing and mixed cropping;
> (d) the use of resistant varieties and strategic planting dates.

In addition, permitted methods of pest and disease control include mechanical controls using traps, barriers and sound; herbal, homoeopathic and biodynamic sprays; waterglass (sodium silicate); bicarbonate of soda; soft soap; steam sterilisation; biological control with naturally-occurring organisms, and conventionally grown seed. If these do not provide sufficient protection, the Standards also contain a number of 'restricted' methods. These include use of a number of plant-based pesticides.

This does not mean that organic farmers never have problems with pests. They do, and then they resort to methods of pest control which sometimes include the use of the small number of plant-based pesticides cleared for use by certified organic producers. But organic farmers vastly reduce the need for pest control, thus reducing the need for pesticides.

First, organic farmers do not control pests where none occur. They do not practice 'insurance spraying' (which means spraying in case a pest or disease should occur) which sometimes does more harm than good by killing off the predators of pests as well.

Second, they do not use 'cosmetic spraying' (applying pesticide to make produce look better). People buying organic food would rather have, say, harmless skin scabs on apples than synthetic pesticide residues within the apples.

Third, organic farmers use non-chemical methods of pest control wherever possible, as described below. This means that when they do have to resort to pesticides, such as pyrethrum and derris, in a genuine emergency, the pesticides have a proportionately greater effect than when the land is routinely soaked in agrochemicals.

Non-chemical Pest Control

On an organic farm, pests and weeds are controlled by a number of non-chemical methods. These can be divided into a number of convenient sub-categories, including:

- rotation
- biological control
- cultivation control
- companion planting
- barriers and traps
- timing
- crop breeding

Rotation

Rotating crops does not just serve to maintain soil fertility, but plays a key role in breaking pest cycles and keeping crops healthy. A large number of pests, fungi, moulds and diseases are crop-specific (they attack only certain species or types of plant). By rotating the crops, pests and diseases which can survive for some time in the soil are 'starved out' and have usually disappeared by the time their host species is planted in the area again.

For example, carrot root fly, which attacks carrots and is economically very important, is usually controlled by a standard four-year rotation. Indeed, the practice of planting the same crops in a field year after year has greatly increased the use of pesticides in conventional agriculture.

Biological Control

Biological control falls into two main types; first the encouragement of wildlife back onto an organic farm to provide natural control over pest species, and second the introduction of specific pest predators to control outbreaks.

Natural biological control is achieved by eliminating pesticide use, increasing the groundcover on a farm or garden (such as hedgerows, field edges, copses etc.) and perhaps specifically encouraging certain birds.

Farmers in both Britain and Malaysia have found that it is more effective to control rats and mice by building nest boxes for barn owls than by laying down rat poisons which sometimes kill the owls as well.[78]

Introducing species specifically to control pests is often more effective, but is difficult to achieve. The predator must be very pest-specific (it must not be capable of preying on everything else in the area as well) so careful tests are needed before it can be used. Biological control organisms are often brought from other countries, so they must be able to survive in the host country's climate. Ideally, a biological control organism will control a pest, then die out once it has disappeared. They work best when there is an introduced pest (when a foreign pest has become established and is being very successful precisely because it has no natural predators). Many biological controls are known. For example *Encarsia formosa*, used against the whitefly, and *Trichogramma* used against the corn borer. There have been some spectacular successes with biological control, but this is by no means a solution to every pest problem.[79]

Some of the major biological control agents commercially available are bacteria used to control specific pests such as thrips or aphids. Others control caterpillars, whitefly and mealy bug.

Cultivation Control

Cultivation control means mechanical removal of weeds and, sometimes, of animal pests as well. On a garden or market garden scale, much of this weeding is carried out manually, which increases labour requirements. Sometimes it is just as effective to remove pests (like slugs) by hand.

On a farm scale, mechanical weeders are usually brought into use. These include tractor drawn-weeders, and in recent years several new designs have been developed to allow careful removal of weeds without damaging the crop. Alternatively, flaming is sometimes used. This is the application of a flame directly onto weeds, which kills the exposed part of the weed plant but unfortunately often leaves the roots untouched so that it can grow again. Once weeds have been removed for several years running or, in the case of larger plants such as bracken, repeatedly cut, the seeds in the soil are progressively reduced and the overall problem declines.

Companion Planting

Companion planting means planting species next to each other in a way which helps protect one, or both. Companion planting works in a number of ways:

- by attracting pest predators. For example, planting buckwheat attracts hoverflies, some of which kill smaller insects.
- as a way of disguising plants from pest species which hunt by scent. For example, marigolds are strong-smelling and, if planted around the edge of a vegetable patch or in between rows of vegetables, can confuse species which hunt by scent.
- by helping block fungi, moulds and other diseases which travel through the soil to attack crop species.

Some gardeners and market gardeners use companion planting a great deal, others do not believe that it is effective.

Barriers and Traps

Some pests can be eliminated by a variety of physical barriers, some of which are routinely used by both organic and many chemical farmers. Examples include:

- grease bands to prevent pests climbing up trees or tall crops. Grease bands are frequently used on fruit trees to prevent attack by those pests which reach fruit by climbing up the trunk.
- mulches are layers of organic matter (compost, manure or even paper) laid around crops and covering the soil in a carpet. These help prevent attack by pests which reach roots by burrowing through the soil, but can sometimes act in the opposite way by providing cover for pests such as slugs and snails. Mulches also inhibit the growth of weeds.
- non-organic carpets, including plastic sheeting, are used to prevent weed growth but can, again, attract pests which need dark, damp conditions.
- many physical barriers are used against slugs and snails, including ash, soot and other dry substances which will tend to desiccate (dry out) the molluscs. None of these are foolproof, and have to be constantly renewed, making these methods fairly labour-intensive.

There are also crop-specific barriers, including those used for protecting carrots from carrot root fly, or brassicas from attack by certain soil-burrowing insects. Some of these are commercially available from garden suppliers, others are reliant on information from organic gardening books. Many are quite labour-intensive, so may be more suitable for the home garden or small market garden rather than a farm-scale operation.

Timing

Timing of planting is also important. Planting early or late to avoid a main pest outbreak can avoid the need for control altogether, but this must be balanced against other factors, including weather, the presence of other crops, etc.

Crop Breeding

Although most crop breeding has not been aimed at minimizing pesticide use, the organic farmer has made some gains from research over the last few years. Particular improvements have been made from crops which are more resistant to attack by moulds and fungi, reducing the need for chemical control.

Pesticides Used in an Organic System

In addition, when pesticides are used, they are restricted to a few plant-based pesticides which are relatively non-toxic to mammalian and bird species and not persistent. Not all plant-based pesticides are non-toxic; for example nicotine is highly poisonous, and is not allowed for use by commercial organic growers. Among the pesticides cleared for use under the Soil Association Standards for Organic Agriculture are:

- cuprous oxide, in certain circumstances as a fungicide;
- derris (also known as rotenone) derived from a plant-based product;
- metaldehyde strip;
- potassium permanganate;
- pyrethrum (only the naturally-occurring variety, and some synthetic pyrethroids are more toxic);
- quassia (the bark of a tropical tree which has insecticidal properties);
- sulphur;
- tar oil.

In addition, the Soil Association Standards recommend the use of new, less toxic pesticides for sheep farmers to carry out the mandatory sheep-dipping. None of these chemicals is allowed for routine use. Pesticides in an organic system should only be used for occasional, specific purposes, such as combating a sudden pest outbreak. Nor should the pesticides be regarded as benign. They are still potentially hazardous to both users and wildlife, but they offer a far safer alternative than most of those currently in use.

Is Organic Growing a Viable Option?

According to conventional agricultural strategists, organic farming has, at best, a minority role to play in the future of agriculture. At the moment, in Britain, it is certainly very small. For example, at the time of writing (December 1990) there are 26,000 acres under the Soil Association Symbol Scheme, roughly the same amount in conversion, and about the same amount in different organic symbol schemes, including the co-operative Organic Farmers and Growers.

This adds up to less than one per cent of Britain's farmland. Seventy per cent of the organic food sold in Britain is currently imported, and imports are making up the increasing shortfall between supply and demand. There are good reasons to believe that the increase in demand will continue, and several opinion polls and market strategies predict organic food occupying about 15–20 per cent of Britain's food market by the year 2000.

The Soil Association is currently running a '20% by 2000' campaign, aimed at converting a fifth of Britain's farmland to organic methods by the end of the century. This figure has not been chosen at random, but is based on the anticipated demand for organic food.

It is also roughly the area of agricultural land which will change fairly radically over the next decade, come what may. These changes are being brought about by the designation of conservation areas, protected by legislation, including Environmentally Sensitive Areas; introduction of extensification programmes to tackle food surpluses; the introduction of Nitrate-Sensitive Areas to protect water quality; and the likelihood of future regulation over farming methods in the areas of land currently under threat from soil erosion.

All of these areas could benefit from organic farming methods. Organic farming is already used on some nature reserves, and has been identified as suitable for Environmentally Sensitive Areas. Organic methods substantially cut nitrate leaching, and may be applicable in Nitrate-Sensitive Areas; the Department of Environment in Britain is currently funding research on this option, and research in the US has found that soil erosion is substantially reduced by the application of organic methods. These areas add up to *more* than 20 per cent. Many farmers on other land areas are also thinking of converting.

Would we Run Out of Food?

Chemical apologists claim that a switch to organic methods would mean serious food shortages in developed countries, and would add to the malnutrition and starvation problems in areas of the Third World. But research suggests that food shortages would not occur. Modelling undertaken at Aberystwyth University suggests that, at current production levels, a 100 per cent conversion to organic agriculture would result in the changes illustrated in Table 12.1.[80] Because the techniques of organic agriculture are still changing, and yields are continuing to improve, the output would be expected to increase significantly before any theoretical 100 per cent conversion was reached. Extrapolation of these figures implies that a 20 per cent shift to organic agriculture would only have a fairly small impact on food production. Dividing the figures by five indicates that cereals would be reduced by about 6 per cent, grain legumes would go up by 30 per cent, beef production would be reduced by less than 3 per cent; and so on.

Table 12.1: Changes in Output With 100% Conversion to Organic Agriculture in Britain

Land use	Change in output(%)
Cereals	-30.1
Potatoes	-12.9
Sugar beet	-58.1
Grain legumes	+174.2
Oil seed rape	-60.5
Milk	-19.0
Beef	-13.1
Sheep	-18.5

Source: Woodward et al, 1988

These figures pre-suppose a change in diet as well, away from high consumption of sugar, for example. They are in line with changes in eating patterns already discernible, for example the change away from eating large amounts of meat towards a diet with smaller consumption of higher-quality meat.[81]

Will organic farming cost more?

Organic food is sometimes priced a great deal higher than conventional food in shops. Part of the reason for this is higher distribution costs and, perhaps, because some stores are still treating it as a luxury food, and keeping supply and demand about the same. Organic producers get a 20–30 per cent premium above standard market value.

At the moment we all subsidise chemical agriculture very heavily. In 1988 the average family was paying £11 per week to subsidise farming in Britain, both through our own legislation and through the Common Agricultural Policy (CAP). This is not usually added in to calculations of the 'cheapness' of British food.[82]

Figures calculated in 1990 suggest that, at present, Britain's organic farmers receive, proportionately to the land area they farm, less than a hundredth of the support offered to chemical farmers in terms of grants, research funding, advice and price support.[83] Much of the research and development of organic methods has been carried out by independent bodies or charitable trusts. If organic farming is going to increase significantly over the next few years, more funding will undoubtedly be needed from government. This will include funding for research, grant support for farmers wishing to convert, and perhaps some other price-support mechanisms. If environmental costs were included in this system as suggested in the so-called 'Pearce Report', prepared for the Department of the Environment, this would further increase the financial viability of organic farming.[84]

Reduction Policies for Pesticides

Faced with growing public awareness about pesticides, several European countries are developing active pesticide policies to reduce the risk to human health and the environment from pesticide use. A central feature of these policies developing in Denmark, the Netherlands, Sweden and other countries is an overall target for reducing the quantity of pesticide use. All these countries have committed themselves to reductions of 25–50 per cent and more during the 1990s.[85] This has provided an impetus for considering other policy measures and defining exactly what is meant by reduction strategies: is it reducing the total number of pesticides available, the overall sales of pesticides, the quantities of active ingredients used, the area treated or the number of applications?

The reduction proposals have drawn attention to the fact that the number of pesticide applications, and types of certain pesticides, are rising in many areas of agriculture, whereas the total quantity of active ingredients is stable, and even declining in some countries, For example, herbicides in the UK.[86] In many respects it is more difficult to regulate or limit the number of applications in the field than it is to control the total quantity of active ingredients used. A variety of fiscal measures have also been proposed as part of pesticide policies. However, a tax, for example, on pesticide consumption would not necessarily be the best way of reducing application frequencies, although it may have an impact on the quantity used.[87]

There is considerable scope for using the CAP to encourage lower-input/lower-output farming, based on reduced chemical pesticide and fertiliser use, including incentives for farmers and growers to switch to organic production. At present an EC standard for organic produce is being negotiated. A satisfactory standard would make it much easier, not only to trade in organic produce, but to direct EC subsidies specifically at organic farmers and growers.

There is also fierce debate in the EC over whether agricultural and horticultural produce should be labelled informing the consumer that it has been treated with synthetic chemical pesticides. There is considerable industry resistance to these proposals and, to date, no decisions have been made.[88]

Some tentative moves have been made within the CAP to encourage low-input production. For example, there is the current scheme whereby Member States have to introduce extensification incentives. The current EC concept of extensification could be broadened so that aid is available not only to organic farmers and growers but to those adopting what the Dutch and Germans call 'integrated agriculture', which relates to reduced pesticide and fertiliser consumption, but not their total elimination. Such schemes are moving very slowly because of the absence of agreed labelling schemes, and limited government support.

The Swedish Action Programme

Starting in 1986, the Swedish government introduced a policy to reduce the use of agricultural pesticide by 50 per cent within five years.[89] Agriculture accounts for approximately one-third of total Swedish pesticide use, the remainder being used in amenity and industrial applications. Pesticide use in forests was stopped in the early 1980s. By the end of the 1980s significant progress had been made with the reduction programme. Pesticide usage had decreased to about 3,000 tonnes of active ingredient per year, equivalent to a 35 per cent reduction in the amount used annually between 1981–85. The full 50 per cent reduction target is likely to be achieved by 1995.[90] The reduction programme uses a variety of measures to achieve its aims, and farmers and growers are not compulsorily required to cut down on pesticide use.[91] The measures include:

- a sales tax (20 per cent) and environmental levy (10 per cent) on pesticides and fertilisers
- certification of approval of spray equipment (from 1991)
- use of lower dose rates
- mandatory training for professional pesticide users
- introduction of more stringent risk/benefit analysis for old and new products
- improved data on long-term health risks and potential effects on the environment
- introduction of stricter review procedures including demonstration of a need for the product
- prevention of the use of chemicals when other equally economical methods of control are available
- strengthening the agricultural advisory service

All 'old' pesticides are being systematically reviewed. Pesticides banned as a result of this safety review include the herbicides atrazine and bromacil, the insecticide lindane and mercury-based seed dressings.[92]

The Swedish government organisations responsible for implementing the reduction programme are the National Chemicals Inspectorate, the National Board of Agriculture, and the Environmental Protection Agency (EPA). According to the EPA, the fall in usage has been most marked with herbicides, which accounted for 75 per cent of pesticide use in agriculture between 1981–85.[93] The EPA also notes that the reduction programme has helped increase farm profitability.[94] The Swedish government is thought to be considering introducing a more comprehensive reduction programme to encompass other areas of pesticide use.

The Danish Pesticide Reduction Policy and Programme

In 1985, the Danish government announced a programme to reduce the use of pesticides in agriculture by 50 per cent by 1997, compared with average annual consumption over the period 1981–85.[95]

The Danish action plan began in 1987 and, according to the Danish National Environmental Research Institute (NERI), the interim aim of a 25 per cent reduction by early 1990 had already been achieved by 1988.[96] NERI also noted that, while volume usage of fungicide and insecticide active ingredients had fallen, herbicide and plant growth regulator use was not being reduced in line with government targets.[97] Following safety reviews of older pesticides, the Danish government had, by the end of 1989, banned a total of 32 pesticide active ingredients.[98]

An active research programme supports the reduction programme and gives a scientific base and authority to reduction measures and techniques. The Danish Institute of Weed Control is researching reduced herbicide use based on the twin strategies of developing methods for non-chemical weed control, and the use of 'factor-adjusted dosages'.[99]

In relation to factor-adjusted doses, field experiments in cereals have shown that the herbicide dose can often be reduced by one-third to two-thirds of the recommended dose without any significant loss of efficacy and yield.[100] An advisory computer data base on weeds and herbicide dose rates is being developed for weed control in spring barley and winter wheat. The aim is to allow farmers to adjust the herbicide dose according to weed species, growth stages and even climatic conditions.[101]

The Netherlands Reduction Policy and Programme

As part of a wider National Environment Plan, the Netherlands government aims to reduce pesticide use by 50 per cent by 1995, with a 70 per cent reduction by the year 2000, based on 1990 levels of usage.[102] The pesticide programme focuses on the reduction of intensive, environmentally damaging agricultural production and the promotion of organic farming. The Netherlands currently has the most intensive agricultural system in Europe with some 22,000 tonnes of pesticide active ingredient being used each year.[103] The rate of application per hectare is approximately four times that of the UK.[104]

The reduction programme calls for a tax on pesticides, and a ban on between one-quarter to one-third of the 320 active ingredients currently approved, especially those linked to water pollution.[105] Government expenditure on environmentally friendly farming is to be at least doubled by 1995.[106]

The UK Approach

In 1989 29,000 tonnes of pesticide active ingredient were applied in England and Wales alone. There is, however, no government reduction policy or programme in the UK similar to the ones that are being developed in other European countries. The 1990 White Paper on the environment – 'This Common Inheritance' – contains just one sentence advocating 'minimum' use of pesticides, but the decision on reduction is clearly left to the user.[107] This begs the question whether market forces alone – in the absence of any central, government strategy – will improve pest management and reduce pesticide use.

It is time the UK government drew up a reduction policy and programme, building on the knowledge and experience of its European counterparts. Authoritative voices are now calling for such policies. For example, the 1990 British Medical Association (BMA) report on pesticides called for a national pesticide policy coordinated by the government.[108] The BMA report stated: 'a pesticides policy would feature a time-tabled reduction of pesticides, as featured in the national pesticides policies of the Netherlands, Germany, and Scandinavia.'[109]

In reply, the UK government has denied it lacks a strategic policy for pesticides.[110]

Conclusion

The problems of pesticide use lead naturally to the development of non-chemical alternatives based on IPM systems, organic farming, and the development of pesticide reduction policies. This chapter has given an overview of current developments in these areas. The UK is still some way behind other European countries and the US in developing an overall strategy to reduce pesticide use. However it is clear that more work is needed to develop and implement alternatives in *all* countries.

13

Conclusions

When we started this book we thought it would be no more than a practical handbook, a sort of route map on pesticides. We recognised that there was a need to explain what pesticides were, what they did, and how they might harm people. It soon became evident, however, that this would not be enough. What was needed, in addition to a jobbing 'do-it-yourself' manual on pesticides, was a book which discussed the law; described what toxicity tests ought to be done; explained how information about risks is used by government and industry, and which provided workers with sufficient information to protect their own health. We realised that readers ought to know how pesticides are marketed in developing countries – often with damaging consequences.

There are alternative ways of growing crops, or controlling insects, that do not depend on dowsing everything with chemicals. Of equal importance are the efforts of farmers to control pests using far smaller quantities of pesticide in conjunction with nature's own predators. Managing pests in this integrated way reduces the quantity of chemicals put on the land, drastically curtails the pollution caused by leaching into water courses and helps curtail the problem of resistance to chemicals that has bedevilled so many public health programmes.

We hope we have achieved our aim. If we have, you will almost certainly agree that changes are necessary in the way pesticides are regulated and used. At present the information which determines whether a pesticide can be marketed is the sole preserve of the company which has developed the pesticide. The Ministry of Agriculture, Fisheries and Food (MAFF) obtains this information when reviewing pesticides, but does not make it public. This is not good enough. We believe it must be accessible if the public is to have confidence in the review system. Until it is, we remain sceptical about the whole process of review. MAFF performs a dual role, with responsibility for maximizing food production, and promoting standards of food production, which in practice means it frequently promotes the interests of the agro-chemical industry at the expense of low-input farming strategies. We do not believe that MAFF can discharge both responsibilities even-handedly, and would feel happier if the Health and Safety Executive

290

(HSE), with its specific remit for ensuring safe working practices, were to make decisions about pesticide safety.

We still have a long way to go in the UK before we have a national policy on pesticides that addresses these inadequacies. Developing countries face a more difficult job to secure a safe living and working environment. Pesticides banned in the West and Japan can still be exported to poorer developing countries. This trade in lethal chemicals has to stop and the prior informed consent agreement (p. 200) will help to do this. Consumer pressure on food manufacturers to eliminate pesticide residues may also help to slow this practice. Workers in the chemical and agricultural industries could help too by insisting that the working practices they have secured in the West also apply in the Third World.

Campaigns

Campaigning groups have brought about many of the improvements in pesticide use achieved to date. Ever since Rachel Carson wrote *The Silent Spring* in 1962,[1] organisations and individuals have fought against the use and abuse of toxic chemicals, spray drift, contaminated food and wildlife hazards. Many of the pesticide controls which exist today have only come about because of the pressure from campaigns. Actions have been local, national and international – often guided by the environmental maxim, 'Think globally, act locally.'

The village of Kimpton in Hertfordshire suffered so regularly from spray drift incidents from crop-spraying planes that in 1985 a group of some hundred local people signed a petition, saying:

'AERIAL CROP SPRAYING CAN DAMAGE YOUR HEALTH BUT CARRIES NO GOVERNMENT HEALTH WARNING!'[2]

The villagers complained about being sprayed directly while walking along public roads; not being informed about aerial spraying; and being told that spraying was taking place on the wrong day. Their actions were important because local people often find it difficult to complain about pesticides when those organising the spraying are their neighbours. Kimpton residents helped persuade the government to tighten aerial spraying regulations, and drew attention to the pesticide issue throughout the county.

Actions do not have to be well-funded or elaborate. Members of the Sussex Green Party have drawn attention to the issue of spray drift within the County by organising public meetings, with speakers from industry and the environmental lobby.[3] On each occasion the local radio stations gave interviews with the protesters and broadcast news about the problem throughout the day before the meeting, and local newspapers carried articles about what was said in the talks.

Local groups often find that support and publicity comes more easily when they are backed by official figures (like their MPs) or national organisations. Friends of the Earth (FoE) has played an active role in drawing public attention to the dangers of pesticides, and in lobbying for improvements. Important FoE action includes organising a poster advertising campaign drawing attention to pesticide residues in food, and publishing two volumes of 'Pesticide Incidents Reports' documenting experiences with pesticide contamination of people's health and environment throughout the UK.[4, 5] The incident reports have demonstrated that concerns about pesticides are very wide-ranging and do not just come from a few 'environmentalists and trouble makers'.

When 17 dead herons were found near Evesham in 1986, the local Friends of the Earth group discovered that the birds all had extremely high levels of DDT and other banned pesticides in their body fat: investigations implicated two local suppliers as the source. When FoE members posed as customers and asked to buy DDT, they were sold as much as they wanted.[6] FoE immediately circulated a press release naming the suppliers, and wrote a report on the incident, which received great publicity locally and nationally. FoE's prompt action stopped these particular supplies. More importantly, it showed that even after pesticides are banned they can remain available 'under the counter' and continue their environmental damage illegally and unrecognised.

The Ramblers' Association has long campaigned against use of irritating or toxic chemicals next to public rights of way, and is currently lobbying the government to introduce legislation on this issue. Many public footpaths are so badly maintained that walkers inevitably rub against vegetation as they pass by.

Research is important, and key areas are often ignored by official bodies. Over the last few years, there has been an increasing number of non-governmental organisations prepared to fund the studies they need to assess pesticide hazards. The London Food Commission has done a great deal to raise consumer awareness on safe foods. A 1986 study highlighted issues which many people find worrying – the incidence of pesticide residues in food; the lack of adequate regulations and testing procedures, and the known or suspected health effects connected with an extremely wide range of agrochemicals.[7] Making information accessible to the non-scientist, the report showed that the then current practices and regulations were inadequate for controlling residue levels on foods.

Growers with land bordering a sprayed area can suffer from spray drift and other contamination which ruins crops. This is a very real menace for organic farmers. Accordingly, the Soil Association, an organic growing organisation, has campaigned against spray drift through reports, parliamentary lobbying and by providing information on the legal rights of people suffering from spray drift. Although spray drift remains an important problem, legislation introduced in 1985 has increased controls, and these small advances only came about because the Soil Association, along with many other groups and individuals, protested so loudly that the government and farming lobby could no longer afford to ignore them.

Proper control and regulation of pesticides is a prime concern of trades unions, which have recognised that control of pesticides is an issue which affects the workplace, the public and the environment. In the early 1980s, unions with an interest in pesticides formed the informal Trade Union Pesticide Group, after publication of a book aimed at highlighting the hazards of 2,4,5-T, *Portrait of a Poison*.[8] The three main user unions in the UK – the General and Municipal Boilermakers' Union (GMB), the National Union of Public Employees (NUPE) and the Transport and General Workers' Union (TGWU) – together produced two *Hidden Peril* reports, including a survey of health effects amongst workers using pesticides.[9]

In London local safety representatives from the GMB, NUPE and the TGWU have successfully lobbied several local authorities to introduce stricter controls on pesticide use, including a ban on pesticides recognised as carcinogenic. Having agreed a ban on the more hazardous substances, the unions and councils set up a joint working party, the Dangerous Substances Working Party, to develop a comprehensive policy for pesticide use. These improvements came about because NUPE members in Waltham Forest organised a successful ban on Rassapron, the BP brand-name herbicide containing aminotriazole, a recognised carcinogen, and banned in a number of other countries, including Sweden.

At the other end of the scale, when a large, international organisation like Oxfam takes up the pesticide issue, it can help make major changes in policy throughout the world. David Bull's influential book on pesticides, *A Growing Problem*, was published by Oxfam.[10] Throughout the world groups are documenting pesticide hazards to health and the environment, and in the early 1980s groups from over 40 countries formed the Pesticides Action Network (PAN). A Regional PAN Coordinator in each continent helps planning on international issues, though groups are entirely independent. Most publish informative newsletters and journals – in many Third World countries they provide the only material warning of the hazards of pesticide use – and carry out educational and campaigning work at local and national levels. At an international level, it was largely due to lobbying by PAN groups which brought about the provision for Prior Informed Consent (see Chapters 9 and 11) in the FAO and UNEP codes governing pesticide exports.

Although many campaigning groups have become increasingly involved in pesticide issues over the past few years, they are reacting to, and backed up by, the actions of hundreds of individuals throughout the country. These people are writing letters to the local papers; complaining to their MPs; working within their local council; tackling their bosses at work, and creating an atmosphere where real changes and improvements have a chance of coming about. They have shown that the agro-chemical lobby is anything but invincible, and that real changes can be achieved if enough people are prepared to fight. To achieve further improvements we need a proper policy on pesticides and we set out below the minimum requirements for such a programme.

Pesticide Policy

The UK should have a more coherent Pesticides Policy based on the overall strategy of *reducing* the use of pesticides wherever possible, in the interests of human health and the environment.

A national Pesticides Policy would aim to:

- minimize pesticide use in agriculture and other areas for the benefit of human and environmental health;
- promote and fund research and development of Integrated Pest Management systems (including biological control), of which pesticide use is only one aspect;
- promote and fund alternative methods of agricultural production especially in relation to low-input and organic farming systems;
- the system reviewing the toxicity of pesticides must be more open, as in the US, and based on a precautionary approach.

A Pesticides Policy would be based on properly researched and conducted social, economic and environmental audits, weighing the merits of chemical control against risks to human and environmental health. For such a policy to operate effectively there would need to be greater freedom of access to the toxicity data on pesticides held by government.

Until such a Pesticides Policy is drawn up and implemented, following the example of countries like the Netherlands, Denmark, Sweden and Germany, a proper assessment of the future role of pesticides in food production and public health is impossible.

Recommendations

A Pesticides Policy is the first priority, within which the other conclusions which follow will operate.

Pesticide Law and Enforcement

The current UK regulatory system for pesticides has five major areas of weakness:

(1) the conflict of interest inherent in MAFF trying to regulate food production as well as pesticide safety;

(2) the lack of enforcement of pesticide laws due to lack of health and safety inspectors and resources;
(3) the lack of any legally empowered Safety Representatives in agriculture, horticulture and forestry in line with other areas of industry;
(4) the lack of an independent toxicity testing scheme;
(5) the lack of independent and public representation.

Conflict of Interest

Although the Advisory Committee on Pesticides (ACP) has an important role in vetting pesticides, the main Secretariat for all agricultural and horticultural pesticides is MAFF. The HSE provides the Secretariat for wood preservatives and other non-agricultural uses. In practice this gives undue influence to MAFF, whose main role is food production. In the US, faced with a similar dilemma, the government removed pesticide control from the US Department of Agriculture and gave it to the Environmental Protection Agency. The ideal would be to create a similar agency in the UK which would have overall responsibility for safety, and not just of pesticides.

Recommendation: In the UK, given current regulatory arrangements, responsibility for pesticide approval should be transferred to HSE, away from MAFF.

Enforcement

Lack of enforcement of pesticide laws is a major weakness. The HSE has too few inspectors (especially Agricultural and Factory Inspectors) to make regular visits to factories and farms, let alone cope with all the extra work generated by new pesticide laws. The main responsibility for enforcing pesticide laws falls on the HSE Agricultural Inspectorate. Visits to farms and horticultural premises have decreased in the 1980s as the number of inspectors has been reduced. According to figures by the Institution of Professional and Managerial Staffs (IPMS):

- labour-employing farms can expect a visit from an inspector every 5 to 8 years.
- self-employed farmers can be large farmers and major pesticide users. Their use of pesticides can pose a risk not only to themselves but to the public and environment. These farmers can expect a routine visit, or any form of health and safety inspection, once every 30 years.[11]

Recommendation: The number of HSE Agricultural and Factory Inspectors must be increased by at least 100.

Safety Representatives

British agricultural, horticultural and forestry workers have no legal right to appoint safety representatives to help protect their own health and safety or that of the public. Unlike other areas of industry, where safety representatives are permitted by law, agricultural and allied workers are exempted from the Safety Representatives and Safety Committee Regulations 1977, made under the Health and Safety at Work Act.

Recommendation: Agricultural horticultural and forestry workers should enjoy the same legal health and safety rights as other sections of the workforce.

Independent Toxicity Testing Scheme

Recommendation: There should be an independent toxicity testing scheme for pesticides, financed by the manufacturers, which would be able to investigate particular aspects of toxicity and provide advice to enforcement and regulatory agencies.

Representation

Recommendation: Public representation in the approval and review process is needed. There should be formal representation on the Advisory Committee on Pesticides (ACP) (or its equivalent) and Scientific Sub-committee by organisations representing consumers, workers, environmentalists and industry.

Information

Recommendations:

(1) full and open access to health and safety data including the basic toxicological data on which pesticide approval is granted by the ACP and its Scientific Sub-Committee;
(2) pollution information on pesticides should also be publicly available.

Incident Reporting and Monitoring

Recommendations:

(1) A national Pesticide Incidents Monitoring Scheme should be established to centralize and coordinate action on pesticide incidents. Given the current regulatory arrangements, it should be run by the HSE with other departments and agencies (such as the Department of Health) involved. The monitoring scheme's primary aim must be to reduce the number of incidents, therefore an enforcement agency must have the lead role.
(2) A comprehensive system, with adequate resources, should be established to monitor pesticide residues in food and water. The system should be independent of MAFF.

Application

Recommendations:

(1) A comprehensive programme of research, testing, and development of spraying systems and other application techniques which are safer to the operator, and environment, is an important priority.

(2) Application hazards should be controlled by the use of technical and engineering controls, as required by the COSHH Regulations, and *not*, as the industry has done in the past, by over-reliance on personal protective equipment as the main means of operator protection.

Disposal

Recommendations:

(1) Research is needed into improved means of disposal for concentrated and dilute pesticide, and for pesticide containers.

(2) Manufacturers should introduce and operate a system of 'reverse chain' disposal for revoked pesticides (where the approval is withdrawn or lapses), unwanted pesticide concentrate, as well as washed, empty containers. If the pesticide industry provided such a service and took back its waste this would have considerable environmental benefits.

Public Research and Advice

Recommendation: More public research and advice is needed. This would range from providing information to the public along US Environmental Protection Agency lines, to scientific research into specific antidotes for acute poisoning. More epidemiological research of exposed populations is also needed.

Training

Recommendations:

(1) Medical training for GPs and other medical and support staff should be improved. Occupational and environmental health should be part of the medical curriculum and a core item in any vocational and refresher training.

(2) There should be mandatory training and refresher training for all professional pesticide users, regardless of age or experience.

(3) The minimum age for anyone using a pesticide should be 18.

Impact on Employment and Communities

Research is needed on programmes to re-train, and reallocate, pesticide workers no longer involved in pesticide manufacture or use. This would include support and help for their local communities as well. *Recommendation:* Job areas requiring investigation are pesticide manufacture, employment opportunities in conservation, and sustainable agriculture.

Export Control

Recommendations: The UNEP and FAO codes relating to the export of hazardous substances, including pesticides, based on principles such as Prior Informed Consent, should be introduced. These recommendations should be incorporated into national laws.

Product liability

Recommendations:

(1) Manufacturers should operate a 'cradle to grave' philosophy for their products, and be more accountable for chemicals in the chain from production through to safe disposal or re-use.
(2) Environmental, health, community and consumer organisations need greater access to manufacturers' information on pesticides.

Labelling

Recommendations:

(1) Labelling of chemicals – industrial or household – must be improved to include better information on human and environmental hazards. The phrase 'harmful to the environment' should be included on new 'eco-labels', and the technical and engineering methods to control the chemical(s) must be clearly spelt out.
(2) Crops and produce treated with pesticides should be labelled 'treated with pesticides.'

Health Surveillance

Recommendations:

(1) Long-term, large-scale epidemiological studies should be carried out to determine the full extent of the pesticide problem affecting workers and the public.
(2) All pesticide users should have regular medical check-ups linked to the provision of a proper occupational health service for industry, including agriculture.
(3) Medical and exposure records should be kept for all pesticide users.

Work Environment Fund

Recommendations:

(1) A Work Environment Fund, financed by a 0.1 per cent payroll levy on all employers, should be set up. The fund would be used to finance the required health studies in relation to all forms of toxic substances
(2) Control of the fund by employers, unions and the HSE would ensure its use on practical research that would be disseminated to industry and services, as in Scandinavia.
(3) The fund would be used to help finance group Occupational Health and Safety services for small employers, as in Scandinavia.

Abbreviations

ACP	Advisory Committee on Pesticides	DAFS	Department of Agriculture and Fisheries, Scotland
ADAS	Agricultural Development and Advisory Service	DoE	Department of the Environment
ADI	Acceptable Daily Intake	EC	European Community/European Economic Community
AI	Agricultural Inspectorate		
BAA	British Agrochemicals Association	EHD	Environmental Health Department
BCPC	British Crop Protection Council, independent group, mainly consisting of people involved in the pesticide trade	EHO	Environmental Health Officer
		EPA	Environment Protection Agency (US)
		FAO	United Nations Food and Agriculture Organisation
BMA	British Medical Association	FDA	Food and Drug Administration (US)
BWPA	British Wood Preserving Association	FEPA	Food and Environment Protection Act 1985
CAP	Common Agricultural Policy	FoE	Friends of the Earth
CDA	Controlled Droplet Application	GIFAP	International Group of National Associations of Pesticide Manufacturers
CDFA	California Department of Food and Agriculture		
CIMAH	Control of Industrial and Major Accident Hazards Regulations	GMB	General and Municipal Workers' Union
		HDRA	Henry Doubleday Research Association
COPR	Control of Pesticides Regulations (under the Food and Environmental Protection Act 1985)	HMIP	Her Majesty's Inspectorate of Pollution
		HMSO	Her Majesty's Stationery Office (government publications)
COSHH	Control of Substances Hazardous to Health Regulations 1988, from HSE	HSC	Health and Safety Commission

HSE	Health and Safety Executive
IOCU	International Organisation of Consumer Unions
IPC	Integrated Pest Control
IPM	Integrated Pest Management
IPMS	Institution of Professionals, Managers and Specialists
MAC	Maximum Admissible Concentrations
MAFF	Ministry of Agriculture, Fisheries and Food
MRL	Maximum Residue Limit
NFU	National Farmers Union
NOEL	No-effect level
NRA	National Rivers Authority
NRDC	Natural Resources Defense Council (US)
NUPE	National Union of Public Employees
OECD	Organisation for Economic Cooperation and Development
PAN	Pesticide Action Network
PIC	Prior Informed Consent (provisions under the UNEP/FAO Codes)
PPE	Personal Protective Equipment
PSPS	Pesticide Safety Precaution Scheme (now defunct)
RSPB	Royal Society for the Protection of Birds
TBTO	Tributyltin Oxide
TLV	Threshold Limit Value
TGWU	Transport and General Workers Union
UNEP	United Nations Environment Programme
WHO	World Health Organisation
WOADS	Welsh Office of Agriculture and Development Service
WPPR	Working Party on Pesticides Residues

Glossary

This glossary contains words used in this book, as well as some common terms which you may come across in technical literature about pesticides.

a.c. aqueous concentrate

a.i. active ingredient

acaricide pesticide which kills spiders and mites

acetylcholine a compound released at the nerve ending, used in the transmission of the nerve impulse

acetylcholine esterase an enzyme that promotes the breakdown of acetylcholine

active ingredient that part of the pesticide which actually kills the pest

adjuvant thickening agent added to pesticides to reduce drift; e.g. gums, cellulose and polymers

aerial spraying applying pesticides by fixed-wing aircraft or helicopters

algicides fungicides controlling algae and other slime-type organisms

anticholinesterase substance that inhibits action of cholinesterase, especially in people sensitive to anticholinesterase compounds (about 15 per cent of the population)

aphicide pesticide which controls aphids

ataxia loss of control over the movements of the voluntary muscles used in walking; sometimes occurs as a result of acute pesticide poisoning

atropine an anti-cholinergic drug

avicides chemicals to control birds

bactericides pesticides controlling bacteria

biocide pesticide which kills a range of insects, animals, etc.

biological control regulation of plant and animal pests by means of encouraging/actively introducing their natural predators

biotechnology genetic manipulation of living matter to produce new plants, animals, material etc; large potential for crop protection purposes

bipyridyls a group of chemicals including the herbicides diquat and paraquat, which are used in the form of dichloride, dibromide or dimethosulphate salt

blow dispersal of granular pesticides away from intended target area by high winds

boom long sprayer attached behind tractor, through which pesticide is applied

carbamates insecticides made from an ester of carbamic acid, which have anti-cholinesterase properties

carcinogenic cancer-inducing

302

carrier liquid or solid added to pesticide to dilute it to help application

chelating agents drugs used to aid the elimination of metallic poisons from the body

chemosterilants chemicals which prevent insect reproduction

cholinesterase an enzyme found in blood plasma that breaks down acetylcholine; this helps regulate the activity of nerve impulses and is necessary for proper nerve functioning

chronic effect long-term health damage

contact herbicide a herbicide which kills primarily by contact with plant tissue rather than as a result of translocation; only the portion of the plant contacted is directly affected

cyanosis blueness of the skin caused by insufficient oxygenation of the blood, sometimes caused by pesticide poisoning

defoliants herbicides which kill green leaves and vegetation

degradability a pesticide's ability to break down (often in the soil) into less harmful by-products

dessicants chemicals which speed the drying of crops, making harvest easier

disinfectant especially applied to fungicides which clean or disinfect the surface of a seed or plant, usually prior to sowing (e.g. as a seed dressing)

diuresis increased production of urine, a symptom of some pesticide poisoning

d.p. dispersible powder

droplet drift drift of pesticide droplets away from target area

dyspnoea difficulty in breathing

e.c. emulsifiable concentrate

emulsion a fluid formed by the suspension of one liquid in another; especially a preparation of an oily substance suspended in a watery liquid – helps pesticides like difenzoquat to stay on foliage

enzyme a substance which has a specific action in promoting a biological change

erythrocytes red blood cells

esters derivatives of acids made by replacing hydrogen with alkyl radicals

fibrillation a small, involuntary, muscular contraction (e.g. a tic); symptom of some forms of pesticide poisoning

flash point lowest temperature at which a material forms a flammable vapour/air mixture in standard conditions

foliar acting a pesticide which is active when applied to foliage

formulation active ingredient plus any other chemicals needed (e.g.: wetting agent, dilutant etc.)

f.p. freezing point

fungicide pesticide which kills fungi

gastric lavage washing out the stomach to remove toxic materials

harvest interval the minimum number of days required by law between the final pesticide application and harvesting the crop

herbicide pesticide which kills plants

hormone weedkiller phenoxy-acid herbicides

insecticide pesticide which kills insects

integrated pest management (IPM) system of pest control which uses an integrated mixture of control measures, including pesticides, time of planting, biological control, use of crop rotations, new crop varieties, etc.; almost always results in far less pesticide being used

keratitis inflammation of the cornea of the eye; one of the commonest side effects of being caught in pesticide spray

larvicide pesticide which controls larval pests

LC 50 concentration required to kill 50 per cent of test animals in water; common measure of toxicity

LD 50 dose required to kill 50 per cent of test animals

Limit of detection lowest concentration of pesticide that can be detected, e.g. on crops or in blood

lipid any group of organic substances (e.g. fat) that does not dissolve in water; general term used for oils, fats and waxes found in living tissue

molluscicide pesticide which kills snails and slugs

m.p. melting point

Maximum residue level (MRL) maximum permitted concentration of a pesticide residue in or on a food crop

μ m micrometre or micron

mutagen induces gene mutations passed on to the next generation

myasthenia gravis muscular weakness, symptom of pesticide poisoning

n.a.d. number average diameter

nematicide pesticide which kills nematodes (eelworms)

nephrotoxicity a toxic action on the kidney

neurotoxin chemical which can damage nerve tissue

n.m.d. number mean diameter

non-selective (or total) a pesticide which kills all species, harmful or otherwise; herbicides are sometimes called total herbicides (although there are attempts to drop this term), while total insecticides are usually known as 'non-selective'

Occupational Exposure Limits see Threshold Limit Values

oligospermic low sperm density (less than 40 million sperms per ml of semen) caused by some pesticides

opisthotonus spasms in back muscles caused by body arching, could be a side effect of pesticide poisoning which causes convulsions

organic farming farming system which does not use artificial pesticides at all

organochlorine pesticide (also known as chlorinated hydrocarbons), one of the hydrocarbons with a certain number of chlorine atoms which give it insecticidal properties; also among the most persistent and toxic groups of pesticide, including DDT

organophosphorus ester of phosphoric acid with insecticidal pesticide properties, acting mainly on anticholinesterase enzyme; also referred to as organophosphates

paraesthesia numbness and tingling; symptom of some types of pesticide poisoning

percutaneous ability of chemical to cause injury as a toxicity result of skin penetration or absorption

peripheral neuropathy a condition of the arm and leg nerves; resulting in loss of function and sensation

peristalsis normal automatic muscular movement of the intestine

persistence length of time a chemical remains in the environment

pesticide general term used to describe all insecticides, herbicides, fungicides as well as miscellaneous chemicals

pesticide resistance/weed resistance resistance developed by some pest species to specific pesticides, rendering them worthless or less effective

pheremones synthetic chemicals which mimic natural insect attractants (also called pheremones) and attract insects into traps or into sprayed areas

post-em post-emergence

ppmv parts per million by volume

pre-em pre-emergence

protectant pesticides (usually fungicides) which coat the plant leaf and prevent fungal/bacterial spore germination and penetration; they only work if applied before a disease is present

pyrexia fever condition

Q application rate (litres/hectare) at dose rate producing toxicity

q application rate (litres per square metre) in stated species

residual a soil-acting or applied pesticide which remains active in the soil for a period of time, which may be weeks, months, or even years

residue small proportion of pesticide left behind on food or in water, and liable to be consumed or to damage wildlife

resistance ability to withstand pesticides, sometimes developed in certain strains of a pest, which quickly spread as others are killed off

r.h. relative humidity

rodenticide pesticide which kills rodents (usually rat poisons)

safener chemical reducing the potential of pesticides to damage crop species as well as pests

s.c. suspension concentrate

seed treatment pesticide for coating or impregnating seeds, often against fungal attack

smokes pesticides burnt in confined spaces, releasing poisons into the air

soil-acting a pesticide which is active within the soil, controlling soil creatures, germinating weeds and fungal/bacterial micro-organisms

solvent chemicals , often toxic in their own right, used in the formulation of pesticides

s.p. soluble powder

spasticity excessive tension of the muscles characterised by rigidity and increased reflexes

spray drift general term for droplet drift, vapour drift and blow

surfactant wetting agent, e.g. a substance such as a detergent which reduces surface tension

systemic refers to pesticides which remain active within the plant, moving around in sap, fluids etc. and controlling pests and diseases at different parts of the plant

TD toxic dose (lowest observed)

teratogen induces birth defects

Threshold Limit Value (TLV) values set which are believed to be the maximum airborne concentration that someone can be repeatedly exposed to on a daily basis without adverse health effects; also referred to as Occupational Exposure Limits; set for some pesticides

toxicity ability of a substance to cause damage to tissue

translocated a pesticide which is moved within the plant and can affect parts of the plant remote from the point of application, resulting in pest control

ULV ultra-low volume spraying

UV ultra-violet

v.a.d. volume average diameter (of spray drop)

vapour drift drift of pesticide which has evaporated after landing on the crop or soil

v.m.d. volume mean diameter (of spray droplet)

v.p. vapour pressure

wettable powder pesticide in powder form, used by mixing with water

wetting agent chemical which makes the formulation 'wetter' and hence more liable to stick to the target.

wood preservative pesticide used to control wood boring insects, fungi etc.

w.p. wettable powder

w.s.c. water-soluble concentrate

Contacts

Statutory Organisations

Advisory Committee on Pesticides (ACP)
c/o Ministry of Agriculture, Fisheries and Food (see below)
Agricultural and Food Research Council (AFRC)
160 Great Portland Street, London W1N 6DT, tel 071-580-6655
Agricultural Development and Advisory Service (ADAS)
c/o Ministry of Agriculture, Fisheries and Food, Whitehall Place, London SW1A
  2HA, tel 071-270-3000
Department of Agriculture and Fisheries for Scotland: (DAFS)
  Chesser House, 500 Gorgie Street, Edinburgh EH11 3AW, tel 031-443-
  4020
Department of Agriculture for Northern Ireland
Dundonald House, Upper Newtownards Road, Belfast BT4 3SB, tel 0232-650-
  111
HM Agricultural Inspectorate: Health and Safety Executive
Daniel House, Trinity Road, Bootle, Merseyside L20 3QZ, tel 051-951-4000
Ministry of Agriculture, Fisheries and Food
Pesticides Safety Division, Ergon House, c/o. Nobel House, 17 Smith Square,
  London SW1P 3JR, tel 071-270-3000
Plant Breeding Institute (PBI)
Maris Lane, Trumpington, Cambridge CB2 2LQ, tel 0223-840-411

Agricultural and Agrochemical Industry Groups

British Agrochemicals Association (BAA)
4 Lincoln Court, Lincoln Road, Peterborough PE1 2RP, tel 0733-49225
British Crop Protection Council (BCPC)
144–150 London Road, Croydon, Surrey CR0 2TD, tel 081-681-6851
National Farmers' Union (NFU)
Agriculture House, 25–31 Knightsbridge, London SW1X 7NJ, tel 071-235-5077
National Federation of Agricultural Pest Control Societies
Agriculture House, 25–31 Knightsbridge, London SW1X 7NJ, tel 071-235-8440
National Institute of Agricultural Engineering
Wrest Park, Silsoe, Bedford, MK45 4HS, tel 0525-60000
Royal Agricultural Society of England (RASE)
National Agriculture Centre, Stoneleigh, Kenilworth, Warwickshire CV8
  2LZ, tel 0203-555-100

Non-governmental Organisations

Farming and Wildlife Advisory Group (FWAG)
The Lodge, Sandy, Bedfordshire SG19 2DL, tel 0767-80551
Friends of the Earth (FoE)
26–28 Underwood Street, London N1 7JQ, tel 071-490-1555
Green Alliance
49 Wellington Street, London, WC2N 4HS, tel 071-836-0341
Greenpeace
Greenpeace House, Canonbury Villas, London N1 2PN, tel 071-354-5100
Parents for Safe Food
c/o National Food Alliance, 102 Gloucester Place, London W1H 3DA, tel 071-935-2099
Pesticides Action Network
an international network of groups which research and campaign on pesticide issues. There is a regional office for each part of the world; in Europe the address is 23 Beehive Place, London SW9 7QR.
Soil Association
86 Colston Street, Bristol BS1 5BB, tel 0272-290-661
The Pesticides Trust: 23 Beehive Place, London SW9 7QR, tel 071-274-8895

Trades Unions

Bakers, Food and Allied Workers' Union
Stanborough House, Great North Road, Stanborough, Welwyn Garden City, Hertfordshire AL8 7TA, tel 07072-60150
GMB (General Municipal and Boilermakers' Union)
22–24 Worple Road, London SW19 4DD.
Institution of Professionals, Managers and Specialists (IPMS)
75–79 York Road, London SE1 7AQ, tel 071-928-9951
Manufacturing Science Finance (MSF)
79 Camden Road, London NW1 9ES, tel 071-267-4422
National Union of Public Employees (NUPE)
Civid House, 20 Grand Depot Road, London SE18 6SF, tel 081-854-2244.
Transport and General Workers' Union (TGWU)
Transport House, Smith Square, London SW1P 3JB, tel 071-828-7788

International Organisations

Commonwealth Institute of Biological Control
Imperial College, Silwood Park, Ascot, Berkshire SL5 7PX, tel 0990-28426
International Register of Potentially Toxic Chemicals
(part of the UN Environment Programme) Palais des Nations, 1211 Geneva 10, tel. 41 (22) 798-84-00
Pesticide Action Network (PAN Europe)
23 Beehive Place, London SW9 7QR.
United Nations Food and Agricultural Organisation
Via delle Terme di Caracalla, 00100, Rome, Italy, tel 39 (6) 5797-1
Worldwide Fund for Nature (WWF)
CH-1196 Gland, Switzerland, tel (022)-64-71-81

Books About Pesticides

Books on pesticides vary in quality and price. We have selected some which we think will be helpful.

Bull, David, *A Growing Problem: Pesticides and the Third World Poor*, Oxfam, 1982.
Carson, Rachel, *The Silent Spring*, Penguin, 1962.
Clutterbuck, Dr C. and Lang, Dr Tim, *P is for Pesticides*, Random House, 1991.
Cook, Judith and Kaufman, Chris, *Portrait of a Poison: the 2,4,5-T Story*, Pluto Press, 1982, (out of print).
Codex Alimentarius Commission, *Recommended International Maximum Limits for Pesticide Residues*, Food and Agriculture Organisation, Rome, (published periodically).
Dudley, Nigel, *This Poisoned Earth: the Truth about Pesticides*, Piatkus Press, 1987.
Hay, Alastair, *The Chemical Scythe: lessons of 2,4,5-T and dioxin*, Plenum Press, 1982.
Matthews, Graham, *Pesticide Application Methods*, Longman, 1982.
Morgan, D.P., *Recognition and Management of Pesticide Poisonings*, Environmental Protection Agency, Washington DC, 1982.
Pesticides Trust, *Pesticides, Policies and People*, London, 1991.
Royal Society of Chemistry, *European Directory of Agrochemical Products* (Four Vols: 1. Fungicides, 2. Herbicides. 3. Insecticides and Acaricides, and 4. Plant growth regulators, etc.),1986.
Royal Society of Chemistry, *Agrochemicals Handbook*, 1987.
Sax, N.I., *Dangerous Properties of Industrial Materials* (3 volumes), van Nostrand Reinhold, 1988.
United Nations, *Consolidated list of products whose consumption and/or sale have been banned, withdrawn, severely restricted or not approved by governments*, UN, New York, 1988.
Van den Bosch, Robert, *The Pesticide Conspiracy*, Doubleday, New York, 1978.
Watterson, Andrew, *Pesticide Users' Health and Safety Handbook*, Gower Technical, 1988.
Weir, David, *The Bhopal Syndrome: Pesticide Manufacture and the Third World*, International Organisation of Consumer Unions, Panang, 1986.
Weir, David and Schapiro, Mark, *Circle of Poison: Pesticides and People in a Hungry World*, Institute of Food and Development Policy, San Francisco, 1981.
World Health Organisation, *Guidelines to the use of the WHO recommended classification of pesticides by hazards*, WHO, Geneva, 1982.

References

Chapter 2

1. International Group of National Associations of Manufacturers of Agrochemical Products (GIFAP), *Pesticide Residues in Food*, Bruxelles, Belgium, 1984.
2. British Agrochemicals Association, *Pesticides in Perspective: why use pesticides?* Peterborough, UK, undated leaflet.
3. ICI Agrochemicals, *Food for Thought*, Farnham, UK 1990.
4. Ministry of Agriculture, Fisheries and Food (MAFF), *Pesticides and Food: A balanced view*, London, UK, undated booklet.
5. Arnold, E. *Infection Control*, British Medical Association Guide. London, 1989.
6. Ibid.
7. London Hazards Centre. *Toxic Treatments: Wood preservative hazards at work and in the home*, London Hazards Centre Handbook, 1989.
8. Watterson, A. *Pesticide Users' Health and Safety Handbook*, Gower Technical, Aldershot, UK, 1988, p. 315.
9. Ibid., p. 261.
10. United States Environmental Protection Agency: Office of Pesticides and Toxic Substances, *Citizen's Guide to Pesticides*, Washington DC, 1987.
11. HSE/MAFF, 'Labelling of timber treatment products', *The Pesticides Register*, No. 12, 1989, p. 1.
12. British Wood Preserving Association, *Preserving Confidence in Timber*, London, undated leaflet.
13. Ordish, G. *The Constant Pest: A short history of pests and their control*, Scientific Book Club, 1976.
14. British Agrochemicals Association, *Annual Review and Handbook*, Peterborough, UK, 1989/90.
15. US Environmental Protection Agency, 'The Economics of Pesticides', in *A Consumer's Guide to Safer Pesticide Use*, Washington DC, 1987, p. 3.
16. Erlichman, J. *Gluttons for Punishment*, Penguin Special, London, 1986.
17. British Agrochemicals Association, *Annual Report and Handbook*, Peterborough, UK, 1988/89.
18. The figures in this paragraph are taken from: Agricultural Development Service, *Arable Farm Crops 1988: Pesticide usage survey report 78*, Reference Book 578, MAFF Publications, London, 1990.
19. Ministry of Agriculture, Fisheries and Food, *A Guide to Veterinary Pesticides*, Reference Book 245, HMSO, London, 1984.
20. Mills, S. 'Salmon farming's unsavoury side: Sterner controls are needed to protect the environment from fish farms', *New Scientist*, 29 April 1989, pp. 58–60.

21. Forestry Commission, *Provisional Code of Practice for the Use of Pesticides in Forestry*, Occasional Paper 21, Edinburgh, UK, 1989.
22. Cook, J. and Kaufman, C. *Portrait of a Poison: the 2,4,5-T Story*, Pluto Press, London, 1982.
23. Ministry of Agriculture, Agricultural Development and Advisory Service (ADAS), *Pesticide Usage Survey Report, 1987*, HMSO, London, 1987.
24. 'Locusts Thrive as Pesticide Ban Stays', *New Scientist*, 31 March 1988, p. 17.
25. Friends of the Earth, *An Investigation of Pesticide Pollution in Drinking Water in England and Wales*, Special Report London, 1988.
26. British Agrochemicals Association, *Chemicals and the Amenity User*, Peterborough, UK, undated leaflet.
27. The figures in this paragraph are taken from Environmental Data Services, 'Survey paves way for controls on non-agricultural herbicides', *ENDS Report 191*, December 1990, p. 8.
28. Transport and General Workers' Union, Textile Group, *Chemicals and Dyestuff Briefing Service*, London, 1989.
29. Fox, B. 'Mothballs Mothballed', *New Scientist*, 5 August 1989, p. 34.
30. US Environmental Protection Agency, *Child-Resistant Packages for Pesticides*, Washington DC, 1985.
31. Arnold, E. *Infection Control*, British Medical Association Guide, London, 1989.
32. HSE/MAFF, 'Review of tributyltin oxide (TBTO) wood preservatives and surface biocides', *The Pesticides Register*, No. 5, 1990, p. 1.
33. British Agrochemicals Association, *Directory of Garden Chemicals 1989/90*, Peterborough UK, 1989.
34. The Soil Association, *How Does Your Garden Grow?*, Bristol, UK, 1986.
35. 'UK Import and Export Statistics, 1987', in *Report on Pesticide Production and Export in the UK*, prepared by Transnationals Information Centre, London, for Greenpeace International, January 1989, unpublished.
36. UN International Programme on Chemical Safety, *Paraquat and Diquat*, Environmental Health Criteria 39, World Health Organization, Geneva, 1984.
37. Transnationals Information Centre London, *Report on Pesticide Production and Export in the UK*, unpublished, January 1989.
38. Drew, B. UK General Manager of Rhone Poulenc, quoted in *Agchem Journal*, British Agrochemical Association, September/October 1988.
39. Pimental, D. et al. 'Pesticides, Insects in Foods and Cosmetic Standards', *Bio Science*, Volume 27, No. 3, 1977, pp. 178–85.
40. Friends of the Earth, *Pesticides Briefing Sheet*, London, 1987.
41. Greenpeace USA, *The International Trade in Wastes*, (4th edition), 1989.
42. The Ramblers' Association, *Pesticides: Waging war on our countryside*, London, 1988.
43. TGWU/NUPE/GMB unions, *Pesticides: The Hidden Peril*, London, 1987.
44. International Organisation of Consumers' Unions, Pesticide Action Network, *Problem Pesticides: Pesticide Problems*, (2nd Edition), Penang, Malaysia, 1988.
45. Graham J. *Pesticides: A Consumer View*, Paper at the Rural Seminar, Wye College, 1989.
46. Bull, D. *A Growing Problem: Pesticides and the Third World*, Oxfam, Oxford, UK, 1982.
47. 'Farming vote spurns green agriculture', *The Guardian*, 15 February 1990.
48. British Organic Farmers' Association, *Organic Farming: An Option for the Nineties*, Bristol, 1990.
49. British Medical Association, *Pesticides, Chemicals and Health Report*, London, in process of publication.
50. Pesticides Trust, *The Pesticides Trust: An Introduction*, London, 1989.
51. Pesticide Action Network, *Breaking the Circle of Poison*, undated leaflet.
52. Glotfelty, D.E. et al. 'Pesticides in Fog', *Nature*, Volume 325, 1987, pp. 602–5.

53. Savage, E.P. et al. 'National study of chlorinated hydrocarbon insecticide residues in human milk', *American Journal of Epidemiology*, Volume 113, 1981, p. 413.

54. Barnet, R.W. et al. 'Organochlorine pesticide residues in human milk samples from women living in Northwest and Northeast Mississipi', *Pesticide Monitor*, Volume 13, 1979, pp. 47–51.

55. Le Marchand et al. 'Trends in birth defects for a Hawaiian population exposed to heptachlor and for the United States', *Archives of Environmental Health*, Volume 41, 1986, pp. 145–8.

56. US Environmental Protection Agency, *A Consumer's Guide to Safer Pesticide Use*, Washington DC, 1987, p. 3.

57. European Institute for Water, 'Report on the Seminar on the EEC Directive 80/778 on the quality of water intended for human consumption', 1988.

58. Ministry of Agriculture, Fisheries and Food, *Working Party on Pesticides Residues report 1985–88*, HMSO, London, 1989.

59. Health and Safety Executive (HSE), Statement made by the HSE Chief Inspector of Agriculture at 'The Killing Fields' press conference, London, May 1988.

60. Moses, M. 'Pesticide related health problems and farmworkers', *AAOHN Journal*, Volume 37, No. 3, 1989, pp. 115–36.

61. UK Parliamentary Select Committee on Agriculture, *Report on the Effects of Pesticides on Human Health*, 1989.

62. Moses, M. *A field survey of pesticide-related working conditions in four locations in the US and Canada*, Pesticide Action Network International, 1988.

63. Health and Safety Executive, *Pesticide Manufacturing Survey*, Bootle, verbal report, 1989.

64. Hay, A. *The Chemical Scythe: Lessons of 2,4,5-T and Dioxin*, Plenum Press, London, 1983.

65. 'Brussels drops need for lethal animal tests', *New Scientist*, 7 October 1989, p. 21.

66. Institution of Professional, Managerial and Scientific Staff, *Health and Safety: Alternative Report*, London, 1989.

67. 'In the US, enforcement of rules on pesticides is spotty and anaemic', *The Wall Street Journal*, Volume VI, No. 253, 26 January 1989.

68. Ibid.

69. Green, M.B. et al. *Chemicals for Crop Protection and Pest Control*, Pergamon International Library, 1987.

70. Royal Commission on Environmental Pollution, *Agriculture and Pollution*, 7th Report, HMSO, London, 1979.

71. Van Den Bosch, R. *The Pesticide Conspiracy*, Prism Press, Dorchester, UK, 1978.

72. Swedish National Board of Agriculture, *Swedish action programme to reduce the risks to health and the environment from pesticide use in agriculture*, 1988.

Chapter 3

1. Sittig, M. (ed.), *Pesticide Manufacturing and Toxic Materials Control Encyclopedia*, Noyes Data Corporation, USA, 1980, pp. 5–10.

2. British Crop Protection Council, *Weed Control Handbook: Principles*, (7th edition), Blackwell Scientific Publications, London, 1982.

3. Dreisbach, R.H. *Handbook of Poisoning: Prevention, Diagnosis and Treatment*, Lange Medical Publications, California, 1989.

4. MAFF, *Data requirements for approval under the Control of Pesticides Regulations*, Appendix 8, London, October 1986.

5. Health and Safety Executive/MAFF, *Pesticides 1990: Pesticides approved under the Control of Pesticides Regulations 1986*, HMSO, London, 1990 (published annually).
6. Canadian Safety Council, *Pesticides Data Sheet*, Ontario, Canada, 1981.
7. Frankel, M. *Chemical Risk: a workers' guide to chemical hazards and data sheets*, Pluto Press, London, 1982.
8. Health and Safety Executive, *Substances for use at work: the provision of information*, Booklet HS (G) 27 (rev.), HMSO, London, 1988.
9. Health and Safety Executive, *Articles and substances used at work*, HSE Leaflet, IND (G) 1(L) Rev., London, 1987.
10. Health and Safety Executive, *COSHH Assessments*, HMSO, London, 1988.
11. Transport and General Workers Union, *How to obtain health and safety information*, TGWU, London, 1989.
12. British Crop Protection Council, *Pest and Disease Control Handbook*, N. Scopes (ed.), Blackwell Scientific Publications, London, 1982.
13. Worthing C.S. (ed.),*The Pesticides Manual: a world compendium*, (8th edition), British Crop Protection Council, Bracknell, UK, 1987.
14. Royal Society of Chemistry, *The Agrochemicals Handbook*, Nottingham, 1987.
15. Audus, L.J. 'Herbicides: physiology, biochemistry', in *Physiology, Biochemistry and Ecology*, Vol. I, Academic Press, London, 1976.
16. Ashton, et al. *Mode of action of herbicides*, (2nd ed), Wiley-Interscience, London, 1981.
17. Hayes, W.J. *Pesticides Studied in Man*. Williams and Wilkins, London, 1982.
18. International Labour Office, *Encyclopaedia of occupational health and safety* Volume 2, (3rd revised edition), Geneva, 1983, pp. 1,616–46.
19. Environmental Protection Agency (EPA), *Recognition and Management of Pesticide Poisonings*, (4th edition), Washington DC, 1989.
20. Ibid.
21. Ibid.
22. World Health Organisation, *The WHO Recommended Classification of Pesticides by Hazard and Guidelines to Classification 1988–89*, WHO, Geneva.
23. Hay, A. 'How the chemical industry could clean up its act', *New Scientist*, London, 12 February 1987, pp. 63–4.

Chapter 4

1. Pesticides: Code of Practice for the Safe use of Pesticides on Farms and Holdings, prepared jointly by the Ministry of Agriculture, Fisheries and Food, the Health and Safety Commission and the Department of the Environment, HMSO, 1990. This Code combines guidance under both Part III of the Food and Environment Protection Act 1985 and the Health and Safety at Work Act, 1974.
2. Graham-Bryce, I.J. 'Crop protection; a consideration of the effectiveness and disadvantages of current methods and of the scope for improvement', *Philosophic Transactions of the Royal Society of London*, Series B 281, 1977, pp. 163–79.
3. Norby, A. and Skuterad, R. 'The effects of boom height, working pressure and windspeed on spraydrift', *Weed Research*, 14, 1985, pp. 385–95.
4. Study for the HSE, carried out by the Ergonomics Department, Loughborough University, and reported to the Trade Union Pesticide Group in 1988.
5. British Agrochemicals Association Ltd, *Spray Operator Safety Study*, Report of a study carried out in conjunction with The Robens Institute of Occupational Health and Safety, University of Surrey, 1983.

6. Rose, C. *Pesticides: The First Incidents Report*, Friends of the Earth, London, 1985. Evidence presented to parliament by FoE.
7. Anon., *Newcastle-Upon-Tyne Journal*, June and August, 1985.
8. Dudley, N. 'Safety Never Assured: The case against aerial spraying', *Food and Environment Report No. 2*, The Soil Association, Haughley, Suffolk, 1985.
9. Civil Aviation Authority, *Information on Requirements to be met by Applicants and Holders of the Aerial Application Certificate*, available from CAA, Printing and Publication Section, Greville House, 37 Gratton Road, Cheltenham, Gloucestershire, GL50 2BN, UK. See also *The UK Pesticide Guide*, from the British Crop Protection Council, published annually, which contains detailed information on legal requirements about aerial application, and a list of products.
10. *Guidelines for the Use of Herbicides on Weed in or near Watercourses and Lakes*, Booklet B2076, Ministry of Agriculture, Fisheries and Food, updated regularly, £2.50.
11. Thorpe, V. and Dudley, N. *Pall of Poison*, The Soil Association, Bristol, UK, 1984.
12. Dodd, P. 'Drifting into Danger, Part 2', *British Farmer and Stockbreeder*, 29 March 1980.
13. Chapman, E. The Pesticide Exposure Group of Sufferers, personal communication.
14. Sotherton, N.W., Rands, M.R.W., Moreby, S.J. 'Comparisons of herbicide treated and untreated headlands for the survival of game and wildlife', *British Crop Protection Council Conference: Weeds*, pp. 221–998, BCPC Publications, 1985.
15. These and other examples are collected in: Dudley, N. *Safety Never Assured*.
16. HSE, AIAC, 'Synopsis of AFRC Research into Spray Drift', HSE ChemAG 90/5, 1990.
17. Ibid.
18. Cross, J.V. et al. *Studies of Spray Drift*, ADAS, Wye, UK, 1988.
19. Rose, C. *Pesticides: The First Incidents Report*.
20. *Storage of Approved Pesticides: Guidance for farmers and other Professional Users*, HSE Guidance Note CS19, HMSO, London, 1988.
21. Edling, C. et. al. 'New Methods for Applying synthetic pyrethroids when planting conifer seedlings: symptoms and exposure relationships', *Annals of Occupational Hygiene*, Volume 29, No. 3, 1985, pp. 421–7.
22. HSE, *Protective Clothing for Use with Pesticides*, Leaflet A26 75M, reprinted 1987.
23. Wakes Miller, Dr. C. 'Heat Exhaustion in Tractor Drivers', *Occupational Health*, August 1982, pp. 361–6.

Chapter 5

1. Council on Scientific Affairs, American Medical Association, 'Cancer risk of pesticides in agricultural workers', *Journal of the American Medical Association*, Volume 260, 1988, 959–66.
2. Barthel, E. 'Increased risk of lung cancer in pesticide exposed male agricultural workers', *Journal of toxicology and environmental health*, Volume 8, 1981, pp. 1,027–40.
3. Nelson, W.C., Likins, M.H., Mackey, J., Newill, V.A., Finklea, J.F. and Hammer, D.I. 'Mortality among orchard workers exposed to lead arsenate spray: a cohort study', *Journal of Chronic Diseases*, Volume 26, 1973, pp. 105–18.
4. Wiklund, K.G., Daling, J.R., Allard, J. and Weiss, N.S. 'Respiratory cancer among orchardists in Washington State 1968–1980', *Journal of Occupational Medicine*, Volume 30, 1988, pp. 561–4.

5. Luchtrath, H. 'The consequences of chronic arsenic poisoning among Moselle wine growers', *Journal of Cancer and Respiratory Clinical Oncology*, Volume 105, 1983, pp. 173–82.

6. International Agency for Research and Cancer, *Monographs on the Evaluation of Carcinogenic Risks to Humans*, Supplement 7, Lyons, France, 1987.

7. Council on Scientific Affairs, American Medical Association, 'Cancer Risk of Pesticides in Agricultural Workers', *Journal of the American Medical Association*, Volume 260, 1988, pp. 959–66.

8. Ibid, p. 964.

9. Ditraglia, D., Brown, D P., Karta, T. and Iverson N. 'Mortality Study of Workers employed at Organo-Chlorine Pesticide Manufacturing Plants', *Scandinavian Journal of Work and of Environmental Health*, Volume 7, 1981, pp. 140–6 .

10. Ribbens, P.H. 'Mortality study of industrial workers exposed to aldrin, dieldrin and endrin', *Internal Archives of Occupational and Environmental Health*, Volume 56, 1985, pp. 75–9.

11. Council on Scientific Affairs, American Medical Association, 'Cancer. Risk of Pesticides in Agricultural Workers', *Journal of the American Medical Association*, Volume 260, 1988, pp. 959–66.

12. Axelson, O. and Sundell, L. 'Herbicide exposure, mortality and tumour incidence. An epidemiological investigation on Swedish railroad workers', *Scandinavian Journal of Work and Environmental Health*, Volume 2, 1974, pp. 21–8.

13. Council on Scientific Affairs, American Medical Association, 'Cancer Risk of Pesticides in Agricultural Workers', *Journal of the American Medical Association*, Volume 260, 1988, pp. 959–66.

14. Ibid.

15. Ibid.

16. Donna, A., Crosignani, P., Robutti, F., Betta, P.G., Bocca, R., Mariani, N., Ferrario, F., Fissi, R. and Berrino F. 'Triazine herbicides and ovarian epithileal neoplasms', *Scandinavian Journal of Work and Environmental Health*, Volume 15, 1989, pp. 47–53.

17. Borzsonyi, M., Torok, G., Pinter, A., et al. 'Agriculturally related carcinogenic risk', *International Agency for Research on Cancer Scientific Publications*, Volume 56, 1984, pp. 465–86.

18. Council on Scientific Affairs, American Medical Association, 'Cancer Risk of Pesticides in Agricultural Workers', *Journal of the American Medical Association*, Volume 260, 1988, pp. 959–66.

19. Moses, M. 'Diseases associated with exposure to chemical substances: pesticides', in Mexey-Resenau, *Public Health and Preventive Medicine*, Last, J.M. (ed.), (Edition 12), Appleton Century Crofts, East Norwalk, Connecticutt, 1986.

20. Council on Scientific Affairs, American Medical Association, 'Cancer Risk of Pesticides in Agricultural Workers', *Journal of the American Medical Association*, Volume 260, 1988, pp. 959–66.

21. Ibid.

22. Ibid.

23. Wang, H.H. and MacMahon, B. 'Mortality of workers employed in the manufacture of chlordane and heptachlor', *Journal of Occupational Medicine*, Volume 21, 1979, pp. 745–8.

24. Shindell, S. and Ulrich, S. 'Mortality of workers employed in the manufacture of chlordane: an update', *Journal of Occupational Medicine*, Volume 28, 1986, pp. 497–501.

25. Ditraglia, D., Brown, D.P., NameKata, T. and Iverson, N. 'Mortality study of workers employed at organochlorine pesticide manufacturing plants', *Scandinavian Journal of Work and Environmental Health*, Volume 7, 1981, pp. 140–6.

26. Wang, H.H. and MacMahon, B. 'Mortality of pesticide applicators', *Journal of Occupational Medicine*, Volume 21, 1979, pp. 741–4.

27. MacMahon, B., Monson, R.R., Wang, H.H. and Zheng, T. 'A second follow up of mortality in a cohort of pesticide applicators', *Journal of Occupational Medicine*, Volume 30, 1988, pp. 429–32.

28. Blair, A,. Grauman, D.J., Lubin, J.H. and Fraumeni, J.F. Jnr. 'Lung cancer and other causes of death among lice and pesticide applicators', *Journal of the National Cancer Institute*, Volume 71, 1983, pp. 31–7.

29. 'Chlorophenoxy herbicides', in *IARC Monographs on the Evaluation of Carcinogenic Risks to Humans. Overall evaluations of carcinogenicity. An Updating of IARC Monographs Volumes 1–42*, Supplement 7, 1987, pp. 156–60.

30. National Toxicology Programme, 'Carcinogenesis bioassay of 2,3,7,8-tetra-chlorodibenzo-p-dioxin (CAS No1746-0-1-6) in Osborne-Mendel Rats and B6C3F mice (Gavage study)', *Technical Report Series No.209*, National Institutes of Health Publication No. 82–1,765, Bethesda, Maryland, 1982.

31. Kociba, R.J., Keyes, D.G., Beyer, J.E., Carreon, R.M., Wade, C.E., Dittenber, D.A., Kalnins, R.P., Frauson, L.E., Park, C.N., Barnard, S.D., Hummel, R.A. and Humiston, C.G. 'Results of a 2 year chronic toxicity and onco-genicity study of 2,3,7,8 tetrochlorodibenzo-p-dioxin in rats', *Toxicology and Applied Pharmacology*, Volume 46, 1978, pp. 279–303.

32. Hardell, L. and Sandstrom, A. 'Case control study: soft tissue sarcomas and exposure phenoxy-acetic acids or chlorophenol', *British Journal of Cancer*, Volume 39, 1979, pp. 711–7.

33. Eriksson, M., Hardell, L., Berg, N.O., Moller, T. and Axelson, O. 'Soft tissue sarcomas and exposure to chemical substances: A case-referent study', *British Journal of Industrial Medicine*, Volume 38, 1981, pp. 27–33.

34. Hardell, L., Eriksson, M., Lenner, P. and Lundgren, E. 'Malignant lymphoma and exposure to chemicals, especially organic solvents, chlorophenols and phenoxy acids: a case control study', *British Journal of Cancer*, Volume 43, 1981, pp. 169–76.

35. Smith, A.H., Pearce, N.E., Fisher, D.O., Giles, H.J., Teague, C.A. and Howard, J.K. 'Soft tissue sarcoma and exposure to phenoxy herbicides and chlorophenols', in New Zealand. *Journal of the National Cancer Institute*, Volume 73, 1984, pp. 1,111–7.

36. Pearce, N.E., Smith, A.H., Howard, J.K., Sheppard, R.A., Giles, H.J. and Teague, E.A. 'Non-Hodgkin's lymphoma and exposure to phenoxy-herbicides, chloro-phenols, fencing work and meat work employment: a case control study', *British Journal of Industrial Medicine*, Volume 43, 1986, pp. 75–83.

37. Pearce, N.E., Sheppard, R.A., Smith, A.H. and Teague, C.A. 'Non-Hodgkin's lymphoma and farming: an expanded case control study', *International Journal of Cancer*, Volume 39, 1987, pp. 155–61.

38. Wiklund, K., Dich, J. and Holm, L.E. 'Risk of malignant lymphoma in Swedish pesticide appliers', *British Journal of Cancer*, Volume 56, 1987, pp. 505–8.

39. Wiklund, K., Lindefors, D.M. and Holm, L.E. 'Risk of malignant lymphoma in Swedish agricultural and forestry workers', *British Journal of Industrial Medicine*, Volume 45, 1988, pp. 19–24.

40. Coggon, D., Pannett, B., Winter, P., Acheson, D.E. and Bonsall, J.L. 'Mortality of workers exposed to 2-methyl-4-chlorophenoxyacetic acid', *Scandinavian Journal of Work and Environmental Health*, Volume 12, 1986, pp. 448–54.

41. Hoar, S.K., Blair, A., Holmes, S.F., Boysen, C.D., Robel, R.J., Hoover, R. and Fraumeni, J.F. 'Agricultural herbicide use and risk of lymphoma and soft tissue sarcoma', *Journal of the American Medical Association*, Volume 256, 1986, pp. 1141–7.

42. Hoar-Zahn, S.H., Blair, A., Holmes, F.F., Boysen, C.D. and Robel, R.J. 'A case reference study of soft tissue sarcoma and Hodgkins disease', *Scandinavian Journal of Work and Environmental Health*, Volume 14, 1988, pp. 224–30.

43. 'DDT', in *IARC Monographs on the Evaluation of Carcinogenic Risk to Humans. Overall Evalutions of Carcinogenicity and Updating of IARC monographs*, Volumes 1–42, Supplement 7, 1987, pp. 186–9.

44. Council on Scientific Affairs, American Medical Association, 'Cancer Risk of Pesticides in Agricultural Workers', *Journal of the American Medical Association*, Volume 260, 1988, pp. 959–66.

45. Borzsonyi, M., Torok, G., Pinter, A., et al. *Agriculturally related carcinogenic risk*, International Agency for Research on Cancer, Scientific Publications, Volume 56, 1984, pp. 465–86.

46. Ibid.

47. Council on Scientific Affairs, American Medical Association, 'Cancer Risk of Pesticides in Agricultural Workers', *Journal of the American Medical Association*, Volume 260, 1988, pp. 959–66.

48. 'DDT' in *IARC Monographs on the Evaluation of Carcinogenic Risk to Humans*, Supplement 7, 1987, pp. 186–9.

49. Hogstedt, C. and Westerlund, B. 'Cohort studies on causes of death of forestry workers with and without exposure to phenoxy-acid preparations', (Swedish) *Lakartidningen*, Volume 77, 1980, pp. 1828–31.

50. Ditraglia, D., Brown, D.P., NameKata, T. and Iverson, N. 'Mortality study of workers employed at organochlorine pesticide manufacturing plants', *Scandinavian Journal of Work and Enviromental Health*, Volume 7, 1981, pp. 140–6.

51. Wong, O., Brocker, W., Davis, H.V. and Nagle, G.S. 'Mortality of workers potentially exposed to organic and inorganic brominated chemicals, DBCP, Tris, PBB, and DDT', *British Journal of Industrial Medicine*, Volume 41, 1984, pp. 15–124.

52. Ibid.

53. Bartel, E. 'Increased risk of lung cancer in pesticide exposed male agricultural workers', *Journal of Toxicology and Environmental Health*, Volume 8, 1981, pp. 1,027–40.

54. Wang, H.H. and MacMahon, B. 'Mortality of pesticide applicators', *Journal of Occupational Medicine*, Volume 21, 1979, pp. 741–4.

55. Blair, A., Grauman, D.J., Lubin, J.H. and Fraumeni, J.F. Jnr. 'Lung Cancer and other causes of death among licensed pesticide applicators', *Journal of the National Cancer Institute*, Volume 71, 1983, pp. 31–7.

56. Wiklund, K.G.K. 'Respiratory cancer among orchardists in Washington State, 1968–1980 (abstract). Dissertation in *Abstracts International*, Volume 44, 1983, p. 128b.

57. '1,2-di-bromo-3-chloropropane', *IARC Monographs on the Evaluation of Carcinogenic Risks to Humans*, Supplement 7, 1987, pp. 191–2.

58. Ibid.

59. Hearn, S., Ott, M.G., Kolesar, R.C. and Cook, R.R. 'Mortality experience of employees with occupational exposure to DBCP', *Journal of Occupational Medicine*, Volume 39, 1984, pp. 49–55.

60. Wong, O,. Rocker, W., Davis, H.V. and Nagle, G.S. 'Mortality of workers potentially exposed to organic and inorganic brominated chemicals, DBCP Tris, PBB and DDT', *British Journal of Industrial Medicine*, Volume 41, 1984, pp. 15–24.

61. 'Ethylene dibromide', in *IARC Monographs on the Evaluation of Carcinogenic Risks to Humans*, Supplement 7, 1987, pp. 204–5.

62. Council on Scientific Affairs, American Medical Association, 'Cancer Risk of Pesticides in Agricultural Workers', *Journal of the American Medical Association*, Volume 260, 1988, pp. 959–66.

63. '1,3-5-dichloropropene (technical grade)', in *IARC Monographs on the Evaluation of Carcinogenic Risks to Humans*, Supplement 7, 1987, pp. 195–6.

64. Markovitz, A. and Crosby, W.H. 'Chemical carcinogenesis: a soil fumigant 1,3- dichloropropene, as possible cause of haematological malignancy', *Archives of Internal Medicine*, Volume 144, 1984, pp. 1,409–11.

65. Council on Scientific Affairs, American Medical Association, 'Cancer Risk of Pesticides in Agricultural Workers', *Journal of the American Medical Association*, Volume 260, 1988, pp. 959–66.

66. 'Dimethylcarbamoyl Chloride', in *IARC Monographs on the Evaluation of Carcinogenic Risk to Humans*, Supplement 7, 1987, pp. 199–200.

67. Council on Scientific Affairs, American Medical Association, 'Cancer Risk of Pesticides in Agricultural Workers', *Journal of the American Medical Association*, Volume 260, 1988, pp. 959–66.

68. Peters, H.A., Gocmen, A., Cripps, D.J., Bryan, G.T. and Dogramaci, I. 'Epidemiology of hexachlorobenzene-induced porphyria in Turkey. Clerical and laboratory follow up after 25 years', *Archives of Neurology*, Volume 39, 1989, pp. 744–49.

69. Wang, G.M. 'Evaluation of pesticides which pose carcinogenicity potential in animal testing', *Regulatory Toxicology and Pharmacology*, Volume 4, 1984, pp. 361–71.

70. Chadwick, R.W., Copeland, M.F., Wolff, G.L., Stead, A.G., Mole, M.L. and Whitehouse, D.A. 'Saturation of lindane metabolism in chronically treated (YS x VY) F_1 hybrid mice', *Journal of Toxicology and Environmental Health*, Volumes 20, 1987, pp. 411–34.

71. Ruch, R.J., Klaunig, J.E. and Pereira, M.A. 'Selective resistance to cytotoxic agents in hepatocytes isolated from partially hepatectomized and neoplastic mouse liver', *Cancer Letters*, Volume 26, 1985, pp. 295–301.

72. Council on Scientific Affairs, American Medical Association, 'Cancer Risk of Pesticides in Agricultural Workers', *Journal of the American Medical Association*, Volume 260, 1988, pp. 959–66.

73. Ibid.

74. Ibid.

75. Ibid.

76. Ibid.

77. Ibid.

78. Ashby, J. and Tennant, R.W. 'Chemical structure, salmonella mutagenicity and extent of carcinogenicity as indicators of genotoxic carcinogenesis among 222 chemicals tested in rodents by the US NCI/NTP', *Mutation Research*, Volume 204, 1988, pp. 17–115.

79. Ibid.

80. Ibid.

81. Ibid.

82. Ibid.

83. Ibid.

84. Paddle, G.M. and Parker, D.G.J. 'A cohort study of employees on a plant manufacturing 4,4 prime-Bipyridyl', *Scandinavian Journal of Work and Environmental Health 1990*, (in press).

85. Bowra, G.T., Duffield, D.P., Osborne, A.J. and Purchase, I.F.H. 'Pre-malignant and neoplastic skin lesions associated with occupational exposure to "tarry" byproducts during manufacture of 4,4 Bipyridyl', *British Journal of Industrial Medicine*, Volume 39, 1982, pp. 76–81.

86. Council on Scientific Affairs, American Medical Association, 'Cancer Risk of Pesticides in Agricultural Workers, *Journal of the American Medical Association*, Volume 260, 1988, pp. 959–66.

87. Ibid.

88. Ibid.

89. World Health Organisation, *IARC Monographs on the Evaluation of Carcinogenic Risk to Humans*, Supplement 7, 1987.

90. World Health Organisation, 'Permethrin Health and Safety Guide. IPCS Health and Safety Guide no. 33. Companion volume to Environmental Health Criteria 94: Permethrin Geneva, 1989.

91. Cantor, K.P. 'Farming and Mortality from non-Hodgkins lymphoma:a case control study', *International Journal of Cancer*, Volume 29, 1982, pp. 239–47.

92. Cantor, K.P. and Blair, A. 'Farming and Mortality from multiple myeloma: a case control study with the use of death certificates', *Journal of the National Cancer Institute*, Volume 72, 1984, pp. 251–5.

93. Burmeister, L.F., Everett, G.D., Van Lier, S.F. and Isacson, P. 'Selected cancer mortality and farm practices in IOWA', *American Journal of Epidemiology*, Volume 118, 1983, p. 72.

94. Steineck, G. and Wiklund, K. 'Multiple myelomas in Swedish agricultural workers', *International Journal of Epidemiology*, Volume 15, 1986, pp. 321–5.

95. Blair, A. and Thomas, T.L. 'Leukaemia among Nebraska farmers: a death certificate study', *American Journal of Epidemiology*, Volume 110, 1979, pp. 264–73.

96. Burmeister, L.F., Van Lier, S.F. and Isacson, P. 'Leukeamia and farm practices in Iowa', *American Journal of Epidemiology*, Volume 115, 1982, pp. 720–28.

97. Pearce, N.E., Sheppard, R.A., Howard, J.K., Fraser, J. and Lilley, B.M. 'Leukaemia among New Zealand Agricultural Workers', *American Journal of Epidemiology*, Volume 124, 1986, pp. 402–9.

98. Corrao, G., Calleri, M., Carle, F., Russo, R., Bosia, S. and Piccioni, P. 'Cancer risk in a cohort of licenced pesticide users', *Scandinavian Journal of Work and Environmental Health*, Volume 15, 1989, pp. 203–9.

99. Rafnsson, V. and Gnunnarsdottir, R. 'Mortality among farmers in Iceland', *International Journal of Epidemiology*, Volume 18, 1989, pp. 146–51.

100. Schumacher, M.C. and Delzell, E. 'A death certificate case control study of non-Hodgkin's lymphoma and occupation in men in North Carolina', *American Journal of Industrial Medicine*, 1988, pp 317–30.

101. Barthel, E. 'Retrospective cohort study on cancer frequency and pesticide exposed male pest control workers', *Zeischeift Erkrant Amt. Org. g*, Volume 166, 1986, pp. 62–8.

102. Armijo, B., Orellana, M., Medina, E., Coulson, A.H., Sayre, J.W. and Detels, R. 'Epidemiology of gastric cancer in Chile: 1 case control study', *International Journal of Epidemiology*, Volume 10, 1981, pp. 53–6.

103. Gallagher, R.P., Threlfal, W.J., Jeffries, E., Band, P.R., Spinelli, J. and Coldman, A.J. 'Cancer and aplastic anaemia and British Columbia farmers', *Journal of the National Cancer Institute*, Volume 72, 1984, pp. 1,311–5.

104. Wiklund, K., Dich, J., Holm, L.E. and Eklund, G. 'Risk of cancer in pesticide applicators in Swedish agriculture', *British Journal of Industrial Medicine*, Volume 46, 1989, pp. 809–14.

105. Musicco, M., Filippini, G., Bordo, B.M., Melotto, A., Morello, G. and Berrino, F. 'Gliomas and occupational exposure to carcinogens: a case control study', *American Journal of Epidemiology*, Volume 116, 1982, pp. 782–90.

106. Musicco, M., Sant, M., Molinari, S., Filippini, G., Gatta, G. and Berrino, F. 'A case control study of brain gliomas and occupational exposure to chemical carcinogens: the risk to farmers', *American Journal of Epidemiology*, Volume 128, 1988, pp. 778–85.

107. Milham, S. and Hesser, J.E. 'Hodgkins disease in woodworkers', *The Lancet*, Volume 2, 1967, pp. 136–7.

108. Acheson, E.D., Hadfield, E.H., Macbeth, R.G. *The Lancet*, Volume 1, 1967, p. 311.

109. Grufferman, S., Dong, T. and Cole, P. 'Occupation and Hodgkins' disease', *Journal of National Cancer Institute*, Volume 57, 1976, pp. 1,193–5.

110. Miller, B.A., Blaier, A.E., Raynor, H.L., Stewart, P.A., Hoar, Zahm, S. and Fraumeni, J.F. 'Cancer and other mortality patterns among United States

furniture workers', *British Journal of Industrial Medicine*, Volume 46, 1989, pp. 508–15.

111. Alavanja, M.C.R., Blair, A., Merkle, S., Teske, J., Eaton, B., Reid, B. 'Mortality among forest and soil conservationists', *Archives of Environmental Health*, Volume 44, 1989, pp. 94–101.

112. Bryant, D.H. 'Asthma due to insecticide sensitivity', *Australian and New Zealand Journal of Medicine*, Volume 15, 1985, pp. 65–8.

113. Lisi, P., Caraffini, S. and Assalve, D. 'Irritation and sensitization potential of pesticides', *Contact Dermatitis*, Volume 17, 1987, pp. 212–8.

114. Spencer, M.C. 'Herbicide Dermatitis', *Journal of the American Medical Association*, Volume 198, 1966, pp. 1,307–8.

115. Laschi-Loquerie, A., Eyraud, A., Morisset, D., Sanou, A., Tachon, P., Veysseyre, C. and Descotes, J. 'Influence of heavy metals on the resistance of mice towards infection', *Immunopharmacology and Immunotoxicology*. Volume 9, 1987, pp. 235–41.

116. Exan, J.H. 'The immunotoxicity of selected environmental chemicals, pesticides and heavy metals. Chemical Regulation of Immunity', in *Veterinary Medicine*, 1984, pp. 355–68.

117. Wharton, M.D., Krauss, R.M., Marshall, S., et al. 'Infertility in male pesticide workers', *Lancet*, Volume 2, 1977, pp. 1259–61.

118. Wharton, M.D., Milby, T.H., Krauss, R.M., et al. 'Testicular function in DBCP exposed pesticide workers', *Journal of Occupational Medicine*, Volume 21, 1979, pp. 161–6.

119. Wharton, M.D., Milby, T.H. 'Recovery of testicular function among DBCP workers', *Journal of Occupational Medicine*, Volume 22, 1980, pp. 177–9.

120. Eaton, M., Schenker, M., Wharton, M.D., Samuels, S., Perkins, C. and Overstreet, J. '7 year follow up of workers exposed to 1,2-dibromo-3-chloro-propane', *Journal of Occupational Medicine*, Volume 28, 1986, pp. 1,145–50.

121. Wong, O, Morgan, R.W., Wharton, M.D. 'An epidemiologic surveillance programme for evaluating occupational reproductive hazards', *American Journal of Industrial Medicine*, Volume 7, 1985, pp. 295–306.

122. Wong, O., Utidjian, H.M. and Karten, V.S. 'Retrospective evaluation of reproductive performance of workers exposed to ethylene di-bromide (EDB)', *Journal of Occupational Medicine*, Volume 21, 1979, pp. 98–102.

123. Cooke, J. and Kaufman, C. *Portrait of a Poison. The 2,4,5-T story*, Pluto Press, London, 1982.

124. Hay, A. *The Chemical Scythe: Lessons of 2,4,5-T and dioxin*, Plenum Press, London and New York, 1982.

125. Westing, A.H. *Reproductive Epidemiology: an overview in Herbicides in War. The long-term ecological and human consequences.*, Westing, A.A. (ed.), Taylor and Francis, London and Philadelphia, 1984, pp. 141–9.

126. Schuck, P.H. *Agent Orange on Trial. Mass Toxic Disasters in the Courts*, The Belknap Press of Harvard University Press, Cambridge, Mass. and London, 1986.

127. Department of Health and Social Security, 'Pesticide poisoning: notes for guidance of medical practitioners', HMSO, 1983, pp. 33–5.

128. Hayes, W. *Pesticides Studied in man*, Williams and Wilkins, Baltimore, and London, 1982, pp. 466–71.

129. Bidstrup, P.L., Bonnell, J.A.L. and Harvey D.G. 'Prevention of acute dinitro-ortho-cresol (DNOC) poisoning', *Lancet*, 1, 1952, pp. 794–5.

130. Edson, E.F. 'Applied toxicology of pesticides', *Pharmacology Journal*, Volume 185, pp. 361–7.

131. Nehez, M., Selypes, A., Paldy A., and Berencsi G. 'The mutagenic effect of a dinitro-o-cresol containing pesticide on mice germ cells', *Ecotoxicology and Environmental Safety*, Volume 2, 1977, pp. 401–5.

132. *IARC Monographs Overall Evaluation of carcinogenicity. An updating of IARC monographs Vols 1–42*, Supplement 7, 1987.

133. Hayes, W. *Pesticides Studied in Man*, 1982. pp. 471–3.

134. Department of Health and Social Security, 'Pesticide poisoning: notes for guidance of medical practitioners', HMSO, London, 1983, pp. 36–7.

135. Conso, F., Meel, P., Pouzoulet, C., Efthymiou, M.L., Gervais, P. and Gaultier, M. 'Toxicité aigue chez l'homme des derivés halogènes de l'hydroxybenzonitrile (ioxynil, bromoxynil)', *Archiv. Maladie Prof. Societé Médicine Hygiène, Travail*, Volume 38, 1977, pp. 674–7.

136. Gordon, D. 'How dangerous is pentachlorophenol?' *The Medical Journal of Australia*, Volume 2, 1956, pp. 485–8.

137. Menon, J.A. 'Tropical hazards associated with the use of pentachlorophenol', *British Medical Journal*, Volume 1, 1958, pp. 1,156–8.

138. Wood. S., Rom, W., White, G.L. and Logan, D.C. 'Pentachlorophenol poisoning', *Journal of Occupational Medicine*, Volume 25, 1983, pp. 527–30.

139. Gray, R.E., Gilliland, R.D., Smith, E.E., Lockard, V.G. and Hume, A.S. 'Pentachlorophenol intoxication: Report of a fatal case, with comments on the clinical course and pathologic anatomy', *Archives of Environmental Health*, Volume 40, 1985, pp. 161–4.

140. World Health Organisation, 'Pentachlorophenol poisoning in Bangkok. Information circular on the toxicity of pesticides to man', No 8, 1962, p. 9.

141. Hassan, A.B., Seligmann, H. and Bassan, H.M. 'Intravascular haemolysis induced by pentachlorphenol', *British Medical Journal*, Volume 291, 1985, pp. 21–2.

142. Roberts, H.J. 'Aplastic Anaemia and Red Cell aplasia due to pentachlorophenol', *Southern Medical Journal*, Volume 76, 1983, pp. 45–8.

143. Ashby, J. and Tennant, R.W. 'Definitive relationships among chemical structure, carcinogenicity and mutagenicity for 301 chemicals tested by the US NTP', *Mutation Research*, Volume 257, 1991, pp. 229–308.

144. Hardell, L., Eriksson, M., Lenner, P. and Lundgcen, E. 'Malignant lymphoma and exposure to chemicals, especially organic solvents, chlorophenols and phenoxy acids: a case control study', *British Journal of Cancer*, Volume 43, 1981, pp. 169–76.

145. Liv, P.T. and Morgan D.B. 'Comparative toxicity and biotransformation of lindane in C57 DL/6 and DBA/2 mice', *Life Sciences*, Volume 39, 1986, pp. 1237–44.

146. Hayes, 'DDT/Chlorinated Hydrocarbon Insecticides' in *Pesticides Studied in Man*, 1982, pp. 172–205.

147. Hayes, 'Aldrin' in *Pesticides Studied in Man*, p. 235.

148. Paul, A.H. 'Dieldrin poisoning', *New Zealand Medical Journal*, Volume 58, 1959, p. 393.

149. Hayes W. 'Dieldrin/Chlorinated Hydrocarbon Insecticides', in *Pesticides Studied in Man*, 1982, p. 243.

150. Terviez, G., Dimitrova, N. and Rusev, P. 'Forensic medical and forensic chemical study of acute lethal poisonings with thiodan', *Folia Medica*, Volume 16, 1974, pp. 325–9.

151. Ely, T.D., Macfarlane, J.W., Galen, W.P and Hine, C.H. 'Convulsions in Thiodan Workers; a preliminary report', *Journal of Occupational Medicine*, Volume 9, 1967, pp. 35–7.

152. Hayes, 'Endosulphan/Chlorinated Hydrocarbon Insecticides', in *Pesticides Studied in Man*, 1987, p. 253.

153. Curley, A. and Garretson, L.K. 'Acute chlordane poisoning', *Archives of Environmental Health*, Volume 18, 1969, pp. 211–5.

154. Derbes, V.J., Dent, J.H., Forrest, W.W. and Johnson, M.F. 'Fatal chlordane poisoning', *Journal of the American Medical Association*, Volume 158, 1955, pp. 1,367–9.

155. Furie, B. and Trubowitz, S. 'Insecticides and blood dyscrasias: Chlordane exposure and self-limited refractory megaloblastic anaemia', *Journal of the American Medical Association*, Volume 235, 1976, pp. 1,720–2.

156. Infante, P.F., Epstein S.S. and Newton, W.A. 'Blood dyscrasias and childhood tumours and exposure to chlordane and heptachlor. Scandinavian', *Journal of Work and Environmental Health*, Volume 4, 1978, pp. 137–50.

157. Wassermann, M. Iliescu, S., Mandric, G. and Horvath, P. 'Toxic hazards during DDT and BHC – spraying of forests against Lymantria monacha', *American Medical Association Archives of Industrial Health*, Volume 21, 1960, pp. 503–8.

158. 'American Medical Association Council on Pharmacy and Chemistry Report on: Health hazards of electric vapourising devices for insecticides', *Journal of the American Medical Association*, Volume 149, 1952, pp. 367–9.

159. Morgan, D.P., Stockdale, E.M., Roberts, R.J. and Walter, A.W. 'Anaemia Associated with exposure to Lindane', *Archives of Environmental Health*, Volume 35, 1980, pp. 307–10.

160. Loge, P.J. 'Aplastic anaemia following exposure to benzene hexachloride (Lindane)', *Journal of the American Medical Association*, Volume 193, 1965, pp. 104–8.

161. West, I. 'Lindane and Haematologic Reactions', *Archives of Environmental Health*, Volume 15, 1967, pp. 97–101.

162. Woodliffe, H.J., Conno, P.M. and Scopa, J. 'Aplastic anaemia associated with insecticides', *The Medical Journal of Australia*, Volume 1, 1964, pp. 628–9.

163. Champlin, R.E. and Gale, R.P. 'Aplastic anaemia', *Medicine International*, 1987, p. 1,748.

164. Hayes, 'Benzene Hexachloride/Chlorinated Hydrocarbon Insecticides', in *Pesticides Studied in Man*, 1982, pp. 211–28.

165. Wang. G.M. 'Evaluation of Peticides which pose carcinogenicity potential', in *Animal Testing Regulatory Toxicology and Pharmacology*, Volume 4, 1984, pp. 361–71.

166. Ruch, R.J. and Klaunig, J.E. 'Antioxidant prevention of tumour promoter induced inhibition of mouse hepatocyte intercullar communication', *Cancer Letters*, Volume 33, 1986, pp. 137–50.

167. Montesano, R., Cabral, J.R.P. and Wilbourn, J.D. 'Environmental Carcinogens: Using pesticides and Nitrosamines as paradigms', *Annals of New York Academy of Sciences*, Volume 534, 1988, pp. 67–73.

168. Chadwick, R.W., Copeland, M.F., Wolff, G.L., Stead, A.G. Mole, M.L. and Whitehouse, D.A. 'Saturation of Lindane metabolism in chronically treated (YS x VY) F1 hybrid mice', *Journal of Toxicology and Environmental Health*, Volume 20, 1987, pp. 411–34.

169. Ruch, R.J. and Klaunig J.E. 'Antioxidant prevention of tumour promoter induced inhibition of mouse hepatocyte intercellular communication', *Cancer Letters*, Volume 33, 1986, pp. 137–50.

170. Hayes, 'Dichlorvos', in *Pesticides Studied in Man*, 1982, p. 349.

171. Kleinman, G.D. 'Occupational disease in California attributed to pesticides and agricultural chemicals', *Archives of Environmental Health*, Volume 1, 1960, pp. 118–24.

172. Michelson, R. 'Insecticide poisoning: A case of suicide with systox', *Tiddsskr. Nor. Laegetoren*, Volume 78, 1958, pp. 356–7.

173. Hayes, 'Demeton', in *Pesticides Studied in Man*, 1982, p. 390.

174. Hayes, 'Parathion', in *Pesticides Studied in Man*, pp. 379–85.

175. Cole, D.C., McConnell, R., Murray, D.L. and Anton F.P. 'Pesticide Illness Surveillance: The Nicaraguan Experience', *Pan American Health Bulletin*, Volume 22, 1988, pp. 119–32.

176. Frils, H. 'Parathion Mortality in Denmark', *Bibl. Laeg.*, Volume 158, 1966, pp. 137–99.

177. Kidd, J.G. and Langworthy, O.R. 'Jake paralysis. Paralysis following ingestion of Jamaica ginger extract adulterated with tri-ortho-cresyl phosphate', *Bulletin Johns Hopkins Hospital*, No. 52, 1933, pp. 39–66.

178. Morgan, J.P. and Penvich, P. 'Jamaica ginger paralysis. Forty-seven year follow-up', *Archives of Neurology*, Volume 35, 1978, pp. 530–2.

179. Montesano, R., Eabral, J.R.P. and Wilbourn, J.D. 'Environmental Carcinogens in Living in a Chemical World', C. Maltoni and I. J. Selikoff (eds), *Annals of the New York Academy of Sciences*, Volume 534, 1988, pp. 67–73.

180. Department of Health and Social Security, 'Pesticide Poisoning: Notes for the guidance of medical practitioners', HMSO, London, 1983 p. 45.

181. Hayes, 'Carbamate Pesticides' in *Pesticides Studied in Man*, 1982, pp. 436–62.

182. Farago, A. 'Suicidal, fatal Sevin (1-naphthyl-N-Methyl Carbamate) poisoning', *Archives Toxikologie*, Volume 24, 1969, pp. 309–15.

183. Aaronson, M., Ford, S.A., Goes, E.A, Savage, E.P., Wheeler, H.W., Gibbons, G. and Stoesz, P.A. 'Suspected carbamate intoxications Nebraska', *Morbidity Mortality Weekly Report*, Volume 28, 1979, pp. 133–4.

184. Hayes, 'Carbamate Pesticides', in *Pesticides Studied in Man*, 1982, pp. 436–62.

185. 'Pesticide Poisoning: Notes for the guidance of medical practitioners', 1983, p. 48.

186. Hayes, 'Synthetic Organic Rodenticides', in *Pesticides Studied in Man*, 1982, pp. 494–519.

187. 'Poisoning cases in 1960', *Pharmaceutical Journal*, Volume 189, 1962, pp. 453–5.

188. Hayes, 'Inorganic and Organometal Pesticides', in *Pesticides Studied in Man*, 1982, pp. 1–74.

189. Goldwater, L.J., Ladd, A.C., Berkhout, P.G. and Jacobs, M.B. 'Acute exposure to phenylmercuric acetate', *Journal of Occupational Medicine*, Volume 6, 1964, pp. 227–8.

190. 'Pesticide Poisoning: Notes for the guidance of medical practitioners', 1983, pp. 50–2.

191. Bloom, G., Lundgren, K.D. and Swensson, A. 'Exposure and hazards from organic mercury compounds in connection with seed dressing on small farms', *Nord. Hyg. Tidsks.*, Volume 36, 1955, pp. 110–7.

192. Engleson, G. and Herner, 'T. Alkyl mercury poisoning', *Acta Paediatrica*, Volume 41, 1952, pp. 289–94.

193. Maghazaji, H.I. 'Psychiatric aspects of methylmercury poisoning', *Journal of Neurology Neurosurgery and Psychiatry*, Volume 37, 1974, pp. 954–8.

194. Amin-Zaki, L., Majeed, M.A., Clarkson, T.W. and Greenwood, M.R. 'Methylmercury poisoning in Iraqi children: Clinical observations over two years', *British Medical Journal*, Volume 1, 1978, pp. 613–6.

195. Amin-Zaki, L., Elhassani, S., Majeed, M.A., Clarkson, T.W., Doherty, R.A. and Greenwood, M.R. 'Studies of infants postnatally exposed to methyl-mercury', *Journal of Pediatrics*, Volume 85, 1974, pp. 81–4.

196. Skerfing, S.B. and Copplestone J.F. 'Poisoning caused by the consumption of organomercury-dressed seed in Iraq', *Bulletin WHO*, Volume 54, 1976, pp. 111–2.

197. 'Pesticide Poisoning: Notes for the guidance of medical practitioners', 1983, pp. 53–4.

198. Hayes, 'Inorganic and Organometal Pesticides', in *Pesticides Studied in Man*, 1982, pp. 1–74.

199. Hayes, 'Inorganic and Organometal Pesticides', in *Pesticides Studied in Man*, 1982, pp. 1–74.

200. Chugh, K.S., Singhal, P.C. and Sharma, B.K. 'Methemoglobinemia in acute copper sulphate poisoning', *Annals of Internal Medicine*, Volume 82, 1975, pp. 226–7.

201. Chugh, K.S.,Singhal, P.C., Sharma, B.K., et al., 'Acute renal failure following copper sulphate intoxication', *Postgraduate Medical Journal*, Volume 53, 1977, pp. 18–23.

202. Fazakas, I.G. 'Lethal copper sulphate poisoning following eating of sprayed grapes', *Acta Medica*, Leg. Seoc. (Liege), Volume 17, 1964, pp. 129–38.

203. Ulner, D. 'Disturbances in Trace Element Metabolism', in *Harrisons Prinicples of Internak Medicine*, R.G. Petersdorf et al. (ed.), McGraw-Hill, London, 1985, pp. 470–2.

204. Hayes, 'Inorganic and Organometal Pesticides', in *Pesticides Studied in Man*, 1982, pp. 1–74.

205. Pimentel, J.C. and Marques, F. 'Vineyard sprayers' lung: A new occupational disease', *Thorax*, Volume 24, 1969, 678–88.

206. Pimentel, J.C. and Menezes, A.P. 'Liver granulomas containing copper in vineyard sprayer's lung. A new etiology of hepatic granulomatosis', *American Reiew of Respiratory Diseases*, Volume 111, 1975, pp. 189–95.

207. Pimentel, J.C. and Menezes, A.P. 'Liver damage in vineyard fungicide applicators', *Journal of Medicine* (Lisbon), Volume 90, 1976, pp. 409–15.

208. Koestler, A. and Wechsler, W. 'Induction of neurogenic and non-neurogenic neoplasma by feeding precursors of methyl and ethylnitrosourea to adult rats', *Journal of Neuropathology and Experimental Neurology*, Volume 33, 1974, p. 178.

209. Musicco, M., Sant, M., Molinari, S., Filippini, G., Gatta, G. and Berrino, F. 'A case-control study of brain gliomas and occupational exposure to chemical carcinogens: the risk to farmers, *American Journal of Epidemiology*, Volume 128, 1988, pp. 778–85.

210. 'Pesticide Poisoning: Notes for the Guidance of Medical Practitioners', 1983, pp. 55–6.

211. Hayes, 'Arsenical Pesticides: Inorganic and Organometal Pesticides' in *Pesticides Studied in Man*, 1987, pp. 1–74.

212. Ibid.

213. Sommers, S.C. and McManus, R.G. 'Multiple arsenical cancers of skin and internal organs', *Cancer*, Volume 6, 1953, p. 347.

214. International Agency for Research on Cancer (IARC), *IARC Monographs on the Evaluation of Carcinogenic Risks to Humans. Overall Evaluation of Carcinogenicity. An Updating of IARC Monographs Volumes 1–42*, Supplement 7, Lyons, France, 1987.

215. World Health Organisation, IARC, Monograph 2, 1973.

216. Ott, G.O., Holder, B.B. and Gordon, H.L. 'Respiratory Cancer and Occupational exposure to arsenicals', *Archives of Environmental Health 29*, 1974, pp. 250–5.

217. Hayes, 'Pesticides derived from plants and other organisms', in *Pesticides Studied in Man*, 1982, pp. 75–111.

218. He, F., Sun, J., Han, K., Wu, Y., Yao, P., Wang, S. and Liv, L. 'Effects of pyrethroid insecticides on subjects engaged in packaging pyrethroids', *British Journal of Industrial Medicine*, Volume 45, 1988, pp. 548–51.

219. 'Pesticide poisoning: Notes for the guidance of medical practitioners, 1983, pp. 59–63.

220. Hayes, 'Herbicides', in *Pesticides Studied in Man*, 1982, pp. 520–71.

221. Okhubo, S., Kanazawa, Y., Tachikawa, H., Hayashi, S.,Komatsuda, H., Hirata, M. and Watanuku,T. 'Findings in two autopsies performed after fatal acute paraquat dichloride poisoning', *Journal of Japanese Association of Rural Medicine*, Volume 24, 1975, pp. 460–1.

222. Harley, J.B., Grinspam, S. and Root, R.K. 'Paraquat suicide in a young woman: Results of therapy directed against the superoxide radical', *Yale Journal of Biological Medicine*, Volume 50, 1977, pp. 481–8.

223. Fitzgerald, G.R., Barniville, G. Black, J., Silke, B., Carmody, M. and O'Dwyer, W.F. 'Occupational paraquat poisoning', *Quarterly Journal of Medicine*, Volume 46, 1977, pp. 561–2.

224. De Alwis, L.B.L. and Salgado, M.S.L. 'Agrochemical poisoning in Sri Lanka', *Forensic Science International*, Volume 36, 1988, pp. 81–9.

225. Barbeau, A., 'Etiology of Parkinsons Disease: A research Strategy', *Canadian Journal of Neurological Science*, Volume 11, 1984, pp. 24–8.

226. Bocchett, A. and Corsini, G.U. 'Parkinson's disease and pesticides' *Lancet II*, 1986, p. 1,163.

227. Rajput., A.H., Uitti, R.J., Stern, W., Laverty, W., O'Donnell, K., O'Donnell, D., Yuen, W.K. and Dua, A. 'Geography, Drinking Water Chemistry, Pesticides and Herbicides, and Etiology of Parkinson's Disease', *Canadian Journal of Neurological Sciences*, Volume 14, 1987, pp. 414–8.

228. Bowra, G.T., Duffield, D.P., Osborn, A.J. and Purchase, I.F.H. 'Premalignant and neoplastic skin lesions associated with occupational exposure to 'tarry' byproduct during manufacture of 4,4'-bipyridyl', *British Journal of Industrial Medicine*, Volume 39, 1982, pp. 76–81.

229. 'Pesticide Poisoning: Notes for the guidance of medical practitioners', 1983, p. 64.

230. Hay, A. 'Dioxin: the 10 year battle that began with Agent Orange', *Nature*, Volume 278, 1979, pp. 108–9.

231. United States Environmental Protection Agency, 'Health Assessment Document for Polychlorinated Dibenzo-p-dioxins', EPA/600/8-84/014F, September 1985.

232. Kociba, R.J., Kayes, D.G., Beyer, J.E., Carreon, R.M., Wade, C.E., Dittenber, D.A., Kalnins, R.P., Frauson, L.E., Park, C.M., Barnard, S.D., Hummel, R.A. and Humbiston, C.G. 'Results of a two-year chronic toxicity and onco-genicity study of 2,3,7,8-tetrachlorodibenzo-p-dioxin in rats', *Toxicology and Applied Pharmacology*, Volume 46, 1978, pp. 279–303.

233. Courtney, D.D. and Moore, J.A. 'Teratology studies with 2,4,5,-trichlorophe-noxyacetic acid and 2,3,7,8- tetrachlorodibenzo-p-dioxin', *Toxicology and Applied Pharmacology*, Volume 20, 1971, pp. 396–403.

234. Hay, A. 'Bill of Health', *Nature* (London), Volume 289, 1981, pp. 4–5.

235. Hardell, L., and Sandstrom, A. 'Case-control study: soft tissue sarcomas and exposure to phenoxyacetic acids and chlorophenols', *British Journal of Cancer*, Volume 39, 1979, pp. 711–7.

236. Hardell, L., Eriksson, M., Lenner, P. and Lundgren, E. 'Malignant lymphoma and exposure to chemicals, especially organic solvents chlorophenols and phenoxyacids: a case control study', *British Journal of Cancer*, Volume 43, 1981, pp. 169–76.

237. Smith, A.H., Pearce, M.E., Fisher, D.O., Giles, H.J., Teague, C.A., and Howard, J.K. 'Soft tissue sarcoma and exposure to phenoxy herbicides and chlorophe-nols in New Zealand', *Journal of the National Cancer Institute*, Volume 73, 1984, p. 1,111.

238. Hoar, S.K., Blair A., Holmes, F.F., Boysen, C.D., Rubel, R.J., Hoover, R. and Fraumeni, Jr. J.F. 'Agriculture herbicide use and risk of lymphoma and soft tissue sarcoma', *Journal of the American Medical Association*, Volume 256, 1986, pp. 1,141–7.

239. Ibid.

240. Hardell, L. and Eriksson, M. 'The association between soft tissue sarcomas and exposure to phenoxyacetic acids', *Cancer*, Volume 62, 1988, pp. 652–6.

241. Hayes, 'Inorganic and organometal pesticides', in *Pesticides Studied in Man*, 1982, pp. 1–74.

242. Ibid.

243. Lebreton, R. 'Etude toxicologique de l'étain et de ses dérivés', *Tours. Imp. Mame.*, France, 1962.

244. Vos, J.G., Deklerk, A., Kranjc, E.I., Kriuzinga, W., Van Ommen, B. and Rozing, J. 'Toxicity of Bis(tri-n-butyltin) oxide in the rat', *Toxicology and Applied Pharmacology*, Volume 75, 1984, pp. 363–86, 387–408.

245. 'Pesticide poisoning: Notes for the guidance of medical prac-titioners', p. 67.

246. Franke, F.E. and Thomas, J.E. 'The treatment of acute nicotine poisoning', *Journal of the American Medical Association*, Volume 106, 1936, pp. 507–12.

247. Kennedy, D. 'Notes on the casualties with poisons in New Zealand', *World Health Organisation Information Circular on the Toxicity of Pesticides to Man*, No. 8, 1962, p. 5.

248. Haines, K.A. and Davis, L.G. 'Protection of workers applying nicotine alkaloid as a concentrated mist spray', *Journal of Ecomomic Entomology*, Volume 41, 1948, p. 513.

249. Faulkner, J.M. 'Nicotine poisoning by absorption through the skin', *Journal of the American Medical Association*, Volume 100, 1933, pp. 1,664–5.

250. 'Pesticide Poisoning: Notes for the guidance of medical practitioners', p. 68.

251. Hayes, 'Fungicides and related compounds', in *Pesticides Studied in Man*, 1982, pp. 578–622.

252. Montesano, R., Cabral, J.R.P. and Wilbourn, J.D. 'Environmental Carcinogens', in *Living in a chemical world*, C. Maltoni and I.J. Selikoff (eds), *Annals of the New York Academy of Sciences*, Volume 534, 1988, pp. 67–73.

253. Erlichman, J. 'Crop spray test for cancer risk', *The Guardian* 17 July 1989.

254. 'Pesticide Poisoning: Notes for the guidance of medical practitioners', pp. 70–1.

255. Schlicher, J.E. and Beat, V.B. 'Dermatitis resulting from herbicide use – a Case Study', *Journal of the Iowa Medical Society*, Volume 62, 1972, pp. 419–20.

256. Hayes, 'Herbicides', in *Pesticide Poisoning in Man*, 1982, pp. 520–77.

257. Ibid.

258. Axelson, O. and Sundell, L. 'Herbicide exposure, mortality and tumor incidence. An epidemiological investigation on Swedish railroad workers', *Work and Environmental Health*, Volume 11, 1974, pp. 21–8.

259. Axelson, O. 'Pesticides and Cancer Risks in Agriculture', *Medical Oncology and Tumor Pharmacotherapy*, Volume 4, 1987, pp. 207–17.

260. World Health Organisation, International Agency for Research on Cancer, *Monographs on the Evaluation of Carcinogenic Risk to Humans*, Supplement 7, 1987, pp. 92–3.

261. 'Pesticide Poisoning: Notes for the guidance of medical practitioners', 1983, pp. 72–3.

262. Timperman, J. and Maes, R. 'Suicidal poisoning by sodium chlorate. A report of three cases', *Journal of Forensic Medicine*, Volume 13, 1966, pp. 123–9.

263. Oliver, J.S., Smith, H. and Watson, A.A. 'Sodium chlorate poisoning', *Journal of the Forensic Science Society*, Volume 12, 1972, pp. 445–8.

264. Hayes, 'Inorganic and Organometal Pesticides' in *Pesticides Studied in Man*, 1982, pp. 1–74.

265. Sasaki, M., Sugimura, K., Yoshida, M.A. and Abe, S. 'Cytogenetic effects of 60 chemicals on cultured human and chinese hamster cells', *Kromosomo II*, Volume 20, 1980, pp. 574–84.

266. Ishidate, M., Sofuni, T., Yoshi Kawa, K., Hayashi, M., Nohmi, T., Sawada, M. and Matsuoka, A. 'Primary Mutagenicity screening of food additives currently used in Japan', *Food and Chemical Toxicology*, Volume 22, 1984, pp. 623–36.

267. Kurokawa, Y., Aoki, S., Matsushima, Y., Takamura, N., Takayoshi, I. and Hayashi, Y. 'Dose Response Studies on the carcinogenicity of potassium bromate in F344 rats after long-term oral administration', *Journal of the National Cancer Institute 77*, 1986, pp. 977–82.

268. Kurokawa, Y., Aoki, S., Jmazawa, T., Hayashi, Y., Matsushima, Y. and Takamura, N. 'Dose related enhancing effect of potassium bromate on renal tumorigenesis in rats initiated with N-ethyl-M-hydroxyethyl-

nitrosamine', *Japanese Journal of Cancer Research* (GANN), Volume 76, 1985, pp. 583–9.

269. Kurokawa, Y., Takamura, N., Matsuoka, C., Imazawa, T., Matsushima, Y., Onodera, H. and Hayashi, Y. 'Comparative studies on lipid peroxidation in the kidney of rats, mice and hampsters and on the effect of cysteine, glutathione, and diethyl maleate treatment on mortality and nephrotoxicity after administration of potassium bromate', *Journal of the American College of Toxicology*, Volume 6, 1987, pp. 489–501.

270. 'Pesticide Poisoning: Notes for the guidance of medical practitioners', 1983, pp. 72–3.

271. Ibid., pp. 74–5.

272. Friedst, B. and Sterner, M. 'Warfarin intoxication from percutaneous absorption', *Archives of Environmental Health*, Volume 11, 1965, pp. 205–8.

273. Pribilla, O. 'Murder caused by Warfarin', *Archives of Toxicology*, Volume 21, 1966, pp. 235–49.

274. Martin-Bouyer, G., Khanh, M.B., Linh, P.D., Hoa, D.Q., Tuan, L.G., Tourneau, J., Barin, C., Guerbois, H. and Binh, T.V., 'Epidemic of Haemorrhagic disease in Vietnamese Infants caused by Warfarin-contaminated talcs', *The Lancet I*, 1983, pp. 230–2.

275. Hay, A. 'Vietnamese Killer Talcum Powder', 3rd World Review, *The Guardian*, 18 February 1983.

276. Hayes, 'Fumigants and nematocides', in *Pesticides Studied in Man*, 1982, pp. 125–71.

277. Van Oettingen, W.F. 'The Halogenated Aliphatic, Olefinic, cyclic, Aromatic, and Aliphatic-Aromatic Hydrocarbons including the Halogenated Insecticides, Their Toxicity and Potential Dangers', *PHS Publication No. 414*, US Government Printing Office, Washington DC, 1955.

278. Rathus, E.M. and Landy, R.J. 'Methyl bromide poisoning', *British Journal of Industrial Medicine*, Volume 18, 1961, pp. 53–7.

279. 'Pesticide Poisoning: Notes for the guidance of medical practitioners', pp. 76–8.

280. Hayes, 'Fumigants and nematocides', in *Pesticides Studied in Man*, 1982, pp. 125–71.

281. International Agency for Research on Cancer, 'Monographs on the Evaluation of carcinogenic risk of chemicals to man', Volume 1, Lyons, France, 1972.

282. Hayes, 'Fumigants and nematocides', in *Pesticides Studied in Man*, 1982, pp. 125–71.

283. Hubbs, R.S. and Prusmack, J.J. 'Ethylene dichloride poisoning', *Journal of the American Medical Association*, Volume 159, 1955, pp. 673–5.

284. International Agency for Research on Cancer, 'Monographs on the evaluation of carcinogenic risk of chemicals to Man', Volume 1.

285. Hayes, 'Fumigants and nematocides', in *Pesticides Studied in Man*, 1982, pp. 125–71.

286. DeSerres, F.J. and Malling, H.V. 'Genetic analysis of the ad-3 mutants of *Neurospora Crassa* induced by ethylene dibromide – a commonly used pesticide', *Newsletter of the Environmental Mutagen Society*, Volume 3, 1970, pp. 36–7.

287. World Health Organisation, International Agency for Reseach on Cancer, 'Monographs on the Evaluation of Carcinogenic Risk to Man', Supplement 7, 1987, pp. 204–5.

288. Ott, M.G., Scharnweber, H.C. and Langer, R.R. 'Mortality experience of 161 employees exposed to ethylene dibromide in two production units', *British Journal of Industrial Medicine*, Volume 37, 1980, pp. 163–8.

289. Alavanja, M.G.R., Rush, G.A., Stewart, P. and Blair, A. 'Proportionate mortality study of workers in the grain industry', *Journal of the National Cancer Institute*, Volume 78, 1987, pp. 247–52.

290. World Health Organisation, International Agency for Research on Cancer, 'Monographs on the Evaluation of Carcinogenic Risk to Man', Supplement 7, 1987, pp. 204–5.
291. US Occupational Safety and Health Administration, news release 5, October 1983.
292. Goldmann, L.R., Mengle, D., Epstein, D.M., Fredson, D., Kelly, K. and Jackson, R.J. 'Acute symptoms in persons residing near a field treated with the soil fumigants methyl bromide and chloropircrin', *Western Journal of Medicine*, Volume 147, 1987, pp. 95–8.
293. Okada, E., Takahashi, K. and Nakamura, H.A. 'study of chloropicrin intoxication', *Nippon Naika/Gakkai Zasshi*, Volume 59, 1970, pp. 1,214–21.
294. Hayes, 'Fumigants and Nematocides', in *Pesticides Studied in Man*, 1982, pp. 125–71.
295. Prentiss, A.M. *Chemicals in War. A Treatise on chemical warfare*, McGraw Hill, New York, 1937.
296. 'Phosphine', in 'Pesticide Poisoning: Notes for the guidance of medical practitioners', 1983, p. 84.
297. Hayes, 'Fumigants and Nematocides', in *Pesticides Studied in Man*, 1982, pp. 133–5.
298. Heyndrickx, A., Van Petegham, C., Van den Heede, M. and Lauwaert R. 'A double fatality with children due to fumigated wheat', *European Journal of Toxicology*, Volume 9, 1976, pp. 113–8.
299. Hayes, W.J. Jr. and Vaughn, W.K. 'Mortality from pesticides in the United States in 1973 and 1974', *Toxicology and Applied Pharmacology*, Volume 42, 1977, pp. 235–52.
300. Hallerman, W. and Pribilla, O. 'Tödliche Vergiftungen mit Phosphorwasserstoff', *Archiv. Toxikology 19*, 1959, pp. 219–42.
301. Khosla, S.N., Nand, M. and Kumer, P. 'Cardiovascular complications of Aluminium Phosphide poisoning', *Angiology*, April 1988, pp. 355–9.
302. Kabra, S.G. and Narayanan, P. 'Aluminium Phosphide: Worse than Bhopal', *The Lancet I*, 1988, p. 1,333.
303. 'Cyanides' in 'Pesticide Poisoning: notes for the guidance of medical practitioners', pp. 85–6.
304. Castello, C., Giussani, A. and Berti, A. 'Poisoning by hydrocyanic acid; clinical and therapeutic considerations and presentation of a case', *Minerva Anestisiology*, Volume 35, 1969, pp. 437–47.
305. Hayes, W.J. Jr. and Vaughn, W.K. 'Mortality from pesticides in the United States in 1973 and 1974', *Toxicology and Applied Pharmacology*, Volume 42, 1977, pp. 235–52.
306. Hayes, 'Fumigants and Nematocides' in *Pesticides Studied in Man*, 1982, pp. 125–71.

Chapter 6

1. British Agrochemicals Association, *Agchem Newsletter*, Feb/Mar 1987, p. 6.
2. National Consumer Council, *Consumers and the Common Agricultural Policy*, HMSO, London, 1988.
3. Owens, S. *Food Production and Quality*, (seminar report) Institute for Public Policy Research and Friederich Ebert Stiftung, London, June 1989, p. 3.
4. Forum on Food and Health, 'Agricultural chemicals in the food chain', *Journal of the Royal Society of Medicine*, Volume 82, May 1989.
5. 'Pesticides: What price the shiny red apple?', *Good Housekeeping*, London, August 1989, pp. 52–3.
6. Ibid.

7. 'Pesticide Residues: a consumers' guide', *Which,* London, October 1990, p. 562.
8. Environmental Protection Agency, *Pesticides and the Consumer,* Washington DC, Volume 13, No. 4, 1987.
9. Ministry of Agriculture, Fisheries and Food (MAFF), 'Report of the Working Party on Pesticide Residues (WPPR), 1985–88. Food Surveillance Paper No. 25', HMSO, London, 1989.
10. British Agrochemicals Association, *Pesticides and Food,* undated leaflet.
11. International Group of National Associations of Pesticide Manufacturers (GIFAP), *Pesticide Residues in Food,* Brussels, 1984.
12. Ames, B.N. et al. 'Dietary Pesticides (99.99% all natural)', *Proceedings of the US National Academy of Sciences,* Volume 87, 1990, pp. 7,777–81.
13. Ames, B.N. and Gold, L.S. 'Chemical Carcinogenesis: Too many rodent carcinogens', *Proceedings of the US National Academy of Sciences,* Volume 87, 1990, pp. 7,772–6.
14. Marks, J. 'Animal Carcinogen testing challenged', *Science,* Volume 250, 1990, pp. 743–5.
15. Ames, B.N. et al. 'Nature's Chemicals and Synthetic Chemicals: comparative toxicology', *Proceedings of the US National Academy of Sciences,* Volume 87, 1990, pp. 7782–6.
16. Ibid.
17. Environmental Protection Agency, 'Pesticides and Food, Environmental Backgrounder', Washington, December 1989, p. 5.
18. Tarkowski, S. 'The World Health Organisation's Role in the Safety Assessment of Pesticides', WHO, Copenhagen. Undated paper.
19. Environmental Protection Agency, *A Consumers' Guide to Safer Pesticide Use,* Washington, 1987, pp. 5–6.
20. Board of Agriculture, National Resource Council, 'Regulating Pesticides' in *Food, The Delaney Paradox,* Committee on Scientific and Regulatory Issues underlying Pesticide Use, Patterns and Agricultural Innovation, National Academy Press, Washington DC, 1987.
21. Ibid.
22. 'Watermelon drama', *New Scientist,* 11 July 1985, p. 19.
23. Green, M.A. et al. 'An Outbreak of Watermelon-Borne Toxicity', *American Journal of Public Health,* Volume 77, 1987, pp. 1,431–4.
24. Fiore, M.C. et al. 'Chronic Exposure to Aldicarb-Contaminated Groundwater and Human Immune Function', *Environmental Research,* Volume 41, 1986, pp. 633–45.
25. Proudfoot, A.T. 'Pesticides Residues in Food – Are they a Risk to Human Health?', in 'Agricultural Chemicals in the Food Chain, Forum on Food and Health Report (1987)', *Journal of Royal Society of Medicine,* Volume 82, 1989, p. 314.
26. Marchand, W. et al. 'Trends in Birth Defects (on a Hawaiian Population Exposed to Heptachlor and for the United States)', *Archives of Environmental Health,* Volume 41, 1986, pp. 145–8.
27. Proudfoot, 'Pesticides Residues in Food', p. 314.
28. Saxena, M.C. et al. 'The Role of Chlorinated Hydrocarbon Pesticides in Abortion and Premature Labour', *Toxicology,* 1980, Volume 17, pp. 323–33.
29. Union of Agricultural Workers, *Pesticides and Your Food,* Ammo, USA, Volume 25, No. 9.
30. Ratner, D. et al. 'Chronic dietary Anticholinesterase Poisoning', *Journal of Medical Science,* Volume 19, Israel, 1983, pp. 810–4.
31. National Research Council, *Regulating Pesticides in Food,* Washington DC, National Academy Press, 1987.
32. MAFF, WPPR Report, 1985–8, 'Food Surveillance Paper No. 25', HMSO, London, 1989.
33. Ibid.

34. 'Toxic Risk "Hidden" in Cereals', *The Guardian*, 13 March 1989, p. 1.
35. Wilkin, D.R. and Fishwick, F.B. 'Residues of Organophosphorus Pesticides in Wholemeal Flour and Bread Produced from Treated Wheat', *Proceedings of the British Crop Protection Conference, Pest and Diseases*, 1981, pp. 183–8.
36. MAFF, WPPR, 'Food Surveillance Paper 16', 1982–85.
37. MAFF, WPPR, 'Food Surveillance Paper No. 25', 1985–88.
38. 'Pesticides found in Cyprus Farm Crops Exceed EC Standards', *The Guardian*, 19 February 1990.
39. MAFF, WPPR, 'Food Surveillance Paper No. 25'.
40. Ibid.
41. Ibid.
42. Kelsey, T. 'Nerve poison found in supermarket salmon', *The Independent on Sunday*, 21 October 1990, p. 1.
43. 'US fungicide bar puts $1 billion wine imports at risk', *The Times*, 3 April 1990.
44. 'US fungicide prompts safety review', *The Guardian*, 22 December 1989.
45. Natural Resources Defense Council, *Chronology of Daminozide*, Washington DC, 1989
46. MAFF, WPPR, 'Food Surveillance Paper No. 25'.
47. Snell, P.J. and Nicoll K.J. *Pesticide Residues and Food: The case for real control*, London Food Commission, 1986.
48. Erlichman, J. 'Baby food makers told to cut toxins', *The Guardian*, 1989.
49. World Health Organisation, *Chemical Contaminants in Food: Global Situation and Trends*, Food Safety Unit, WHO, Environmental Health Division, Geneva, 1989.
50. Erlichman, J. 'Millions of apples sprayed with "cancer" chemical', *The Guardian*, 12 April 1989, p. 1.
51. Campt, D. 'Daminozide: a case study of a pesticide controversy', *Pesticides and the Consumer*, Volume 13, No. 4, Environmental Protection Agency, Washington DC, 1987, pp. 32–4.
52. London Food Commission, *Pesticides residues in food: the question of Alar*, London, 1989.
53. Health and Safety Executive, MAFF, 'ACP review of daminozide', *The Pesticides Register*, No. 11, London, 1989.
54. National Farmers' Union, 'Safety comes first on pesticides', press release, London, 17 July 1989.
55. Ibid.
56. Brown, D. 'Scientists find deadly dioxins in mothers' milk', *The Sunday Telegraph*, 14 May 1989, p. 1.
57. Ibid.
58. Erlichman, J. 'Crop spray tests for cancer risk', *The Guardian*, 17 August 1989.
59. HSE, MAFF, 'Review of ethylene bisdithiocarbamate (EMDC) fungicides', *The Pesticides Register*, No. 1, 1989, p. 1.
60. Advisory Committee on Pesticides, 'Review of ethylene bisdithiocarbamate (EBDC) fungicides', MAFF news release, 31 January 1990.
61. Natural Resources Defense Council, *Intolerable risk: Pesticides in our children's food*, Washington DC, 1989.
62. Ibid.
63. Friends of the Earth, *Pesticide residues in food worry government experts*, p. 6.
64. Lean, G. 'Highly secret: the poisons we eat', *The Observer*, 16 April 1989, p. 10.
65. MAFF, 'Consultative document on pesticide residues', London, October 1986.
66. MAFF, 'Second Consultative Documents on Pesticide Residues', London, April 1988
67. HSE, MAFF, 'Advisory MRL for tecnazene', *The Pesticide Register*, Issue No. 7, London, 1989, p. 1.

68. EC Directive 4092/1/89 COM 88, 'Proposal for a Council Regulation on the fixing of maximum levels for pesticide residues in and on certain products of plant origin, including fruit and vegetables, and amending Directive 76/895/EEC as regards procedural rules'.
69. 'The Pesticide (Maximum Residue Levels in Food) Regulations 1988. SI 1378, plus Consultative Documents', HMSO, London.
70. British Agrochemicals Association, *Pesticide Residues and Food*, p. 1.
71. MAFF, 'Who Monitors Food', in *Pesticides and Food: a balanced view*, MAFF, London, undated.
72. MAFF, 'UK first country to issue full pesticide monitoring results', press release, London, 27 September 1990, p. 2.
73. Ibid, p. 1.
74. Snell, P.J.I. and Nicol, K.J. 'Pesticide Residues and food: the case for real control', London Food Commission Report, 1986.
75. Ibid.
76. Ibid.
77. Nicolson, R.S. 'Agricultural chemicals in the food chain', *Journal of the Royal Society of Medicine*, Volume 82, May 1989, pp. 313–14.
78. Clutterbuck, C. *Food: a trade union issue*, TUC, London, 1990.
79. Ibid.
80. Ibid.
81. 'Crops destroyed under pesticide safety rules', *The Independent*, London, 5 July 1989.
82. Pesticides Trust, 'California Dreaming: An American comparison', *Pesticides News*, Volume 4, May 1989, p. 4.
83. Ibid.
84. California Department of Food and Agriculture, *Pesticide Residue Annual Report for 1987*, Sacramento, California, 1988.
85. Pesticides Trust, 'California Dreaming: an American comparison', *Pesticides News*, London, Volume 4, May 1989.
86. Pesticides Trust, 'California Dreaming'.
87. National Consumer Council, *The Future of Food Regulation*, London, 1989.
88. Erlichman, J. 'Call for watchdog on food safety', *The Guardian*, London, 1989.
89. Transport and General Workers Union/NUPE/GMB, *Hidden Peril Report*, London, 1988.
90. British Medical Association, *Pesticides, Chemicals and Health Report*, forthcoming.
91. Consumers in the European Community Group, *European Community controls on the use of pesticides in foodstuffs*, London, 1990, p. 1.
92. European Institute for Water, *The EC Directive 80/778 on the quality of water intended for human consumption: pesticides*, Como Seminar, 1988.
93. Department of the Environment, 'Environmental Protection Water Statistics', *Information Bulletin*, London, 28 March 1990, p. 5.
94. Craig, F. and Craig, P. *Britain's Poisoned Water*, Penguin, London, 1989, p. xiii.
95. Pryke, P. 'Minister warns EEC not to prosecute on water', *The Guardian*, 20 July 1989.
96. Environmental Data Services, 'European Water utilities seek stricter pesticide control', *ENDS Report* 172, London, May 1989, p. 6.
97. Environmental Protection Agency, *Environmental Progress and Challenges: EPA's Update*, Washington DC, 1988, pp. 52–6.
98. Croll, B.T. 'Agricultural chemicals in the food chain. 1987 Forum on Food and Health Report', *Journal of the Royal Society of Medicine*, Volume 82, May 1989, p. 313.
99. Friends of the Earth, *An Investigation of Pesticide Pollution in Drinking Water in England and Wales*, Special Report, London, November 1988.

100. Institution of Environmental Health Officers, London Food Study Group, *London's Drinking Water: Summary Report*, London, 1989, p. 4.
101. 'EEC Argues Over Pesticides in Drinking Water.', *ENDS Report* 161, June 1988, p. 13.
102. Environmental Protection Agency, *Environmental Progress and Challenges*, Washington DC, 1988, pp. 50–60.
103. Office of Water, *National Survey of Pesticides in Drinking Wells: Health Advisory Summaries*, Washington DC, January 1989.
104. 'Textile industry moves to clean up effluent discharges', *ENDS Report* 191, December 1990, pp. 7–8.
105. 'EEC Argues over Pesticides in Drinking Water', *ENDS Report* 161, June 1988, p. 14.
106. 'Pesticides found in London's Water Supplies', *The Food Magazine*, October/December 1990, p. 4.
107. 'EEC Argues over Pesticides in Drinking Water', *ENDS Report* 161.
108. Koenig, P. 'Shell to Challenge £50 million clean-up Bill', *The Independent on Sunday*, London, 21 October 1990, p. 3.
109. Ibid.
110. Tarkowski, S. *WHO's Role in the Safety Assessment of Pesticides*, WHO, Copenhagen, undated paper.
111. Ibid.
112. Ibid.
113. Department of the Environment (DoE)/Welsh Office, *Guidance on Safeguarding the Quality of Public Water Supplies*, HMSO, London, 1990.
114. DoE, 'Environmental Protection Act, 1990 – a Landmark in Environmental Legislation', press release, London, 1 November 1990, p. 3
115. National Rivers Authority, *Guardians of the Water Environment*, London, 1989.
116. DoE, *Proposed National Classification Scheme for Dangerous Substances in Groundwater*, press release, London, 31 October 1990.
117. DoE/Welsh Office, *The Water Act, 1989*, London, 1989.
118. European Institute for Water, *Seminar on the EEC Directive 80/778 on the quality of water intended for human consumption: Pesticides. Seminar report and conclusions*, November 1988.
119. Ibid.
120. Ibid.
121. Ibid.
122. Ibid.
123. Ibid.
124. Ibid.
125. Ibid.
126. 'EEC Argues over Pesticides in Drinking Water', *ENDS Report* 161.
127. Ibid.
128. Ibid.
129. Lean, G. 'Ministers rush to Legalise Sewage Dumping in Rivers', *The Observer*, 2 July 1989, p. 3.
130. Ibid.
131. 'European Water Utilities seek Stricter Pesticide Controls', *ENDS Report* 172, May 1989, pp. 5–6.
132. DoE/Welsh Office, *The Water Act*.
133. Ibid.

Chapter 7

1. Ministry of Agriculture Fisheries and Food, *'Investigation of Poisoning of Red Kite'*, news release, London, 26 June 1990.
2. Arden-Clarke, C. *The environmental effects of conventional and organic/biological farming systems in impacts on wildlife and habitat*, World Wildlife Fund, 1988, pp. 1–80.
3. Mellanby, K. 'The Biology of Pollution', *Studies in Biology*, No. 38, 1972, p. 46.
4. Potts, G.R. 'Monitoring changes in the cereal ecosystem', in *Agriculture and the Environment*, D. Jenkins (ed.), ITE Symposium 13, 1984, pp. 128–34.
5. Dewey, S.L. 'Effects of the herbicide, atrazine, on aquatic insect community structure and emergence', *Ecology*, Volume 67, 1986, pp. 148–62.
6. Mellanby, K. 'Pesticides and Pollution', *The Fontana New Naturalist*, Fontana, London, 1970, p. 77.
7. Mellanby, 'The Biology of Pollution', pp. 46–7.
8. Vickerman, G.P. et al. 'Effects of some foliar fungicides on the chrysomelid beetle, gastrophysa polygenu', *Pesticide Science*, Volume 14, 1983.
9. Tomkin, A.D. et al. 'Effects of insecticides and a fungicide on the numbers and biomass of earthworms in pasture', *Bulletin of Environmental Contamination Toxicology*, Volume 12, 1974, pp. 487–92.
10. Stringer, A. et al. 'The effects of benomyl and some related compounds on lumbricus terrestris and other earthworms', *Pesticide Science*, Volume 4, 1973, pp. 165–70.
11. Van Leeuwen, C.J. et al. 'Aquatic toxicological aspects of dithiocarbamates and related compounds', *Aquatic Toxicology*, 1985, pp. 145–64.
12. HSE/MAFF, 'Review of Tributyltin oxide (TBTO) wood preservatives and surface biocides', *The Pesticides Register*, No. 5, 1990.
13. ' Ban on TBT anti-fouling paints should extend to all ships, urge Lords', *ENDS Report* 170, March 1989, p. 25.
14. Ibid.
15. 'Ecology groups cry foul on new paints', *New Scientist*, 12 May 1989, p. 43.
16. Newton, I. 'Monitoring of persistent pesticide residues and their effects on bird populations in Britain since "Silent Spring"', *Proceedings of the Institute of Biology*, 1988, pp. 33–45.
17. Mellanby, 'Pesticides and Pollution', p. 137.
18. Taylor, J.C. et al. 'A short note on the heavy mortality in foxes during the winter 1959–60', *Veterinary Record*, No. 73, 1961, pp. 232–3.
19. Mellanby,'Pesticides and Pollution', p. 140.
20. Van Klingeren, B. et al. 'A study of *nepus europaens* in relation to the use of pesticides in the Polder in the Netherlands', *Journal of Applied Ecology*, 1966, pp. 125–31.
21. Jeffries, D.J. et al. 'The ecology of small mammals in arable fields drilled with winter wheat and the increase in their dieldrin and mercury residues', *Zoological Journal*, No. 171, 1973, pp. 513–39.
22. Mason, C.F. et al. 'Organochlorine residues in British otters', *Bulletin of Environmental Contamination Toxicology*, Volume 36, pp. 656–61.
23. 'Britain's eels extensively contaminated by organochlorines', *ENDS Report* 165, October, 1988, p. 7.
24. Carson, R. *The Silent Spring*, (latest edition), Penguin Books, Middlesex, UK, first published 1962, pp. 40–1
25. Menzle, C.M. 'Fate of pesticides in the environment', *Ann. Ref. Entomol.*, Volume 17.
26. Hardy, A.R. 'The impact of the commercial agricultural use of organophosphorus and carbamate pesticides on British wildlife', *Agriculture and Environment*, D. Jenkins (ed.), 1984, pp. 72–80.

27. Ibid.
28. Moore, N.W. 'The future prospects for wildlife', *Ecological Effects of Pesticides*, Perring, F.H. (ed.), Linnean Society Symposium Series, No. 5, Academic Press, London, 1977, p. 178.
29. Potts, G.R. *The Partridge: pesticides, predation and conservation*, Collins, London, 1986.
30. Law, S. 'Pesticides team up to damage birds', *New Scientist*, London, 21 October 1989, p. 36.
31. Johnston, G. 'Enhancement of malathion toxicity to hybrid red-legged partridge following exposure to prochloraz', *Pesticide Biochemistry and Physiology*, Volume 35, 1989, pp. 107–18.
32. Ibid.
33. Arden-Clarke, *The environmental effects of conventional and organic/biological farming systems*.
34. Nature Conservancy Council, *Focus on bats: their conservation and the law*, NCC, Peterborough, 1989.
35. Mellanby, 'The Biology of Pollution, p. 53.
36. Select Committee on the European Communities, Correspondence with Ministers, House of Lords Sessions 1987–88, 1st Report, 1987, HMSO, London.
37. Foraan, et. al. 'Acute toxicity of aldicarb, aldicarb sulfoxide and aldicarb sulfone to *Daphnia laevis*, *Bulletin of Environmental Contamination Toxicology*, Volume 35, 1985, pp. 546–50.
38. Dudley, N. *This Poisoned Earth: the truth about pesticides*, Piatkus, London, 1987, pp. 102–3.
39. Mellanby, 'The Biology of Pollution', p. 53.
40. Nutall, N. 'Poison Risk to Shellfish', *The Observer*, 1 January 1989.
41. Rudd, R.L. *Pesticides and the Living Landscape*, Faber and Faber, London, 1964.
42. Stevenson, J.H., et al. 'Poisoning of honeybees by Pesticides', *MAFF Report*, Harpenden, UK, 1977, pp. 55–72.
43. Wilkinson, W. et al. 'Safety to honey bees when used in cereals', *British Crop Protection Conference: Pests and Diseases*, 1986, pp. 1,085–92.
44. MAFF, 'The protection of bees: the toxicity to bees of a selection of pesticides', Wallchart, HMSO, London, 1990.
45. Newton, 'Monitoring of pesticide residues', pp. 33–45.
46. 'Dieldrin is poisoning Norfolk's otter', *New Scientist*, London, 14 August 1986, p. 12.
47. Felton, C.L. et al. 'Bird poisoning following the use of warble fly treatment containing famphur', *Verterinary Record*, Volume 108, 1981, p. 440.
48. MAFF, *The illegal poisoning of animals: background notes*, Appendix 2, Pesticide Safety Division, 1990, p. 1.
49. MAFF, 'Poisoning of animals by pesticides', Agricultural Development and Advisory Service, 1990, p. 3.
50. Advisory Committee on Pesticides, *Pesticide poisoning of animals 1988: suspected incidents in Great Britain*, Environmental Panel Report, MAFF Publications, London, 1990.
51. Ibid.
52. Macrae, C. 'Poison bait slaughters rare birds', *The Observer*, 5 June 1988, p. 2.
53. Ibid.
54. Ibid.
55. Ibid.
56. HSE/MAFF, 'Successful prosecutions by MAFF for possession and use of unapproved pesticide', *The Pesticides Register*, No. 6, London, 1990, p. 1.
57. HSE/MAFF, 'Further successful prosecution', *The Pesticides Register*, No. 7, London, 1990, p. 3.

58. Mills, J.A. 'Some observations on the effects of field applications of fensul-fothion and parathion on bird and mammal populations', *Proceedings of New Zealand Ecology Society*, Volume 20, 1973, pp. 66–71.

59. Arden-Clarke, 'The environmental effects of conventional and organic/biological farming systems', pp. 1–80.

60. Ratcliffe, D.A. 'Decrease in eggshell weight in certain birds of prey', *Nature*, No. 215, 1967, pp. 208–10.

61. Newton, 'Monitoring of pesticide residues', pp. 33–45.

62. Moore, N.W. 'Future prospects for wildlife', *Ecological Effects of Pesticides*, p. 179.

63. Rudd, 'Pesticides and the living landscape'.

64. Arden-Clarke, 'The environmental effects of conventional and organic/biological farming systems', pp. 1–80.

65. Ibid.

66. Ibid.

67. 'Banned pesticides pollute European pines', *New Scientist*, 16 September 1989, p. 26.

68. Farming Wildlife and Advisory Group, *A hedgerow code of practice*, FWAG, Sandy, UK. Undated leaflet.

69. Solomon, M.S. 'Fruit and Hops', in *Integrated Pest Management*, Burn, A.J. (ed.), Academic Press, London, 1987, pp. 329–60.

70. Kogan, M. *Ecological Theory and IPM Practice*, Wiley Interscience, 1986.

71. Ibid.

72. Sotherton, N.W. 'The Cereals and Game Birds Research Project: Overcoming the indirect effects of pesticides', in *Britain since Silent Spring*, Institute of Biology, Harding, D. (ed.), London, 1988, pp. 64–70.

73. Potts, *The Partridge*.

74. Green, R.E. 'The feeding ecology and survival of partridge chicks on arable farmland in East Anglia', *Journal of Applied Ecology*, No. 21, 1984, pp. 817–30.

75. Sotherton, 'The Cereals and Game Birds Research Project', p. 69.

76. Harvey, J. '"Don't spray" data upsets agro giant', *Farmers' Weekly*, 25 January, 1985.

77. Sotherton, 'The Cereals and Game Birds Research Project'.

78. Ibid.

79. Rands, M.R.W. and Sotherton, N.W. 'Pesticide use on cereal crops and changes in the abundance of butterflies on arable farmland in England', *Biological Conservation*, Volume 36. 1986, pp. 71–82.

80. Jarvis, R.H. 'The Boxworth Project', in *Britain Since Silent Spring*, pp. 46–55.

81. MAFF, '*Boxworth Project shows environmentally safer pesticide regime for cereals*, news release, 29 June 1989.

82. Ibid.

83. Murphy, K.J. 'Herbicide Resistance in Weeds: A Growing Problem', *Biologist: Journal of the Institute of Biology*, Volume 30, No. 4, September 1983, pp. 211–9.

84. Blake, A. 'Crushing the resistance', *Farmers' Weekly*, 4 November 1988, p. 44.

85. Ibid.

86. Norman, B. 'Fubol-resistant bremia threat to lettuce crops: problem could escalate', *The Grower*, 16 February 1984, pp. 2–3.

87. 'Eyespot resistance is a growing problem', *Big Farm Weekly*, 6 October 1983.

88. Metcalf, R.L. 'Insecticide Resistance' in *Ecological Theory and IPM Practice*, Kogan, M. (ed.), Wiley Interscience, 1986, Chapter 10.

89. Ibid.

90. Solomon, 'Fruit and Hops'.

91. Sax, I. 'International toxicity update: Bees and DDT', *Dangerous Properties of Industrial Materials*, Volume 4, No. 3, 1984.

Chapter 8

1. Dudley, N. *Garden Pesticides*, The Soil Association, Bristol, 1986.
2. Such blunt assertions are normally confined to verbal responses in radio interviews, lectures etc. Printed statements are similarly reassuring. For example, the annual *BAA Directory of Garden Chemicals 1989/90* from the British Agrochemicals Association, has the following statement: 'All products, whether they have a warning symbol or not, can be used safely, provided the instructions and safety precautions are observed.'
3. See Dudley, *Garden Pesticides*, and also Dudley, N. *How Does Your Garden Grow?*, The Soil Association, Bristol, 1986.
4. United Nations Secretariat, 'Consolidated List of Products Whose Consumption and/or Sale have been Banned, Withdrawn, severely Restricted or Not Approved by Governments, first issue revised July 1984, DIESA/WP/1/Rev. 1'.
5. The British government called for a voluntary ban on the sale of ioxynil in shops on 13 June 1985 after evidence emerged that it was a potential teratogen and also affected the thyroid. It remained on sale in many garden shops for some time afterwards, and is sold in a number of formulations for use on farms.
6. *Gardening Which?*, 1988.
7. Earth Resources Research, Unpublished report referred to in 'Is your Environment Making you Ill?', *Which?*, April 1989.
8. Much of the information on bats has been built up by the Nature Conservancy Council in Britain, which also publishes a list of 'bat-friendly' wood preservatives to be used in places where bats might nest. Not all of these are also particularly safe for humans. Contact: Nature Conservancy Council, Northminster House, Northminster Road, Peterborough, Cambs., PE1 1UA, UK.
9. Ibid.
10. London Hazards Centre, *Toxic Treatments, A London Hazards Centre Handbook*, London, 1988.
11. Chandler, A.C. and Read, C.P. *Introduction to Parasitology*, John Wiley, New York, 10th edn. 1961, pp. 621–39.
12. Ibid. p. 625.
13. Ibid, p. 632.
14. Ibarra, J. *To be a Bug Buster*, Community Hygiene Concern report, London, 1990.
15. Ibid.
16. Burgess, I. 'Carbaryl lotions for head lice – new laboratory tests show variations in efficacy', *The Pharmaceutical Journal*, 4 August 1990, pp. 159–61.
17. Ibarra, *Bug Buster*.
18. Ibid.
19. Louse Watch Help Line, *Bug busting 1989 monitoring*, Summary of results, July 1990.
20. Ibarra, *Bug Buster*, p. 19.
21. ADAS, *Bed Bugs: Technical Information*, MAFF, Wildlife and Storage Biology, 1990.
22. The information in this paragraph is taken from: King, F. 'Mind the bugs don't bite', *New Scientist*, 27 January 1990, pp. 51–4.
23. The information on fleas and pesticides is taken from: ADAS, *Fleas: Technical Information*, MAFF, Wildlife and Storage Biology, 1990.
24. HSE/MAFF, *Pesticides 1990*, Reference Book 500, HMSO, London, 1990.
25. Meyer, A.N. *Rodent biology and control*, MAFF/ADAS, Wildlife and Storage Biology, 1990.
26. HSE/MAFF, *Pesticides 1990*.

27. Meyer, *Rodent biology and control*.
28. HSE/Maff, *Pesticides 1990*.
29. Ibid.
30. Meyer, *Rodent biology and control*.
31. Ibid.
32. 'Foresters wage poison war on grey squirrels', *The Independent*, 30 May 1990.

Chapter 9

1. Goulding, R. 'Accidental pesticide poisoning: the toll is high', *International Pest Control*, March/April 1989, p. 32.
2. Pesticides Trust, *The FAO Code: Missing Ingredients*, London, October 1989.
3. Postel, S. 'Defusing the Toxics Threat: Controlling Pesticides and Industrial Waste', *Worldwatch Paper 79*, September 1987.
4. United Nations Food and Agriculture Organisation (FAO), *FAO Trade Yearbook 1984*, Rome, 1985.
5. Weir, D. *The Bhopal Syndrome: Pesticide Manufacturing and the Third World*, International Organisation of Consumer Unions, Penang, 1986.
6. *The FAO Code: Missing Ingredients*, pp. 10–11.
7. Ibid., pp. 12–13.
8. Ibid., pp. 13–14.
9. Ibid., pp. 17–18.
10. World Bank Staff Appraisal Report, *Ghana Cocoa Rehabilitation Project*, Report No. 6818-Gh., October 19 1987, p. 32.
11. *The FAO Code: Missing Ingredients*, p. 16.
12. Ibid., pp. 14–15.
13. Ibid., p. 11.
14. Postel, S. *Defusing the Toxics Threat: Controlling Pesticide and Industrial Waste*, World Water Paper 79, September 1987.
15. Weir, D. *The Bhopal Syndrome*.
16. Goldenman, G. and Rengam, S. *Problem Pesticides, Pesticide Problems*, International Organization of Consumer Unions and Pesticide Action Network, Penang, 1988.
17. Boardman, R. *Pesticides in World Agriculture. The Politics of International Regulation*, Macmillan, London, 1986.
18. Wood, G.A.R. and Lass, *Cocoa*, Longman Scientific and Technical, Singapore, (4th Edn.), 1985.
19. Dethier, V.G. *Morris Plague? Insects and Agriculture*, The Darwin Press, Princeton. N.J., USA, 1976, p. 91.
20. Dahlberg, K. *Beyond the Green Revolution: The Ecology and Politics of Global Agricultural Development*, Plenum Press, New York, 1980, pp. 81–2, 149–66.
21. Shou-Zhen, X. 'Health Effects of Pesticides: A review of epidemiologic research from the perspective of developing nations', *American Journal of Industrial Medicine*, Volume 12, 1987, pp. 269–79.
22. *The FAO Code: Missing Ingredients*, pp. 40–4.
23. Ibid., pp. 45–6.
24. Ibid., p. 47.
25. Ibid., pp. 29–30.
26. Ibid., pp. 35–7.
27. Kabra, S.G. and Narayanan, R. 'Aluminium Phosphide, Worse than Bhopal', *The Lancet*, Volume 1, 11 June 1988, p. 1,333.
28. *The FAO Code: Missing Ingredients*, p. 37.
29. Ibid., pp. 37–8.

30. Amarasingam, R.D. and Lim, A.S. 'Review of Cases of Human Poisonings recorded from 1977–1981', Toxicology Division of Department of Chemistry, Petaling, Haya, 1981.

31. Chan, K. W. and Cheong Izham, K.S. 'Paraquat Poisonings: A clinical and Epidemiological Review of 30 cases', *Medical Journal of Malaysia*, Volume 37, No. 3, 1982.

32. Joyaratnam, J., Lun, K. C. and Phoon, W.O. 'Survey of Acute Poisoning Among Agricultural Workers in Four Asian Countries', *Bulletin of the World Health Organisation*, Volume 65, 1987, pp. 521–7.

33. Ramasamy, S. and Nursiah, M.T.A. 'A survey of pesticide use and associated incidences of poisoning in peninsular Malaysia', *Journal of Plant Protection in the Tropics*, 5(1), 1988, pp. 1–9.

34. *The FAO Code: Missing Ingredients*, pp. 2, 30–1.

35. Ibid., pp. 47–8

36. Corrales, D. 'Problemática de los agroquimicos en el occidente de Nicaragua', in *Instituto Nicaraguense de Recuros Naturales y des Ambiente. Actas des II Seminario Nacional de Recursos Naturales y del Ambiente*, Managua, August 1981, pp. 83–98.

37. Cole, D.C., McConnell, R., Murray, D.L. and Anton, F.P. 'Pesticide Illness Surveillance: The Nicaraguan Experience', *Pan American Health Organisation Bulletin 22(2)*, pp. 119–32.

38. *The FAO Code: Missing Ingredients*, pp. 39–40.

39. Mowbray, D.L. 'Pesticide Poisoning in Papua New Guinea and the South Pacific', *Papua New Guinea Medical Journal*, 29, 1986, p. 141

40. Ibid.

41. Ibid.

42. *The FAO Code: Missing Ingredients*, pp. 49–50.

43. Ibid., p. 40.

44. Loevinsohn, M.E. 'Insecticide use and increased mortality in rural Central Luzan, Philippines', *The Lancet*, Volume 1, 1987, pp. 1,359–62.

45. Thiam, A. and Dieng, A.G. 'Les Pesticides au Senegal: Utilisation, Distribution, Reglementation', ENDA, Senegal, 1988, p. 7.

46. *The FAO Code: Missing Ingredients*, pp. 8–9, 31–2.

47. De Alwis, L.B.L. and Salgado, M.S.L. 'Agrochemical poisoning in Sri Lanka' *Forensic Science International*, 36, 1988, pp. 81–9.

48. *The FAO Code: Missing Ingredients*, pp. 32–3.

49. Edmiston, S. and Maddy, K. T. 'Summary of Illnesses and Injuries Reported in California by Physicians in 1986 as potentially related to Pesticides', *Veterinary and Human Toxicology*, Volume 29 (5), 1987, pp. 391–7.

50. Proudfoot, A.T. and Dougall, H. 'Poisoning Treatment Centre Admissions following acute incidents involving pesticides', *Human Toxicology*, 7, 1988, pp. 255–8.

51. Durham, W.F. and Armstrong, J. F. 'Exposure of workers to pesticides', *Archives of Environmental Health*, Volume 14, 1967, pp. 622–33.

52. Cole, D.C., McConnell, R., Murray D.L. and Anton, F.P. 'Pesticide Illness Surveillance: The Nicaraguan Experience', *Pan American Health Organization Bulletin*, Volume 22 (2), 1988, pp. 119–32.

53. Ibid.

54. Edmiston, S. and Maddy, K.T. 'Summary of Illnesses and Injuries Reported in California by Physicians in 1986 and potentially related to Pesticides', *Veterinary and Human Toxicology*, Volume 29 (5), 1987, pp. 391–7.

55. Ramasamy, S. and Nursiah, M.T.A. 'A survey of pesticide use and associated incidences of poisoning in peninsular Malaysia', *Journal of plant Protection in the Tropics*, 5 (1), 1988, pp. 1–9.

56. Ibid.

57. *The FAO Code: Missing Ingredients*, pp. 39–40.

58. 'Environmental Effects of Major Federal Actions', US Executive Order 12114, 4 January 1979.
59. 'Export of Banned and Significantly Restricted Substances', US Executive Order 122644, 15 January 1981.
60. Engler, R. 'Technology Out of Control', *National*, 27 April 1985, p. 492.
61. Simonian, L. 'Pesticide Use in Mexico:Decades of Abuse', *The Ecologist*, Volume 18, 1988, pp. 82–7.
62. Weir, D. and Shapiro, M. *Circle of Poison, Pesticides and People in a Hungry World*, Institute for Food and Development Policy, San Francisco, 1981.
63. Bull, D. *A Growing Problem. Pesticides and the Third World Poor*, Oxfam, 1982.
64. *International Code of Conduct on the Distribution and Use of Pesticides*, Food and Agriculture Organization of the United Nations, Rome, 1986.
65. Weir and Shapiro, *Circle of Poison, Pesticides and People in a Hungry World*.
66. Bull, *A Growing Problem*.
67. *Monitoring and Reporting the Implementation of the International Code of Conduct on the Use and Distribution of Pesticides. Final Report, October 1987*, Environment Liaison Centre, P.O. Box 72461, Nairobi, Kenya.
68. Goldenman and Rengam, *Problem Pesticides, Pesticide Problems*.
69. *The FAO Code: Missing Ingredients*.
70. *Pesticide News. Newsletter of the Pesticides Trust*, November 1989, pp. 2–3.
71. Hanson, M. *Escape from the Pesticide Treadmill: Alternatives to Pesticides in Developing Countries*, Institute for Consumer Policy Research, Consumers Union, Washington DC, 1987.
72. Department of Agriculture, *Meat and Poultry Inspection 1981. Report of the Secretary of Agriculture to the US Congress*, Food Safety and Inspection Service, USDA, US Printing Office, Washington DC, 1982.
73. Leonard, H. J. *Natural Resources and Economic Development in Central America. A Regional Environmental Profile*, Transaction Books, Oxford, UK, 1985, pp. 152–3.
74. Ibid.
75. Ibid.
76. 'PCBs, PCDDs and PCDFs in Breast Milk: Assessment of Health Risks', World Health Organisation, Copenhagen, Volume 29, 1988.
77. Jenson, A.A. 'Chemical contaminants in human milk', *Residue Review*, Volume 89, 1983, pp. 1–128.
78. Swezey, D.L., Daxl, R.G. and Murray, D.C. 'Nicaragua's revolution in pesticide policy', *Environment*, Volume 28 (1), 1986, pp. 6–9, 29–36.
79. FAO/WHO, *Guide to Codex Maximum limits for pesticide residues* CAC/PRI, Rome, 1978.
80. Olszyna-Marzys, A.E., De Campos, M., Farrar, M.T., and Thomas, M. 'Residues of chlorinated pesticides in human milk in Guatemala', *Bull. of Sanit. Panam*, Volume 74, 1973, pp. 93–107.
81. Zaidi, S.S.A., Bhatnager, V.K., Banerjee, B.D., Balakrishnan, G. and Shah, M.P. 'DDT residues in human milk samples from Delhi, India', *Bulletin of Environmental Contamination and Toxicology*, Volume 42, 1989, pp. 427–30.
82. Hanson, M. *Escape from the Pesticide Treadmill*.
83. Moustafa, F. I. and Georghiou, G. P. 'Extension of resistance to adults as a result of selection in mosquito larvae with temephos, and the cross-resistance to other organophosphorus insecticides', *International Pest Control*, May/June 1989, pp. 61–5.
84. Hanson, *Escape from the Pesticide Treadmill*, p. 22.
85. Folsom, J.W. 'Calcium arsenate as a cause of aphid infestation', *Journal of Economic Entomology*, 20, 1927, pp. 840–3.
86. DeBach, P. 'An insecticidal check method for measuring the efficacy of entomophagous insects', *Journal of Economic Entomology*, 39, 1946, pp. 695–7.

87. DeBach, P. *Biological Control by Natural Economies*, Cambridge University Press, New York 1974, p. 323.
88. Stern, V.M., Smith, R.F., van den Bosch, R. and Hagon, K.S. 'The integrated control concept', *Hilgardia*, 29, 1959, pp. 131–54.
89. Krieger, R.I., Feeny, P.P. and Wilkinson, C.F. 'Detoxification enzymes in the guts of caterpillars: an evolutionary answer to plant defenses', *Science*, Volume 172, 1971, pp. 579–81.
90. Plapp, F.W. Jr. 'The nature, mode of action; and toxicity of insecticides', in *C.R.C. Handbook of Pest Management in Agriculture*, Volume III, Pimentel, D. (ed.), CRC Press, Boca Raton, Florida, 1981, pp. 3–16.
91. Hanson, *Escape from the Pesticide Treadmill*, p. 139.
92. International Rice Research Institute, *Annual Report for 1979*, Los Banos, Philippines, 1980.
93. Kenmore, P.E., Carino, F.O., Perez, C.A., Dyck, V.A. and Gutierrez, A.P. 'Population regulation of the rice brown planthopper (*ilaparvata lugens*) StdD within rice fields in the Philippines', *Journal of Plant Protection of the Tropics*, Volume 1 (1), 1984, pp. 19–37.
94. Hanson, *Escape from the Pesticide Treadmill*, pp. 131–2.
95. Ibid., pp. 31–182.
96. Saffaur, O. 'A new crop of pest controls', *New Scientist*, 14 July 1988, pp. 48–51.
97. Wanigasundara, M. 'Bugs beat Sri Lankan weed problem', *Eastern Review*, 25 March 1989.
98. Hanson, M. *The First Three Years: Implementation of the World Bank Pesticide Guidelines 1985–1988*, Institute for Consumer Policy Research, Consumer Union, San Francisco (in press), p. 11.
99. Ibid., p. 13.
100. Ibid., p. 31.
101. Ibid., p. 25.
102. Ibid., p. 34.
103. Ibid., p. 37.
104. Ibid., pp. 43–6.
105. Ibid., pp. 50–4.
106. Ibid.
107. Gonzalez, G.E.T. – personal communication, 15 May 1989.
108. Da Silva, M.L. – personal communication, 15 May 1989.
109. Aculey, P. – personal communication, 15 May 1989.
110. *Manual de Segurança No Uso de Agrotoxicos*, Fundacentro, Ministerio Do Trabalho, Sao Paulo, Brazil, 1987.
111. Da Silva, M. L. – personal communication, 15 May 1989.
112. Federación Nacional de Campesinos Libres del Ecuador, *Riesgos del Trabajo Agrícola*, Fenacle, Guayaquil, Ecuador, February 1989.
113. *Demise of the Dirty Dozen*, Pesticide Action Network, Pan North America Regional Centre, 1989.
114. *Plaguicidas, 1989*, Calendar produced by FEMACLE, FETLAE, FITPAS and CEOSL, c/o FEMACLE, Guayaquil, Ecuador.
115. Gasser, C.S. and Fraley, R.T. 'Genetically Engineered Plants for Crop Improvement', *Science*, Volume 244, 1989, pp. 1,293–9.
116. Kneitschel D. *Guidado Veneno. Las Plaguicidas y Sus Peligros*, Fundación Friedrich Ebert, Apartado 2691, Santo Domingo, Dominican Republic, 1986.

Chapter 10

1. Aculey, P. – personal communication, 15 May 1989.
2. Ibid.
3. Ibid.
4. World Bank Staff Appraisal Report, Ghana Cocoa Rehabilitation Project, Report No. 6818, Gh., 19 October 1987, p. 32.
5. Ibid., p. 25.
6. Ibid., p. 56.
7. Aculey, P. – personal communication, 15 May 1989.
8. Ibid., 24 September 1988.
9. Ibid.
10. Ibid.
11. Hayes, W. *Copper Sulfate in Pesticides Studied in Man*, Williams and Wilkins, Baltimore and London, 1982, pp. 5–6.
12. House, R. 'The Cocoa Slaves of Brazil', *The Independent Magazine*, 9 September 1989, pp. 22–7.
13. Ibid.
14. Ibid.
15. Ibid.
16. Ibid.
17. Da Cruz, P.F.N. – personal communication, 9 May 1989.
18. Ibid.
19. Ribeiro, J.C. – personal communication, 16 May 1989.
20. Aculey, P. – personal communication, 24 September 1988.
21. Silva, L.C.S. – personal communication, 24 September 1988.
22. Aculey, P. – personal communication, 24 September 1988.
23. 'World Bank Staff Appraisal Report', Ghana Cocoa Rehabilitation Project, Report No. 6818, Gh., October 18 1987, pp. 64, 79.
24. Ibid., p. 78.
25. Vespar, A. – personal communication, 9 May 1989.
26. International Agency for Research on Cancer, *Monographs on the Evaluation of the Carcinogenic Risk of Chemical to Humans: An Updating of IARC Monographs Volumes 1–42*, Supplement 7, Lyons, France, 1987.
27. *IRPTC Bulletin*, Volume 7, No. 3, December 1985.
28. *Se o mato nao deixa seu cacau render*, Gramoxone nele, ICI, Brasil SA.
29. *Manual de aplicaçao do Cobre Sandoz*, Sandoz SA Divisao Agroquimica.
30. Caufield, C. 'Companies defy Brazilian pesticide law', *New Scientist*, 11 August 1983, p. 393.
31. Morales, J. and Silva, L.C.S. – personal communication, 24 September 1988.
32. *Salvador Evening News*, March 1988 (full date unknown); *Evening News*, 12 April 1988.
33. *Salvador Evening News*, 24 April 1988.
34. *The Amsterdam Report on International Conference in Pesticides and Cocoa Production*, Amsterdam, 22–25 September 1988.
35. Ibid.
36. Brinkman H. – personal interview, 25 September 1988.
37. Ibid.
38. Beauchamp, E.T. The Biscuit Cake, Chocolate and Confectionery Alliance – personal communication, 18 April 1989.
39. Richardson D P. 'The Nestle Company Letter', 20 June 1989.
40. Mustapha, E.B. *International Conference on Cocoa*, Ilheus, Brazil, 8–18 May, 1989.
41. Aculey, P. – personal communication, 16 May 1989.
42. Ladeji, J. – personal communication, 16 May 1989.

Chapter 11

1. *The WHO Recommended Classification of Pesticides by Hazard and Guidelines to Classification 1988–89*, WHO, Geneva.
2. *International Code of Conduct on the Distribution and Use of Pesticides*, FAO, Rome, 1986, amended 1990.
3. UNEP, *The London Guidelines for the Exchange of Information on Chemicals in International Trade*, UNEP, Nairobi, amended version, 1989.
4. EC, *Draft proposal for an amendment to Council Regulation No. 1734/88*, 26 June 1990.
5. Pesticides Trust, *Pesticides News*, No. 1, June 1988, Double issue, Nos. 9 and 10, November 1990.
6. *Codex recommended national regulatory practices to facilitate acceptance and use of Codex maximum limits for pesticide residues in food*, FAO, Rome, 1985.
7. *Convention concerning Safety in the Use of Chemicals at Work*, Convention 170, ILO, Geneva, 1990.
8. *Guiding Principles on Information Exchange Related to Export of Banned or Severely Restricted Chemicals*, OECD, Paris.
9. *International glossary of key terms in chemicals control legislation*, OECD, Paris 1982.
10. 'Curtain rises on unscreened chemicals', *New Scientist*, 18 November 1989, p. 25.
11. EC, *Plant Protection Prohibitions Directive 79*, 117/EC.
12. EC, *Directive, Concerning the Placing of EC-authorised Plant Protection Products on the Market*, COM(89) 34 final.
13. HSE ChemAG Committee, verbal report to the Committee, 12 December 1990.
14. *Health and Safety: Control of Substances Hazardous to Health Regulations*, SI 1988, No. 1657, HMSO, London.
15. EC *Directive. Proposal for a Council Regulation on the fixing of maximum levels for pesticide residues in and on certain products of plant origin, including fruit and vegetables*, 4092/1/89 COM(88) 798.
16. EC Directive on the freedom of access to information on the environment.
17. Ibid.
18. Pesticides Trust, 'The FAO Code: Missing Ingredients', PAN Report, London, 1989.
19. MAFF, *UK Pesticides Safety Precautions Scheme*, Pesticides Branch, London, revised 1981.
20. Food and Environment Protection Act (FEPA), 1985, Part 3 Chapter 48, HMSO, London.
21. The Control of Pesticides Regulations (COPR), 1986, SI 1986, No. 1510, HMSO, London.
22. Advisory Committee on Pesticides (ACP), *Annual Report 1988*, HMSO, London.
23. HSE/MAFF, *The Pesticides Register*, Published monthly, HMSO, London.
24. HSE/MAFF, *Pesticides 1990: Pesticides Approved Under COPR 1986*, Reference Book 500 (updated annually), HMSO, London.
25. *Code of Practice for Suppliers of Pesticides to Agriculture, Horticulture and Forestry*, MAFF Publications, London, Ref PB 0091, February 1990.
26. *Health and Safety: Classification, Packaging and Labelling of Dangerous Substances Regulations 1984*, SI 1984, No. 1244, HMSO, London.
27. *Health and Safety: Notification of New Substances Regulations 1982*, SI 1982, No. 1496, HMSO, London.
28. HSE/MAFF, *Pesticides 1990*, Reference Book 500.
29. Ibid.

30. The Medicines Act 1968, Chapter 67, HMSO, London.
31. MAFF, *The Work of the Veterinary Medicines Directorate*, MAFF Publications, London, ref PB 0156, 1990.
32. The Medicines (Labelling) Regulations 1976, SI 1976, No. 1726, HMSO, London.
33. The Health and Safety at Work Act 1974, Chapter 37, HMSO, London.
34. MAFF, *Data Requirements for Approval Under the Control of Pesticides Regulations 1986*, Pesticides Safety Division, London, 1986.
35. MAFF, *Advisory Committee on Pesticides Annual Report 1987*, Appendix 3, HMSO, London.
36. HSE, *COSHH Approved Code of Practice*, COP 29, HMSO, London.
37. MAFF, *Data Requirements for Approval Under the COPR*.
38. Health and Safety Agency (HSA) for Northern Ireland, *Health and Safety at Work (Northern Ireland) Order 1978: the Order outlined*, Leaflet HSA 1, HSA, Belfast, undated.
39. HSA for Northern Ireland, *Where can you obtain advice?*, Leaflet HSA 14, HSA, Belfast, undated.
40. Morrisey H. *Northern Ireland health and safety report*, Transport and General Workers' Union, Belfast, November 1988.
41. HSC/MAFF, *Pesticides: Code of Practice for the safe use of pesticides on farms and holdings*, HMSO, London, 1990.
42. HSC, *COSHH Approved Code of Practice on the safe use of non-agricultural pesticides at work*, HMSO, London, 1991.
43. HSC, *COSHH Approved Code of Practice on control of substances hazardous to health in fumigation operations*, COP 30, HMSO, London, 1989.
44. HSC/MAFF, *Pesticides: Code of Practice for the safe use of pesticides on farms and holdings*.
45. HSC, *COSHH Approved Code of Practice on the safe use of non-agricultural pesticides at work*.
46. Transport and General Workers' Union (TGWU), *Hidden Peril*, Report No. 2, TGWU, London, 1989.
47. HSE, *Training in the use of pesticides*, Leaflet AS 25, revised 2/90 30M, HSE offices.
48. HSC/MAFF, *Pesticides: Code of Practice*.
49. HSC, *Recommendations for training users of non-agricultural pesticides*, HMSO, London, 1990.
50. Agricultural Training Board (ATB), FEPA, *Training and testing in pesticide use in agriculture*, ATB, Beckenham, File No 91.46, November 1989.
51. HSE/MAFF, 'Labelling requirements', *The Pesticides Register*, No. 12, 1989, HMSO, London.
52. BASIS, *A registration scheme for distributors of crop protection products*, London, 1989.
53. MAFF, *Code of Practice for suppliers of pesticides to agriculture*.
54. MAFF, 'New certificate for suppliers of forestry pesticides', Press release, 3 July 1989.
55. HSE, *Storage of approved pesticides: guidance for farmers and other professional users*, Guidance Note CS 19, 1988, HMSO, London.
56. HSE, *Pesticide incidents investigated in 1989/90: a report by HM Agricultural and Factory Inspectorates*, HSE, Bootle, 1990.
57. HSE/MAFF, *Pesticides 1990*, Reference Book 500.
58. *Air Navigation Order 1985*, HMSO, London.
59. Civil Aviation Authority, *The Aerial Application Certificate: information on requirements to be met by applicants and holders*, CAP 44, CAA, Cheltenham, 1989.
60. Health and Safety, *Control of Industrial Major Accident Hazards Regulations 1984*, SI, 1984, No. 1902, HMSO, London.

61. EC, *Major accident hazards of certain industrial activities*, 82/501/EEC as amended.
62. Hay, *The Chemical Scythe.*
63. Health and Safety, The Road Traffic (Dangerous Substances in Road Tankers and Tank Containers) Regulations 1981, SI 1981, No. 1059, HMSO, London.
64. Health and Safety, The Road Traffic (Carriage of Dangerous Substances in Packages etc.) Regulations 1986, SI 1986, No. 1951, HMSO, London.
65. Health and Safety, Dangerous Substances in Harbour Regulations 1987, SI 1987, No. 87, HMSO, London.
66. The Control of Pollution Act 1974, HMSO, London.
67. The Special Waste Regulations 1981, SI 1981, HMSO, London.
68. Agriculture Act 1947, Chapter 48, Section 109(3), HMSO, London.
69. Health and Safety: Genetic Manipulation Regulations 1989, SI 1989, HMSO, London.
70. HSE, *Genetic Manipulation: your responsibilities under the Regulations*, Leaflet IND(G) 86L 1990, HSE offices.
71. The Environment Protection Act 1990, Part 6, HMSO, London.
72. HSC, *Annual Report 1989/90*, HMSO, London, p. 24.
73. Institution of Professionals, Managers and Specialists (IPMS), *Health and Safety: an Alternative Report – the real facts about the work of the HSE in 1989*, IPMS, London, 1990.
74. Goddard, G. 'Occupational accident statistics 1981–85', HSE Economics and Statistics Unit, in *Employment Gazette*, January 1988, pp. 15–21.
75. HSC, *Annual Report 1989/90*.
76. IPMS, *Health and Safety: an Alternative Report 1989*.
77. Ibid.
78. Ibid., p. 15.
79. Ibid., p. 16.
80. HSC, *Annual Report 1989/90*.
81. The Institution of Environmental Health Officers, *Environmental Health Report 1985*, London, 1986.
82. Nicolson, R.S. 'Agricultural Chemicals in the Food Chain' *Journal of the Royal Society of Medicine*, Volume 82, May 1989, pp. 313–14.
83. TGWU, *Hidden Peril*, Report 2.
84. Ibid.
85. 'Programme to review safety of older pesticides gets under way', ENDS Report 170, March 1989.
86. MAFF, 'Review of older agricultural pesticides is making good progress', news release, 1 November 1990.
87. HSE/MAFF, 'Government measures on pesticide approvals, reviews and residue monitoring', *The Pesticides Register*, No. 3, 1990, HMSO, London.
88. MAFF, 'Routine review of pesticides', news release, 16 March 1989.
89. IPMS, *Health and Safety: an Alternative Report*.
90. MAFF, 'Review of older agricultural pesticides is making good progress', news release, 1 November 1990.
91. IPMS, *Health and Safety: an Alternative Report 1989*.
92. Goddard G. *Occupational Accident Statistics 1981-85*.
93. IPMS, *Health and Safety: an Alternative Report 1989*.
94. IPMS, *Health and Safety: an Alternative Report 1988*, p. 16.
95. Ibid.
96. Ibid.
97. Green Alliance, 'Environmentalists and agrochemicals industry join forces to urge improved pesticide control', press release, 7 August 1989
98. HSC, 'Annual Report 1989/90', press release, 13 December 1990.
99. HSC, *Annual Report 1987/88*, HMSO, London.
100. Green Alliance, 'Environmentalists and agrochemical industry join forces'.

101. Ibid.
102. MAFF, Letter from J. Gummer to Green Alliance, 26 September 1989
103. Friends of the Earth, 'How to report a spray incident', *Pesticides Broadsheet*, Spring 1985.
104. The Reporting of Injuries, Diseases and Dangerous Occurrences Regulations 1985, as amended, SI 1985, No. 1457, HMSO, London.
105. *The Environment and Safety Information Act*, HMSO, London, 1988.
106. The Ramblers Association, *Pesticides: waging war on our countryside*, 1988.

Chapter 12

1. Kornberg, Sir H. *Royal Commission on Environmental Pollution, 7th Report: Agriculture and Pollution*, HMSO, London, 1979, p. 68.
2. Solomon, M.G. 'Fruit and Hops', in *Integrated Pest Management*, Burn, A. (ed.), Academic Press, London 1987.
3. *Royal Commission on Environmental Pollution, 7th Report*, p. 49.
4. Carson, *The Silent Spring*, pp. 237–9.
5. British Agrochemicals Association, *Pesticide Resistance. Pesticides in Perspective*, Peterborough, undated.
6. Debach, P. *Biological Control of Natural Enemies*, Cambridge University Press, London, 1974.
7. Ibid.
8. Van Den Bosch, R. et al. *Source Book on Integrated Pest Management*.
9. Sattaur, O. 'A new crop of pest controls', *New Scientist*, 14 July 1988, pp. 48–54.
10. Ibid.
11. 'Integrated mosquito control in India', *Biocontrol News and Information*, Volume 9, No. 1, 1988, CAB (Commonwealth Agricultural Bureau), Wallington, Oxfordshire, UK, p. 2.
12. Ley, S.V. et al. 'Insect antifeedants',*Chemistry in Britain*, January 1990, pp. 31–5
13. Ricard, J.L. et al. 'Biocontrol of pathogenic fungi in wood and trees, with particular emphasis on the use of Trichoderma', *Biocontrol News and Information*, Volume 9, No. 3, 1988, CAB, pp. 133–42.
14. Galden, R.D. 'A capsule history of biological control of weeds', *Biocontrol News and Information*, Volume 9, No. 2, CAB, 1988, pp. 55–61.
15. Templeton, G.E. 'Status of weed control with plant pathogens', in *Biological Control of Weeds with Plant Pathogens*, Charudattan, R. (ed.), John Wiley Interscience, London, 1982.
16. 'Disease Sprayed on Crops Controls Problem Weeds', *Big Farm Weekly*, July 1983, p. 21.
17. 'Water weed control in Papua New Guinea', *Biocontrol News and Information*, Volume 7, No. 1, CAB, 1986, p. 1.
18. Ford, J. 'Beekeepers stung by release of moths', *New Scientist*, 25 August 1988, p. 26.
19. Wilken, D.R., et al, 'Residues of organophosphorus pesticides in wholemeal flour and bread produced from treated wheat', *Proceedings of British Crop Conference, Pests and Diseases*, 1981, pp. 183–8.
20. 'Simple traps keep the grain bugs at bay', *New Scientist*, 28 July 1988, p. 41.
21. Northall, P. 'Revolutionary methods to control Nicaragua's pests', *New Scientist*, 14 July 1988, p. 51.
22. Ibid.
23. Ibid.
24. 'Blandford fly meets its Waterloo', *New Scientist*, 13 May 1989, p. 30.
25. Cherfas, J. 'Enzyme test reveals mechanism of pests' resistance', *New Scientist*, 4 March 1989, p. 30.

26. Sattaur, O. 'Pheromones add a new twist to cotton crops', *New Scientist*, 7 January 1989, p. 35.
27. Ibid.
28. 'Vacuum cleaners suck strawberries free of pests', *New Scientist*, 28 January 1989, p. 41.
29. Pesticides Action Network, 'Fighting pests the natural way', *PAN Europe, 1988.*
30. 'Grow your own pesticide', *New Scientist*, 6 June 1985, p. 10.
31. Pesticides Action Network, 'Fighting pests the natural way'.
32. Grimbley, P.E. – personal communication, Glasshouse Crops Research Institute, March 1985.
33. Payne, C.C. 'Prospects for biological control', in *Britain Since the Silent Spring*, pp. 103–16.
34. Glasshouse Crops Research Institute (GCRI), 'The Biological Control of Cucumber Pests', *Growers Bulletin*, No. 1 (3rd edn), Littlehampton, West Sussex, UK, undated.
35. GCRI, 'Integrated control of tomato plants', *Growers Bulletin*, No. 3, Littlehampton, West Sussex, UK, undated.
36. Ibid.
37. Payne, C.C. 'Prospects for biological control'.
38. Ibid.
39. Sly, J.M. *Pesticide Usage in England and Wales: Orchards 1983*, Preliminary Report 38, MAFF, London, 1985.
40. Solomon, 'Fruit and Hops'.
41. Ibid., pp. 343–6.
42. Ibid., p. 348.
43. Ibid., pp. 352–4.
44. Pain, S. 'Salmon farmers put "cleaner fish" on the payroll, *New Scientist*, 21 October 1989, p. 35.
45. Ibid.
46. 'Pumps deliver a fatal blow to grain pests', *New Scientist*, 24 June 1989, p. 38.
47. Ibid.
48. 'Cooked cockroaches', *New Scientist*, 17 February 1990, p. 35.
49. Ibid.
50. Payne, C.C. 'Prospects for biological control'.
51. 'Ecological effects of pesticides', *Biocontrol News and Information*, Volume 9, No. 2, CAB, 1988, pp. 51–2.
52. Miller, L.K., et al. 'Bacterial, viral and fungal insecticides', *Science*, Volume 219, 1983, pp. 715–21.
53. Ibid.
54. Payne, C.C. 'Prospects for biological control'.
55. Miller, L.K., et al. 'Bacterial, viral and fungal insecticides'.
56. Payne, C.C. 'Prospects for biological control'.
57. Davison, J. 'Plant beneficial bacteria', *Biotechnology*, Volume 6, March 1988, pp. 282–6.
58. 'Ecological effects of pesticides', *Biocontrol News and Information*, Volume 9, No. 2, CAB, 1988, pp. 51–2.
59. Royal Commission on Environmental Pollution, *The release of genetically engineered organisms to the environment*, 13th Report, CM 720, HMSO, London, 1989.
60. Connor, S. 'Genes on the loose', *New Scientist*, London, 26 May 1988, pp. 65–8.
61. Straughan, R. *The genetic manipulation of plants, animals and microbes: the social and ethical issues for consumers*, discussion paper, National Consumer Council, London, 1989.
62. Biotechnology Working Group, *Biotechnology's Bitter Harvest: Herbicide-tolerant crops and the threat to sustainable agriculture*, US, undated.
63. Ibid.

64. 'Approval for first British virus release experiment', *Nature,* London, Volume 320, 6 March 1986, p. 2.
65. The Natural Environment Research Council, Public Notice, NERC Institute of Virology, Oxford, 1 April 1988.
66. 'Gene-spliced pesticide uncorked in Australia', *New Scientist,* 4 March 1989, p. 23.
67. Anderson, I. 'The strawberry field that has halted biotechnology', *New Scientist,* London, 20 February 1986, p. 18.
68. Connor, S. 'Genes on the loose'.
69. 'The unkindest cut of all for Strobel', *New Scientist,* 10 September 1987, p. 26.
70. Balfour, E.B. *The Living Soil,* (1943, second edition 1975) latest edition, Universe Books, New York.
71. Steinbeck, J. *The Grapes of Wrath,* Penguin, Middlesex, various editions.
72. Soil Association, *The Living Earth* (formerly *Mother Earth* and *The Soil Association Quarterly Review*), Suffolk and Bristol.
73. Henry Doubleday Research Association, *HDRA Membership News,* (Quarterly), Ryton Gardens, near Coventry.
74. British Organic Farmers and the Organic Growers' Association, *New Farmer and Grower,* Bristol.
75. European Community Standards.
76. National Farmers' Union Survey.
77. British Organic Standards Committee, *Standards for Organic Agriculture,* The Soil Association, Bristol, latest edition, 1990.
78. Wenz, C. 'New Chemicals Under Fire', *Nature,* London, 28 June 1984.
79. Elm Farm Research Centre, unpublished research note, 1988.
80. Woodward, L., Stopes, C., Lampkin, N., Dudley, N., Arden-Clarke C., and Midmore, P. *The Potential for Developing Organic Agriculture as a Mainstream Policy Option for Reducing Surpluses and Protecting the Environment, A report to the management committee of British Organic Farmers,* The Organic Growers' Association, and The Soil Association, Elm Farm Research Centre, 1988.
81. Ibid.
82. Costs of the CAP per family.
83. Figures calculated by British Organic Farmers (unpublished), as background to *20% of Britain Organic by the Year 2000,* British Organic Farmers, The Organic Growers Association and the Soil Association
84. Pearce, D., Markandya, A. and Barbier, E.B. *Blueprint for a Green Economy, a report for the UK Department of the Environment,* Earthscan Publications, London, 1989.
85. Baldock D. *Reduction strategies for pesticides,* Institute for European Environmental Policy, London, 1990.
86. British Agrochemicals Association, *Annual Report and Handbook 1988/89.*
87. Baldock, D. *Reduction strategies for pesticides.*
88. 'Environmentally-safe label for UK food?', *Agrow World Crop Protection News,* No. 86, 5 May 1989, p. 9.
89. Bernson,V. 'Regulation of pesticides in Sweden', *Brighton Crop Protection Conference: Pests and Diseases,* 1988, p. 1,059.
90. 'Progress of Swedish pesticide reduction programme', *Agrow World Crop Protection News,* London, No. 115, 13 July 1990.
91. Swedish National Board of Agriculture, *Action programme to reduce the risks to health and the environment in the use of pesticides in agriculture,* summary paper, Crop Production Division, June 1988.
92. 'Progress of the Swedish pesticide reduction programme'.
93. Ibid.
94. Ibid.
95. Pesticides Trust, 'Towards a reduction in pesticide use', *Pesticide News,* No. 7, March 1990, pp. 9–11.

96. 'Denmark bans 20 more active ingredients in 1989', *Agrow World Crop Protection News*, No. 119, 14 September 1990, p. 1.
97. Ibid.
98. Ibid.
99. Thonke, K.E. 'Research on pesticide use in Denmark to meet political needs', in *Aspects of Applied Biology*, No. 18, 1988, Institute of Weed Control, Denmark, pp. 327–9.
100. Kudsk, P. 'Experiences with reduced herbicide doses in Denmark and the development of the concept of factor-adjusted doses', in *Proceedings of Brighton Crop Protection Conference: Weeds*, 1989, pp. 545–54.
101. Baandrup, M. et al. 'Three year field experience with an advisory computer system applying factor-adjusted doses', in *Proceedings of Brighton Crop Protection Conference: Weeds*, 1989, pp. 555–60.
102. 'Dutch government's pesticide usage proposals', *Agrow World Crop Protection News*, No. 119, 14 September 1990, p. 1.
103. 'Further news on reduction strategies', *Pesticide News*, No. 9–10, November 1990, p. 17.
104. 'Dutch government's pesticide usage proposals', *Agra Europe*, 31 August 1990, p. 9.
105. 'Dutch plan calls for 50% cut in agrochemicals', *Agrow*, No. 89, 16 June 1989, p. 7.
106. Ibid.
107. UK Government White Paper, 'This Common Inheritance', Cm 1200, HMSO, London, 1990.
108. British Medical Association, *Pesticides, Chemicals and Health*, BMA Report, 1990.
109. Ibid., p. 137.
110. Health and Safety Executive/MAFF, BMA working group on pesticides, chemicals and health. *The Pesticides Register*, No. 10, 1990, p. 1.

Chapter 13

1. Carson, Rachel, *The Silent Spring*, Penguin Books, Middlesex, UK, 1962.
2. Reported in Dudley, N. *Safety Never Assured: The Case Against Aerial Spraying*, The Soil Association, Bristol, UK, 1986.
3. The meetings took place in 1987 and 1988 and involved speakers from the British Agrochemicals Association and the Soil Association.
4. Rose, C. *The Pesticides Incidents Report*, Friends of the Earth, London, 1986.
5. Lees, A. and Blake, L. *Pesticides: The Second Incidents Report*, Friends of the Earth, London, 1988.
6. Anon. *The Avon Herons*, Vale of Evesham Friends of the Earth, UK, 1985.
7. Snell, P. and Nicoll, K. *Pesticide Residues and Food: The case for real control*, The London Food Commission, 1986.
8. Cook, J. and Kaufman, C. *Portrait of a Poison: the 2,4,5-T Story*, Pluto Press, London, 1982.
9. Anon. *Pesticides: The Hidden Peril, a Joint Trade Union Report on Pesticide Usage*, Agricultural and Allied Workers' Trade Group (TGWU), General and Municipal Workers' Union, and National Union of Public Employees, London, 1987.
10. Bull, D. *A Growing Problem: Pesticides and the Third World Poor*, Oxfam, Oxford, UK, 1982.
11. Institute of Professional, Managerial and Scientific Staff, *Health and Safety: An Alternative Report*, London, 1989.

Index

demeton-s-methyl, 45, 146, 218
dengue, 8
Denmark, 4, 32, 92, 137, 225, 252, 285, 287–8
 Institute of Weed Control, 287
 NERI (National Environmental Research Group), 287
Department of Employment, 232, 251
Department of Health, 136, 169, 232, 251, 296
dermatitis, 25, 85, 90, 99, 104–5
derris, 11
 rotenone, 5
di-ethylthiocarbamic acid, 80
diallate, 45
diatomaceous earth, 36
diazinon, 126, 204
dibenzofurans, 203
dichlorvos, 15, 45, 57, 79–80, 85, 91, 122, 148, 269
diclobenil, 57
dieldrin, 8, 15, 16, 44, 72, 74, 89–90, 91, 122, 123, 133, 135, 145–6, 148, 149, 202, 217, 218, 223
diethylamine, 75
diethyltin di-iodide, 103
difenothrin, 18
dimethoate, 79–80, 146
dimethylcarbamoyl chloride, 77
dinitro compounds, 87
dinoseb, 87, 143
dinoterb, 87
dioxins, 75, 101, 123–4, 203, 245
dipropylnitrosoamine, 81
diquat, 44, 57, 100–1
disposal methods, 2, 28, 38, 49, 60–3, 131, 247–8, 297
disulphaton, 45, 198
dithiocarbamates, 45, 72, 104, 144
diuron, 16, 72, 105
DNOC (dinitro-o-cresol), 11, 87, 143
DOE (Department of the Environment), 136, 227, 232, 251, 284
 Pearce Report, 285
Dominican Republic, 207–8
Dow, 21, 22, 197, 218
 Elanco, 20, 274
Draize test, 28
drycleaners, 73
DTI (Department of Trade amd Industry), 232
DuPont, 20, 218, 274
Dutch Elm Disease, 275
Dutch Institute for Working Conditions, 221

Earth Resources Research, 166
EC (European Community), 4, 12, 24, 25, 28, 32, 127–8, 131–4, 140–1, 145, 225, 226, 229–32
 CAP (Common Agricultural Policy), 7, 112, 278, 285, 286
 Drinking Water Directive, 131, 135–40
 Seveso Directive, 245
 Standing Committee on Plant Protection, 230
Ecuador, 174, 178, 198, 207, 210, 223
 Association of Free Farmers, 207
EDB (ethylene dibromide), 76–7, 86, 108–9
Egypt, 176
EHD (Environmental Health Department), 255
EHO (Environmental Health Officer), 128
El Salvador, 94
Electrodyn Sprayer Conveyor Treatment, 65
electrostatic spraying, 55
Elm Farm Research Centre, 277
encephalitis, 8
endosulfan, 44, 90, 197, 218
endrin, 44, 72, 74, 91, 145, 146, 149
enzyme tests, 266
EPA (Environmental Protection Agency), 10, 17, 25, 29–30, 104, 118, 120, 124, 125, 131–2, 167, 251, 275, 295, 297
epichlorohydrin, 77
EPTC, 45
ethoprop, 197
ethyl mercury phosphate, 95
ethylene dichloride, 108
 oxide, 9, 81
ethyltin triiodide, 103
ETU (ethylene thio-urea), 72–3, 125
eye damage, 97, 167, 213, 215

famphur, 149
FAO (Food and Agriculture Organisation), 16, 114, 118, 192, 223, 226–7, 228, 231–2, 293, 298
 Pesticides Code, 191 ff.
Farm and Food Society, 276
farmworkers, 26, 76, 81–3, 98, 101
FDA (Food and Drug Administration), 114
fenitrothion, 15, 197, 198, 204, 270
fensulfothion, 151
fenthion, 85, 204
fentin (triphenyltin), 102–3